Current Topics in Microbiology and Immunology

169

Editors

Bunyaviridae

Edited by D. Kolakofsky

With 34 Figures and 69 Tables

Springer-Verlag
Berlin Heidelberg New York
London Paris Tokyo
Hong Kong Barcelona
Budapest

Daniel Kolakofsky
Professor of Microbiology
University of Geneva
School of Medicine, C.M.U.
Avenue de Champel 9
1211 Geneva 4
Switzerland

ISBN-13: 978-3-642-76020-4 e-ISBN-13: 978-3-642-76018-1
DOI: 10.1007/978-3-642-76018-1

Softcover reprint of the hardcover 1st edition 1991

Library of Congress Catalog Card Number 15-12910

Typesetting: Thomson Press (India) Ltd, New Delhi;

23/3020-543210—Printed on acid-free paper

Preface

The Bunyaviridae represent an extremely diverse family of viruses, which until recently were relatively poorly understood. These viruses have become increasingly important in both human and veterinary medicine, and, most recently, bunyaviruses have been recognized as plant pathogens as well. This book attempts to treat all aspects of their biology, including their natural history, genetics, virion structure, unusual pathway of intracellular assembly, gene structure and the mechanisms of its expression, antigenicity, and pathogenesis, in a single volume. As such, it fills a void in the virology literature. This volume is also timely, as the molecular description of this family is now almost complete. The reader can expect to find the present state of the art on how bunyaviruses are maintained in nature, and how they replicate and sometimes cause disease.

I would like to thank my co-contributors for the time and effort they have invested to make their chapters as complete as possible, and not least of all, for by and large respecting the deadline. I am also grateful to Dick Compans for suggesting this volume and helping to design its contents, and for the many discussions during his stay in Geneva.

DANIEL KOLAKOFSKY

List of Contents

List of Contributors

The Bunyaviridae and Their Genetics—An Overview

C. R. PRINGLE

1 Introduction

This chapter reviews briefly the special features that distinguish members of the family Bunyaviridae from other viruses and surveys current knowledge of the genetic properties of these viruses, the factors determining their pathogenic potential for vertebrates, and their transmissibility by arthropod vectors. Detailed descriptions of the individual members of the Bunyaviridae and their attributes together with comprehensive referencing can be found in the ensuing chapters and elsewhere (ELLIOTT 1990; GONZALEZ-SCARANO and NATHANSON 1989; BISHOP 1986). Only the genetics of the Bunyaviridae and factors affecting the variability of these viruses are accorded more than superficial attention here.

Biological Sciences Department, University of Warwick, Coventry CV4 7AL, UK

Current Topics in Microbiology and Immunology, Vol. 169

1.1 The Distinctive Features and Diversity of Bunyaviruses

Perhaps the single most distinctive feature of the family Bunyaviridae is the remarkable number of named viruses included in this famly (KARABATSOS 1985; BISHOP 1986; CALISHER and KARABATSOS 1988). Almost half (47.4%) of the 530 viruses registered with the American Committee on Arthropod-borne Viruses by the end of 1987 were considered to be bunyaviruses or bunya-like viruses.

The number of confirmed or putative members of the family Bunyaviridae continues to increase, and a more recent survey lists some 342 animal bunyaviruses or bunya-like viruses, now supplemented by a solitary plant bunyavirus (GONZALEZ-SCARANO and NATHANSON 1989). Molecular characterization of the few viruses examined in detail so far suggests that this profusion of names is indeed a true reflection of the genetic diversity of the family. The abundance of names does not reflect an unusual prevalence of bunyaviruses, although in endemic situations bunyaviruses may predominate among arthropod-transmitted viruses in actual frequency. In a survey of arthropod-borne viruses in coastal Ecuador spanning the years 1974–1978, 328 of the 379 viruses isolated from mosquitoes and sentinel hamsters were identified as bunyaviruses, and one bunyavirus alone—the otherwise little renowned Maguari virus—accounted for 64% of the isolations (CALISHER et al. 1983).

The names attributed to bunyaviruses range from the whimsical or enigmatic (Facey's Paddock, Main Drain, Bimbo) to the starkly descriptive (Crimean haemorrhagic fever, Nairobi sheep disease, Rift Valley fever), reflecting the extent to which these individual viruses have impinged on human consciousness and welfare. Apart from their multifarious and sometimes outlandish names the features which distinguish bunyaviruses are the following:

1. Tripartite structure of the genome and the negative sense of the linear single-stranded virion RNA
2. Subdivision of the family predominantly on the basis of coding strategy into six distinct genera (*Bunyavirus, Hantavirus, Nairovirus, Phlebovirus, Uukuvirus,* and *Tospovirus*), which embrace subgeneric groupings of viruses with varying degrees of serological relationship (serogroups)
3. Ambisense encoding of a putative non-structural protein gene in the S genome segment in viruses of the *Phlebovirus*, *Uukuvirus* and *Tospovirus* genera
4. Enveloped pleomorphic particles with four virus-specified proteins: the putative polymerase protein (L), cysteine-rich precursor-derived envelope glycoproteins (G1 and G2, plus G3 in the nairoviruses), and the nucleoprotein (N)
5. Three circular nucleocapsids, comprising in viruses of the *Bunyavirus* genus approximately 2100 molecules of N and 25 molecules of L per particle, all enveloped by a lipid envelope containing approximately 600 molecules of glycoproteins G1 and G2

6. Absence of a matrix protein
7. Maturation by budding from membranes of the Golgi apparatus
8. Close association with specific vertebrate hosts and specific arthropod vectors, with the exception of the phytobunyavirus tomato spotted wilt virus (TSWV), which is transmitted by thrips, and the hantaviruses which do not appear to be arthropod-borne
9. Temporal separation of internal and envelope protein synthesis
10. Priming of synthesis by a cap transfer mechanism analogous to that of influenza virus, except that it occurs in the cytoplasm, and absence of 3′-polyadenylation of mRNA
11. Insensitivity to actinomycin-D and alpha-amanitin, and variable dependence on the presence of an intact functional nucleus
12. Limitation of gene exchange to reassortment of genome sub-units between closely related viruses within serogroups.

1.2 Taxonomic Status

The viruses grouped together as the family Bunyaviridae do not fit easily into the current scheme of classification of RNA viruses, which is based on the form and sense of the genomic RNA sequestered in the extracellular virion. The viruses included in the genera *Bunyavirus, Hantavirus* and *Nairovirus*, at least as far as they have been characterized, can be regarded as conventional negative-strand RNA viruses such as the unsegmented filoviruses, paramyxoviruses and rhabdoviruses, and the 7/8 segmented orthomyxoviruses. In contradistinction the viruses belonging to the *Phlebovirus* and *Uukuvirus* genera and the newly designated *Tospovirus* genus (DE HAAN et al. 1990) resemble the arenaviruses in that the small genome segment (the S RNA) is a linear single-stranded structure of covalently linked negative- and positive-sense half molecules from which separate mRNAs are transcribed. In the Bunyaviridae ambisense encoding of genetic information is restricted to the S RNA and present in three of the six genera only, whereas in the Arenaviridae both sub-units of the genome incorporate ambisense encoded information (SALVATO 1989; SALVATO and SHIMOMAYE 1989; IAPALUCCI et al. 1989). The functional significance of the ambisense coding strategy in the S RNA of bunyaviruses is less obvious than in the arenaviruses where the differential encoding of the membrane glycoprotein and nucleocapsid protein genes in opposite senses in the S RNA may provide temporal separation of the synthesis of internal and external proteins (see Fig. 2). The ambisense S RNA of the bunyaviruses encodes the nucleocapsid protein gene and a putative non-structural protein gene of unknown function. A further distinction may be the differential encapsidation of strands: Full length molecules of both polarities of the S RNA of Uukuniemi virus have been reported to be encapsidated in the ratio of 10:1, whereas only negative-strand molecules of the middle-sized (M) RNA were found in virions (SIMONS et al. 1990).

The features which unite these diverse viruses within a single family are a common morphology, the tripartite genome structure, the absence of a matrix protein, and budding into intracytoplasmic vesicles from the internal membranes of the Golgi apparatus (BISHOP et al. 1980; SCHMALJOHN and DALRYMPLE 1983; DE HANN et al. 1989; ELLIOTT 1990). The taxonomy of the bunyaviruses below the family level has not yet stabilised. The *Bunyavirus*, *Hantavirus*, *Nairovirus*, *Phlebovirus* and *Uukuvirus* genera have been distinguished by the molecular size and coding properties of their genomes (recently reviewed in ELLIOTT 1990) and the possession of common CF antigens and variable neutralization antigens (or vice versa in the case of the phleboviruses). These genera are also distinguished biologically by their modes of transmission; the bunyaviruses are predominantly mosquito borne, the nairoviruses and uukuviruses are tick borne, and the phleboviruses are sandfly and gnat borne. The phytobunyavirus TSWV is transmitted by thrips and is the only segmented genome plant virus with a lipid envelope. The members of the genus *Hantavirus* are exceptional in that they are transmitted by rodents and have no arthropod vector.

Curiously, although the hantaviruses came to prominence because of their association with epidemic haemorrhagic disease in non-Korean troops serving in Korea in the early 1950s, transmission is always directly from rodents, and person-to-person spread of virus has not been reported. Many bunyaviruses are uncharacterized and remain outside the existing taxonomy; it is not yet clear whether they can be accommodated within the existing generic structure, or whether new genera will be required (ZELLER et al. 1989). The recent recognition of the ambisense nature of the S RNA of Uukuniemi virus may signal the end of the ascendancy of the splitters in Bunyaviridae taxonomy and the emergence of the lumpers, since it has recently been proposed that the phleboviruses and the uukuviruses should be subsumed within the one genus (SIMONS et al. 1990).

1.3 Genome Coding Potential and Gene Products

In common with all other negative-stranded RNA viruses the coding potential of the genome is utilized efficiently. In the genome of Bunyamwera virus, the prototype of the family and the only bunyavirus completely sequenced so far, 95.3% of the RNA encodes polypeptides (ELLIOTT 1989a, b; LEES et al. 1986). Figure 1 illustrates diagrammatically the mean sizes of the RNA subunits of the six genera and their coding properties (data from ELLIOTT 1990). Despite the considerable difference in relative size of the corresponding sub-units in different genera, the coding properties are similar. The large-sized (L) RNA of Bunyamwera virus has a single open reading frame (ORF) encoding a protein of 2238 amino acids (M_r 259000). A short ORF is present in the negative-sense strand, but no gene product has been identified (ELLIOTT 1989b). No sequence similarity or other homology to any other viral RNA polymerase has been discerned, suggesting that the bunyaviruses are not a recently evolved group. The L protein of viruses of other genera are comparable in size to that of Bunyamwera virus, and it can be presumed that the L RNA of the hantaviruses,

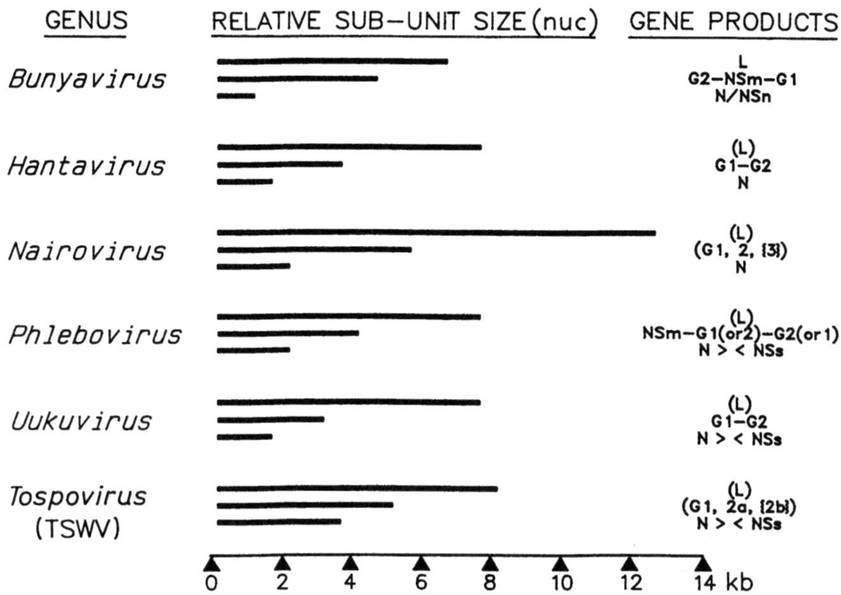

Fig. 1. The tripartite structure of the genome of the Bunyaviridae. Relative sizes of the L, M, and S RNA sub-units and their coding specificities. The sizes of the large (*L*), middle (*M*) and small (*S*) RNAs are represented approximately to scale. The known gene products of each genome sub-unit are listed to the right. From the L RNA: *L*, large (putative polymerase) protein. (The single open reading frame has been confirmed in the *Bunyavirus* genus only). From the M RNA: G1, G2a, G2b, and G3, glycosylated envelope proteins; NSm, a non-structural protein, derived from a single open reading frame, in the order indicated by dashes where known; the internal brackets ({ }) indicate suspected gene products. From the S RNA: N, nucleoprotein; NSs, a non-structural protein; > < indicates ambisense encoding

phleboviruses, uukuviruses and TSWV encode only an L protein. The L RNA of the nairoviruses is considerably larger, and it remains to be established whether additional polypeptides are encoded in this segment.

The M RNA encodes the two (or three in some nairoviruses) envelope glycoproteins of the virion, and in the *Bunyavirus* and *Phlebovirus* genera a non-structural protein in addition. Only one ORF is present in the M RNA of the nine different viruses belonging to the genera *Bunyavirus*, *Hantavirus*, *Phlebovirus* and *Uukuvirus* which have been sequenced so far. The individual polypeptides are thought to be derived by nascent cleavage during translation. Post-translational cleavage of a precursor polypeptide has been reported only for Uukuniemi virus in an in vitro system (Ulmanen et al. 1981). The NSm polypeptide is located terminally in the phleboviruses and internally in the bunyaviruses. As is discussed below the M RNA segment appears to be the major determinant of the pathogenicity and tissue tropism of bunyaviruses.

The S RNA encodes the nucleoprotein and in four of the six genera also a small non-structural protein. The NSs protein is encoded in distinct modes. In the genus *Bunyavirus* the NSs protein is encoded in an overlapping reading frame,

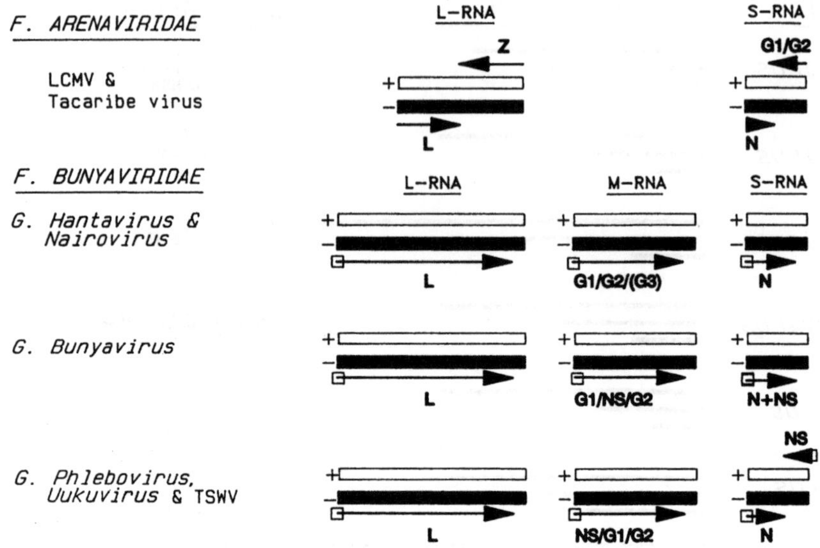

Fig. 2. Comparison of coding strategies in the Bunyaviridae and the Arenaviridae. The negative and complementary positive strands of the genomic sub-units are not drawn to scale. *Arrows*: messenger RNA orientated in the 5′–3′ direction by the *arrowhead*; the encoded gene products are indicated by *letters*. Ambisense encoding of genetic information in non-overlapping reading frames is characteristic of both sub-units of the genome of arenaviruses. The larger sub-unit codes in the negative strand for the L (putative polymerase) protein and in the opposite strand for the Z (putative RNA binding) protein. The smaller sub-unit codes in the negative strand for the N (nucleoprotein), and in the opposite strand for the G1/G2 (envelope) glycoproteins. The bunyaviruses can be separated into three groups; ambisense encoding of genetic information is observed only in the S RNA sub-unit of viruses of the genera *Phlebovirus* and *Uukuvirus* and in TSWV. These viruses are also distinguished from the others by the order of encoding of gene products in the single open reading frame of the M RNA. Viruses in the genus *Bunyavirus* encode a non-structural protein in an overlapping reading frame in the S RNA, and another non-structural protein in a non-overlapping reading frame in the M RNA, whereas viruses in the genera *Hantavirus* and *Nairovirus* encode neither. The bunyavirus mRNAs are capped (□); the symbol "/" indicates derivation from a common precursor; the symbol "+" indicates encoding in overlapping reading frames

and it is now accepted that the N and NSs polypeptides are translated from the same mRNA by separate initiation. A third ORF in the same frame as NSs is present in some (Germiston virus and Maguari virus) but not all members of the genus *Bunyavirus*. However, no gene product has been detected. In the phleboviruses, the uukuviruses and TSWV, however, the N protein is encoded in the complementary (positive-sense) strand as in the other viruses, but the NSs protein is encoded in the negative-sense strand. The N coding sequence is located in the 3′ half of the genome and the NSs coding sequence in the 5′ half, as illustrated diagrammatically in Fig. 2, and the N and NSs proteins are translated from separate subgenomic RNAs. The NSs protein of Rift Valley fever phlebovirus has been reported to be phosphorylated and that of Punta Toro phlebovirus associated with the nucleocapsid, suggesting an involvement in replication or transcription. The NSs protein of Rift Valley fever phlebovirus is unusual in that it has been detected in the nucleus of infected cells. Only one ORF corresponding to the N protein is present in the S RNA of the hantaviruses and the nairoviruses.

1.4 Genome Structure and Replication

The 3′-terminal sequences of the three RNA sub-units are conserved within each genus. Furthermore, the terminal sequences of the genome sub-units of the phleboviruses and uukuviruses are identical, which together with the ambisense nature of their S RNAs emphasizes the close relationship of these two genera.

The 3′ and 5′ ends of each RNA sub-unit exhibit complementarity. At the 3′ terminus residues 1–11, with the exception of 9, are conserved in the L, M and S RNA of Bunyamwera virus and complementary to the corresponding residues at the 5′ terminus. The next 20 or more residues show segment-specific complementarity. A mismatched residue has been found at position 9 in the terminal sequences of all viruses of the genus *Bunyavirus*, but not in those of viruses of other genera, except for a mismatch at position 12 in Hantaan virus.

The circularity of the nucleocapsids is probably a consequence of the complementarity of the termini, and it has been confirmed that the ends of full-length encapsidated strands were in fact base paired (RAJU and KOLAKOFSKY 1989). Only full-length strands of both negative and positive polarity were encapsidated, whereas mRNAs which were capped and truncated at the 3′ end were not. The evidence favours location of the nucleocapsid recognition signal at the 5′ end (RAJU and KOLAKOFSKY 1987).

As in all other negative-strand RNA viruses, primary transcription of the negative-sense genome is mediated by a virion-associated RNA polymerase. This virion-associated enzyme is also endowed with an endonuclease activity and may be involved in mediating the process of mRNA capping in the cytoplasm. However, although in vitro enzyme activity has been demonstrated, primary transcription in the absence of protein synthesis has been difficult to detect in many instances. BELLOCQ et al. (1987) have attempted to resolve this anomaly by proposing that in the absence of ribosome binding and protein synthesis the nascent mRNA and its template interact, preventing progression of the polymerase. This hypothesis was based on the observation that the presence of rabbit reticulocyte lysate in the in vitro polymerase reaction stimulated production of complete transcripts of S RNA in place of the incomplete transcripts normally observed. Concurrent protein synthesis appeared to be necessary to prevent premature termination during synthesis of mRNA on the S RNA template. RAJU et al. (1989) have shown that this requirement was host cell-dependent, and that it could be mimicked in vitro. In mammalian cells translational competence was required to avoid premature termination at nucleotide 175, whereas in C6/36 mosquito cells read-through was independent of concurrent translation. Reconstitution experiments suggested that this translational requirement was mediated by host cell factors present in BHK-21 cells which were absent in C6/36 cells.

Maturation occurs predominantly at smooth membranes in the Golgi complex and consequently is inhibited by monensin. Experiments with vaccinia virus recombinants have shown that targeting to the Golgi apparatus is a property of the glycoproteins alone (MATSUOKA et al. 1988; PENSIERO et al. 1988). The glycoproteins are anchored in the membrane with their amino-termini located externally and their carboxy-termini internally. Virus particles are formed by budding into vesicles which are transported to the cell surface, eventually releasing their contents to the exterior. Characteristically bunyavirus-infected cells retain their gross morphology until late in the infectious cycle, long after most of their contents have been lost.

2 The Genetics of the Bunyaviridae

2.1 Sequence Homologies

Only limited sequence information is available at present, but it is clear that the pattern of sequence relationships closely reflects taxonomic status. In general, viruses classified in different genera show little or no sequence similarity. However, significant homology has been detected between the N proteins (33%–35%) and the G proteins (15%–25%) of the prototype uukuvirus Uukuniemi virus, and the phleboviruses Punta Toro virus and Sicilian sandfly virus (SIMONS et al. 1990). This together with their common adoption of ambisense coding and possession of identical sub-unit terminal sequences indicates that the two genera are more closely related than the others and may not merit separate status as genera. The proteins of TSWV, the other virus exhibiting ambisense encoding in the S RNA, have no discernible similarity with these viruses. A separate genus is justified for TSMV for this reason and on account of its unique host and vector specificity.

Viruses classified within the same genus may exhibit limited sequence homology, and viruses belonging to the same serogroup generally show close similarity. For example, the overall homology of the G2 proteins of four members of the *Bunyavirus* genus is about 66%, that of the NSm proteins about 50%, and that of the G1 proteins about 40%. Comparison of the N proteins of six members of the *Bunyavirus* genus belonging to three serogroups revealed 40% overall similarity, whereas the similarity was 80% for members of the same serogroup (ELLIOTT 1989b). The degree of sequence divergence observed even among viruses of the same serogroup is sufficient to justify the designation of these viruses by individual names.

2.2 Temperature-Sensitive Mutants and Homologous Recombination

Temperature-sensitive (ts) mutations have been employed to investigate gene function, to define the genetic basis of virulence and to delineate the extent of reassortment and the limits to gene exchange. The majority of ts mutants were obtained by mutagenization using either 5-fluorouracil, 5-azacytidine or *N*-methyl-*N*′-nitrosoguanidine. The data of OZDEN and HANNOUN (1978) suggested that 5-fluorouracil was a slightly more effective mutagen than NTG. GENTSCH et al. (1977) isolated spontaneous mutants from snowshoe hare bunyavirus at a frequency of 1.7% and from La Crosse virus at a frequency of 1.0%. However, the frequency of spontaneous mutants in a stock of Maguari virus was higher and estimated at 2.7% (IROEGBU 1981).

Some 200 ts mutants have been isolated from 11 viruses; 10 of these viruses belong to two different serogroups of the genus *Bunyavirus* and the remaining one to the genus *Uukuvirus* (Table 1). Homologous recombination between individual pairs of mutants, subsequently shown to be mediated by sub-unit reassortment (GENTSCH et al. 1977, 1979; IROEGBU and PRINGLE 1981a), has been observed in each virus. The majority of the mutants have been assigned to reassortant groups; i.e. non-ts virus is present only in the progeny from mixed infections with parental viruses classified in different reassortant groups. A significant feature was the consistent recovery of mutants assignable to two reassortant groups rather than the three expected from the tripartite structure of the genome.

Table 1. Bunyavirus temperature-sensitive mutants

Genus	Serogroup	Virus	No. of ts Mutants	Reassortment groups[a]				References[b]
				I	II	III	Unclass	
Bunyavirus	Bunyamwera	Batai	5	1	4	0	0	
		Bunyamwera	8	5	3	0	0	1–5
		Maguari	46	12	31	1	2	
		Germiston	8	2	6	0	0	6
		Guaroa	4	2	2	0	0	7
			9	3	5	1[c]	0	8
	California	La Crosse	20	6	14	0	0	
		Snowshoe hare	48	26	20	0	2	
		Tahyna	20	1	17	0	2	8–12
		Trivittatus	12	4	6	0	2	
		Lumbo	9	2	7	0	0	
			8	1	7	0	0	13
Uukuvirus		Uukuniemi	13	3	2	0	8	14

[a] The homology of the reassortment groups does not extend across serogroup boundaries
[b] [1] IROEGBU and PRINGLE 1981a; [2] IROEGBU and PRINGLE 1981b; [3] PRINGLE and IROEGBU 1982; [4] ELLIOTT et al. 1984; [5] IROEGBU 1981; [6] OZDEN and HANNOUN 1980; [7] PRINGLE, unpublished data; [8] BISHOP 1979; [9] GENTSCH and BISHOP 1976; [10] GENTSCH et al. 1977; [11] GENTSCH et al. 1979; [12] GENTSCH et al. 1980; [13] OZDEN and HANNOUN 1978; [14] GAHMBERG 1984
[c] Tentative

Within the Bunyamwera and California serogroups the equivalence of the reassortant groups has been established by mixed infections with heterologous parental viruses. However reassortment has never been observed between viruses belonging to different serogroups, although the combinations of viruses tested has been rather limited (see below). Consequently the homologies of the reassortant groups listed in Table 1, at least as defined by genetic experiments, does not extend beyond the boundaries of serogroups.

Maguari virus of the Bunyamwera serogroup is the exception: One mutant of the 46 ts mutants isolated from mutagenized wild type Maguari virus, mutant tsMAG23(III), recombined with all members of panels of mutants representing reassortant groups I and II and two putative double I+II mutants, thus identifying the missing third reassortant group (PRINGLE and IROEGBU 1982). Mutant tsMAG23(III) exhibited a prounced host restriction which may explain the dearth of group III mutants. The plaque-forming ability of tsMAG23(III) on BS-C-1 cells was at least a 100-fold lower than on BHK-21 cells and yields were a 1000-fold less, whereas group I and II mutants of Maguari virus and several other viruses did not show such marked host cell-dependent differences.

2.3 Complementation

GENTSCH and BISHOP (1976) demonstrated that mutants in different reassortment groups of snowshoe hare virus were able to complement one another, and likewise OZDEN and HANNOUN (1978) detected intergenic complementation between recombining ts mutants of Lumbo virus. However intergenic complementation was difficult to discriminate from reassortment on account of the high frequency and early occurrence of the latter in the multiplication cycle.

IROEGBU and PRINGLE (1981a, b) detected intragenic complementation between individual pairs of ts mutants of Maguari virus belonging to reassortment group I following mixed infection of BHK-21 cells at high multiplicity of infection. These observations are compatible with the assignment of group I mutants to the S RNA with its three ORFs. Complementation was confined to specific pairs of mutants, however, and the group I mutants could not be grouped on this basis. No intragenic complementation was detected in the case of group II mutants, which suggests in view of the presumptive nascent cleavage of the membrane glycoprotein precursor that the assembly and maturation of bunyavirus virions is strictly compartmentalized.

2.4 Heterologous Reassortment and Gene Assignment

Heterologous crosses of different viruses within the same serogroup have been employed to equate reassortment groups and genome sub-units. RNA fingerprinting and dot hybridization with sub-unit specific cDNA probes, in conjunction with phenotypic analyses by SDS/PAGE of radiolabelled polypeptides and by immunological methods, have been employed extensively to correlate the phenotype and genotype of non-ts reassortant clones present in the progeny from mixed infections with ts mutant parental viruses (GENTSCH and BISHOP 1976; GENTSCH et al. 1977, 1979, 1980; ROZHON et al. 1981; BISHOP et al. 1984; IROEGBU et al. 1981a, b; PRINGLE and IROEGBU 1982; PRINGLE et al. 1984a, b; ENDRES et al. 1989). In general, unequivocal assignments were possible, although some anomalies have persisted. It was observed generally that heterologous reassortment was restricted and not always reciprocal, the restriction increasing with the genetic divergence of the parental viruses.

In the California serogroup, crosses of the closely related snowshoe hare and La Crosse viruses yielded the non-ts reassortants SSH/LAC/LAC[1] and SSH/LAC/SSH when the snowshoe hare parent was a group I ts mutant and the La Crosse parent a group II mutant. These are the only two reassortants that would be expected if the group I mutation were located in the M RNA and the group II mutation in the L RNA. However the same two reassortants were recovered and none of the other four possible reassortant classes, when the parents were wild-type viruses and the progeny screened without temperature selection. No viable reassortants were isolated from the reciprocal cross of a group I La Crosse virus mutant and a group II snowshoe hare virus mutant (GENTSCH et al. 1979).

By a series of ingenious backcrosses GENTSCH et al. (1979) obtained further reassortants. First, group I and II ts mutants were induced in the SSH/LAC/SSH and SSH/LAC/LAC reassortants by random mutagenesis and each in turn backcrossed to the appropriate ts parental virus to yield further non-ts reassortants. The SSH/LAC/SSH group II reassortant backcrossed to the SSH/SSH/SSH group I grandparent yielded a new reassortant having the gene combination LAC/LAC/SSH. In their turn, these new reassortants were mutagenized to obtain group I and group II ts mutants, and these in turn were backcrossed to the original SSH/SSH/SSH and LAC/LAC/LAC parental viruses and further non-ts reassortants obtained and genotyped. The fourth gene combination SSH/SSH/LAC was obtained as a non-ts reassortant in progeny from the cross of a group II SSH/SSH/LAC reassortant and the original SSH/SSH/SSH group I mutant. Overall the data are consistent with the assignment of the group I lesion to the M RNA and the group II lesion to the L RNA.

However several crosses yielded no reassortants whatever, and the two missing gene combinations (LAC/SSH/SSH and LAC/SSH/LAC) were obtained only later in experiments by ROZHON et al. (1981) in which virus concentrated by polyethylene glycol precipitation was used as the inoculum. It could be concluded, therefore, that the asymmetric and restricted patterns of gene segregation observed in these experiments were not due to inherent in-

[1] The genotype is indicated by abbreviations in the order of the L/M/S sub-units: SSH, snowshoe hare; LAC, La Crosse; BUN, Bunyamwera; MAG, Maguari; BAT, Batai; CAR, Caraparu

compatibility of the genome sub-units of these two members of the California serogroup but rather were attributed to phenotypic properties such as reduced cytopathogenicity and aberrant morphogenesis, or compartmentalization of replication within the cell. The interaction between parental genomes appeared to be an early event, at least in homologous crosses (IROEGBU and PRINGLE 1981a); consequently the relative multiplicities of infection of the parental viruses could have an important bearing on the outcome of any particular mixed infection.

A different and more regular form of restriction was observed in the case of heterologous crossing of Bunyamwera group viruses (see below) and also to some extent when more distantly related members of the California group were crossed. The results of crossing group II mutants of Tahyna virus with group I mutants of snowshoe hare virus or trivitattus virus, group I mutants of Tahyna virus with group II mutants of La Crosse virus, and group I mutants of trivitattus virus and group II mutants La Crosse virus, yielded reassortants whose genotypes were, as before, consistent with the assignment of the group I lesion to the M RNA and the group II lesion to the L RNA. There was a marked tendency, however, for the L and S RNA sub-units to cosegregate; 58 of the 65 reassortants characterized derived their L and S RNA from the same parent. The same phenomenon, was observed in the cross of a group I La Crosse virus and a group II Tahyna virus where 9 of 12 reassortants derived their L and S RNA from the same parent; however, this cross was exceptional in that two of the 12 non-ts reassortants had genotypes which were not anticipated on the basis of the group I and II gene assignment predicted from the outcome of the other crosses. This discrepancy was presumed to be the result of an experimental error. Again certain combination of parental viruses (Tahyna group I × snowshoe hare or trivitattus group II, and La Crosse group I × trivitattus group II) yielded no reassortant progeny virus.

Reciprocal crosses of the group I and group II mutants of three Bunyamwera serogroup viruses, on the other hand, did yield reassortants with all possible combinations of genome sub-units (IROEGBU and PRINGLE 1981a, b). In contradistinction to the California serogroup data, however, analysis of the genotype of the reassortants suggested that the group I mutation in Batai, Bunyamwera and Maguari viruses resided in the S RNA and the group II mutation in the M RNA (IROEGBU and PRINGLE 1981a, b; PRINGLE et al. 1984). Furthermore from the genotypic analyses it was evident that segregation did not occur at random in heterologous crosses of these three viruses, and only the M RNA sub-unit appeared to segregate without restriction. This was the case both where temperature selection was employed to isolate non-ts reassortant viruses, and where a high frequency of reassortment allowed direct analysis of unselected progeny virus. Curiously, in the temperature-selected progeny from crosses of ts mutant parents, invariably only one of the two predicted classes of non-ts reassortants was obtained. A similar bias was observed in unselected progeny, and of a total of 279 reassortants characterized all appeared to derive their L and S RNA subunits from the same parent.

A different situation obtained, however, when these reassortant viruses in their turn were used as the parental viruses. For example, all six possible reassortants were obtained at comparable frequencies in unselected progeny from a mixed infection of viruses with the genotypes BUN/MAG/BUN and BAT/BUN/BAT. Consequently, as observed in the case of the California serogroup viruses, the restricted patterns of genome segregation cannot be a consequence of any inherent incompatibility of the RNA sub-units of the different viruses. The molecular basis of these restricted segregations has not been elucidated, but the apparent linkage of the L and S RNA sub-units which encode the virion core and replicative proteins suggests that the restrictions operate at the gene product level. The Bunyamwera serogroup data suggested that once a homologous gene combination has been disrupted by recombination, there is no preferred association of genes or gene products with any selective advantage. If this phenomenon extends beyond the confines of in vitro culture it has evolutionary implications because reassortment seems to be enhanced following the break-up of adapted gene combinations, suggesting that bunyaviruses may have the same potential for rapid and abrupt changes in properties as the type A influenza viruses.

A single ts mutants of Maguari virus defines the third reassortant group predicted on the basis of the tripartite structure of the genome. Because of the phenomenon of restriction encountered in crosses of heterologous viruses it could not be established conclusively that the group III lesion is located in the L RNA. Non-ts reassortants of genotype BUN/MAG/BUN were obtained from the cross of MAGts23 (III) with a group II Bunyamwera virus parent which excluded location of the group III lesion in the M RNA, but the non-ts reassortant class required to assign the group III lesion unequivocally to the L RNA was not obtained because no non-ts progeny was obtained from the critical cross of MAGts23(III) and the group II Bunyamwera parent (PRINGLE and IROEGBU 1982).

An attempt was made to resolve this problem by analysing the genotype of unselected virus from crosses of the various ts mutants and the heterologous wild type virus, employing the ts phenotype as a fourth marker (MURPHY and PRINGLE 1987; HAMPSON 1987). Reassortants of genotype MAG/BUN/BUN/ts$^+$ and MAG/BUN/MAG/ts were obtained from various crosses of group I Maguari virus mutants with wild-type Bunyamwera virus, a result which is consistent with the previous assignment of the group I lesion to the S RNA. Reassortants BUN/MAG/MAG/ts and MAG/MAG/BUN/ts were obtained from several crosses of group II Maguari virus mutants with wild-type Bunyamwera virus, a result which is consistent with the previous assignment of the group II lesion to the M RNA. The cross of MAGts23(III) with Bunyamwera virus wild type yielded the reassortants MAG/BUN/BUN/ts, BUN/MAG/MAG/ts$^+$, MAG/MAG/BUN/ts and BUN/MAG/BUN/ts$^+$, a result which is consistent with assignment of the group III lesion to the L RNA.

The assignments of the group I, II and III lesions to the S RNA, M RNA and L RNA in the Bunyamwera serogroup differ from the California serogroup result, where apparently no N gene mutations and many L gene mutations were obtained. The California serogroup result is more in line with expectation from the relative target sizes of the sub-units and the experience with other negative-standard RNA viruses (PRINGLE 1987, 1990). The reasons for the discordance of the Bunyamwera serogroup data have not been resolved.

In this and previous experiments (IROEGBU and PRINGLE 1981a, b) with Bunyamwera serogroup viruses apparently heterozygous viruses were occasionally present in the progeny. These heterozygous or segregating clones were sometimes difficult to detect since in mixed infections Maguari virus appeared to have a replicative advantage over Bunyamwera virus. Consequently, errors in the assignment of genotype could have occurred occasionally.

The physical nature of these heterozygous viruses has not been determined, but TALMON et al. (1987) have shown by electron microscopy of vitrified specimens that the number of sub-units in different virions may vary. A different but operationally similar problem has been encountered in the genetic analysis of rotaviruses, where it has recently been recognized that the genetic background of the recipient can affect the phenotype conferred by the gene introduced by reassortment (CHEN et al. 1989). The assignment of the group I and III mutations in the Bunyamwera serogroup in particular will not be unequivocal until the L RNA and S RNA of the mutants have been sequenced.

2.5 Heterologous Reassortment and the Concept of Gene Pools

Consideration of the available evidence indicates that reassortment of genome sub-units has only been achieved in vitro and recognized in vivo for viruses belonging to the same serogroup. Figure 3 summarizes diagrammatically the outcome of the in vitro heterologous recombination experiments described above which involved the crossing of ts mutants and determining the phenotypes and genotypes of non-ts reassortants.

A less definitive but more generally applicable procedure to test potential for gene exchange is the following. In the experiments with ts mutants it was established that the M RNA gene products are solely responsible for the induction of neutralizing antibodies (GENTSCH et al. 1980; IROEGBU and PRINGLE 1981). Subsequently, MILLICAN and PORTERFIELD (1982) prepared nine hamster antisera against three parental Bunyamwera serogroup viruses (Batai, Bunyamwera and Maguari) and their six heterologous reassortants and established independently that the specificity of neutralization was strictly determined by the M RNA, whereas antibody-dependent enhancement of infectivity was more cross-reactive. Furthermore non-ts revertants of a group II mutant of Muguari virus often exhibited changed responses to neutralizing antibody with simultaneous alteration of the electrophoretic mobility of the G1 protein (MURPHY and PRINGLE 1987). Therefore, the neutralization of virus infectivity is unequivocally determined by the M RNA. If any virus to which a

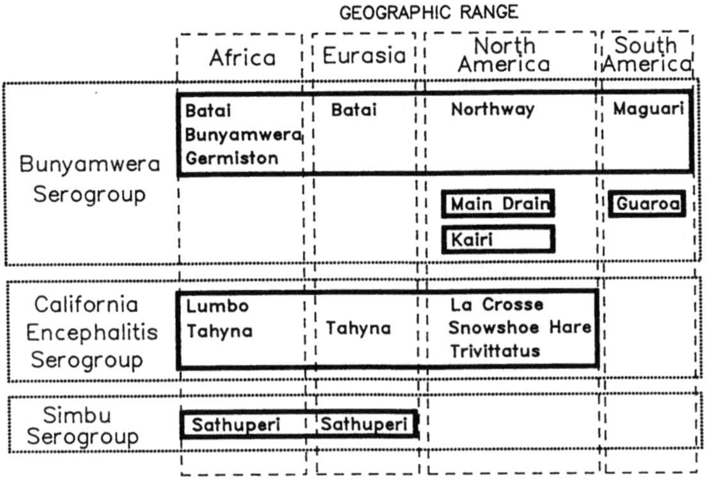

Fig. 3. Patterns of genome sub-unit exchange among members of the genus *Bunyavirus*. The *large rectangles* enclose serologically related viruses. Viruses which can exchange genome sub-units by genetic reassortment and can be regarded as a common gene pool are contained within the same *heavily lined box*. Viruses in different boxes cannot exchange genome sub-units even where they belong to the same serogroup. The viruses are arranged vertically by geographic range; all except Batai, Tahyna, and Sathuperi are restricted to a single continent

specific neutralizing antiserum can be prepared is crossed with a ts mutant parental virus which donates a sub-unit encoding a non-ts G1 protein (e.g. group II California serogroup mutants, and group I and III Bunyamwera serogroup mutants), it should be possible to detect reassortment. Consequently by screening the progeny at the restrictive temperature for the mutant and in the presence of specific antiserum to the other virus, the parental viruses are eliminated and the detection of reassortants favoured.

Figure 3 is supplemented by data obtained in this way. All of the six viruses of the California serogroup which have been studied were able to exchange genome sub-units. However, only five of the eight Bunyamwera serogroup viruses studied were able to exchange sub-units. It is apparent, therefore, that in addition to the restricted segregations observed in the reciprocal ts mutant crosses described previously, some viruses within the same serogroup are less able (or unable) to exchange genome sub-units and are in effect genetically isolated. The pattern of restriction within the Bunyamwera serogroup seems to follow the serological relationships (Hunt and Calisher 1979); the more divergent serologically, the greater is the degree of restriction in reassortment.

In Fig. 3 Guaroa virus has been represented as a Bunyamwera serogroup virus, but it can be regarded as a bridging virus. In complement fixation tests it appears related to the Bunyamwera serogroup, but in neutralization tests it has reactivity similar to California viruses (Whitman and Shope 1962). Although independently isolated ts mutants of Guaroa virus are available (Gentsch et al. 1980; PRINGLE, unpublished), no evidence of reassortment with either Bunyamwera or California serogroup viruses has been reported.

Independent experiments have failed to show evidence of reassortment of sub-units between the Bunyamwera and the California serogroups. A difference in the restrictive temperature appropriate for the two groups of viruses, however, limited the stringency of the screening for non-ts reassortant. Similarly, attempts to obtain reassortment between Sathuperi virus of the Simbu serogroup and viruses of the Bunyamwera and California serogroups were unsuccessful. The same restrictions on reassortment were observed in mixed infections carried out in either BHK-21 or *Aedes albopictus* C6/36 cells (ELLIOTT et al. 1984). The evidence at present suggests, therefore, that the various serogroups of the Bunyaviridae represent separate gene pools, and that variation can arise both by gene capture from foreign viruses and by genetic divergence brought about by ecological isolation.

Figure 3 illustrates that the propensity to undergo reassortment is not related to the known geographical range of these viruses. For instance, Maguari virus which is thought to be limited to South America can exchange genome segments with African, Eurasian and North American Bunyamwera serogroup viruses. Conversely, Northway, Main Drain and Kairi viruses, which are thought to be present only in North America, do not reassort. It must be presumed that the very specific host and vector requirements of these viruses have promoted their genetic isolation. In general the propensity for reassortment closely parallel the serological relationships of bunyaviruses. The pattern of potential gene exchange in the Bunyaviridae is different from that observed in the Orthomyxoviridae, where reassortment appears to be possible among all influenza viruses of the same serological type. However, although non-random reassortment in influenza A virus has been reported (LUBECK et al. 1979), its extent has not been examined systematically on account of the complexity of the analysis where 254 different recombinants are possible rather than the 6 of the bunyaviruses.

2.6 Role of the Vector

Reassortment of genome sub-units can occur in invertebrate cells perhaps more readily than in vertebrate cells (IROEGBU and PRINGLE 1981), and it has been confirmed experimentally that reassortants are generated during mixed infection of mosquitoes (BEATY et al. 1981, 1985). Reassortment occurred at high frequency in *Aedes triseriatus* mosquitoes orally infected with La Crosse virus and Snowshoe hare virus.

Furthermore, replication and reassortment were enhanced in the ovaries of female mosquitoes after multiple blood meals, and the newly generated reassortants were transmitted transovarially to about 10% of their progeny (CHANDLER et al. 1990). These newly evolved reassortants in their turn could be transmitted to a susceptible vertebrate host by oral transmission. In nature, *Aedes triseriatus* mosquitoes feed on a variety of hosts, which increases the chance of multiple infection with different viruses since many California serogroup viruses cocirculate in the same area sharing the same vector. The ovaries of infected mosquitoes appear to be an active site of virus replication and consequently of evolution of new virus genotypes. This coupled with the possibility of transovarial transmission of new genotypes suggests that reassortment in invertebrate vectors is an important mechanism for establishment of new gene combinations.

Presumptive reassortant bunyaviruses have been isolated from mosquitoes in nature (KLIMAS et al. 1981; USHIJIMA et al. 1981). The evidence is circumstantial, but persuasive. For example, USHIJIMA et al. (1981) observed that the oligonucleotide fingerprints of the L RNA and S RNA of Shark River virus and Pahayokee virus, two members of the Patois serogroup, were virtually indistinguishable whereas the fingerprints of their M RNAs were quite distinct. Both these viruses originated from *Culex melanoconium* mosquitoes caught in southern Florida, and it was concluded that these two viruses represented natural recombinant viruses which had acquired their M RNA from different unknown donor viruses. It is possible, however, that the L and S RNAs of the Patois serogroup viruses are more conserved in evolution than the M RNAs, although this was not the pattern that was observed in a survey of variation among natural isolates of La Crosse virus of the California serogroup from different ecological niches (EL SAID et al. 1979; KLIMAS et al. 1981).

It is clear that sub-unit reassortment is a potential and actual driving force in the evolution of bunyaviruses. The role of change by progressive mutational drift, as exemplified by the antigenic drift of influenza viruses, has not been studied systematically. The comparison of natural isolates of La Crosse virus referred to above revealed that the oligonucleotide fingerprints of the L, M, and S RNAs of each isolate were distinguishable, suggesting that variation can arise by progressive or random accumulation of mutational changes. Isolates from other sources, however, have shown little variation; for example, two La Crosse viruses recovered 18 years apart from human brain tissue appeared virtually identical (BISHOP and SHOPE 1979). Furthermore,12 cycles of transovarial transmission through female phlebotamine flies did not induce significant changes in the genome of a phlebovirus (BILSEL et al. 1988). At present the relative importance of genetic drift in the evolution of the Bunyaviridae is difficult to assess since the available data are limited predominantly to comparisons of oligonucleotide fingerprints.

2.7 Deletion Mutants, Interference and Persistent Infection

Deletion mutations located predominantly in the viral polymerase genes are associated with autointerference and persistent infection in all RNA viruses.

Such deletion mutants have been difficult to characterize in bunyaviruses on account of the frequent inability to recover sub-unit RNA in equimolar amounts from purified virus (ELLIOTT 1990). Undiluted passage of Bunyamwera virus in BHK-21 (KACSAK and LYONS 1978) or BS-C-1 cells (IROEGBU and PRINGLE, unpublished data) culminated in reduced yields of infectious virus. In BHK-21 cells this was associated with the appearance of S RNA-containing particles. CUNNINGHAM and SZILAGYI (1987) have detected truncated L RNA genome sub-units in the cytoplasm of infected BHK-21 cells infected with Germiston virus (genus *Bunyavirus*) which resemble more the polymerase gene deletions characteristic of the defective interfering particles of other negative-strand RNA viruses.

Heterologous interference was observed between different members of the Bunyamwera serogroup in dual infection experiments, such that Maguari virus appeared to suppress replication of Bunyamwera virus or Batai virus. This suppression parallelled the cytopathogenicity of the viruses and may have contributed to the biased outcome of reassortment experiments (MURPHY and PRINGLE 1987; HAMPSON 1987).

Persistent infections have been established by propagation of Dugbe virus (genus *Nairovirus*) in pig kidney cells (DAVID-WEST and PORTERFIELD 1974), Toscana virus (genus *Phlebovirus*) in Vero cells (VERANI et al. 1984) and Bunyamwera virus in C6/36 mosquito cells (ELLIOTT and WILKIE 1986). In the latter situation autointerfering particles released from the persistently infected cells appeared to contain intact S RNA but to be deficient in L and M RNA.

Deletion mutations affecting the M RNA have not been reported. MURPHY and PRINGLE (1987) have described non-ts revertants of a group II ts mutant—tsMAG8 (II)—of Maguari virus which exhibited gross shifts in the electrophoretic mobility of the G1 protein. These viruses grew normally, and there was no evidence of extensive deletion of the M RNA. The accelerated mobility of the G1 protein was tentatively considered to be the result of exposure of a cryptic proteolytic cleavage site.

3 Genetic Determinants of Virulence

The biology of the family Bunyaviridae is dominated by the M RNA since this sub-unit encodes the genes concerned in many of the most important interactions with the host. Virulence, host range, tissue tropsim, transmissibility, neutralization, haemagglutination, and membrane fusion are the principal phenotypic properties that have been attributed to M RNA gene products. Definition of the predominant role of the M RNA in determining the disease-producing potential of bunyaviruses has been to a large extent a by-product of the production and characterization of heterologous reassortants.

3.1 Virus Virulence in Vertebrates

As in many other viruses, the virulence of bunyaviruses may become rapidly modified during the course of serial passage in cultured cells. Field isolates of Tahyna virus were generally virulent in suckling mice, but laboratory passage resulted in the rapid appearance of strains with differing in vivo virulence and in vitro cytopathogenicity. Conversely,a laboratory strain remained neurovirulent but exhibited reduced peripheral virulence as a consequence of serial passage in mouse brain (JANSSEN et al. 1984). Tahyna virus passaged serially in cultured mosquito cells at 20 °C replicated less well in vertebrate cells at 37 °C and exhibited reduced virulence for mice by any route of inoculation (MALKOVA and MARHOEL 1976). Clearly, bunyaviruses can adapt rapidly to changed environmental circumstances, and it is likely that this occurs by selection of randomly occurring genetic variation rather than by reassortment of pre-existing variation.

Experiments with California serogroup viruses in laboratory animals have shown that disease production is determined by the M RNA (SHOPE et al. 1981). Viruses of differing virulence for mice were crossed, and it was observed that the disease-producing ability of their progeny segregated according to the parental origin of the M RNA sub-unit. Tahyna virus was avirulent by intracerebral or subcutaneous inoculation into 4-week-old mice, whereas La Crosse and snowshoe hare wild-type viruses and reassortants carrying LAC or SSH M RNA were virulent. Exceptionally, LAC/LAC/LAC and LAC/LAC/SSH reassortants obtained from a cross of tsLAC16(I) and a group II its mutant of genotype SSH/LAC/SSH were significantly less virulent following intracerebral inoculation into mice. It was subsequently established by analysis of further reassortants that the L RNA of mutant tsLAC16(I) carried an extraneous mutation which could modify virulence (ROZHON et al. 1981).

JANSSEN et al. (1986) independently verified these findings and further showed that the S RNA as well as the L RNA could modify virulence. Reassortants derived from the cross of an avirulent strain of Tahyna virus and a virulent La Crosse virus confirmed that in California serogroup viruses the M RNA is the major determinant of peripheral virulence. Peripheral virulence mimics the natural route of transmission from insect vector to vertebrate host and is defined as susceptibility to subcutaneous inoculation. In addition, however, it was apparent that the origin of the L and S RNAs also influenced the level of peripheral virulence displayed by these reassortants. GONZALEZ-SCARANO et al. (1988) were able to further partition the virulence phenotype into four independently varying components (neuroinvasiveness and neurovirulence in mice, and oral transmissibility and intrathoracic susceptibility in mosquitoes), establishing the essentially polygenic nature of the virulence phenotype.

Selection for rapid penetration of BHK-21 cells has resulted in attenuation of alphaviruses (OLMSTED et al. 1984), and following this rationale ENDRES et al. (1989) succeeded in isolating a novel neuroattenuated variant by serial passage of a TAH/LAC/LAC reassortant in BHK-21 cells. The neuroattenuated phenotype (resistance of adult mice to intracerebral inoculation of virus) of this variant is unique among California serogroup viruses. This variant also exhibited decreased neuroinvasiveness following subcutaneous inoculation of suckling mice, replicated less well in neuroblastoma cells than neurovirulent viruses, and was temperature-sensitive for growth in BHK-21 cells at 38.9 °C. Non-ts revertants simultaneously regained neurovirulence and ability to replicate efficiently in neuroblastoma cells. Thus neuroblastoma cells may provide a promising in vitro system for further analysis of the determination of neurovirulence in mice.

To date there has been only one study of virulence determinants outside the California serogroup. BISHOP and colleagues (1984) have described reassortants of two strains of Caraparu virus of the group C serogroup which demonstrated that M RNA gene products determined virulence in this serogroup also. The group C serogroup bunyaviruses have not been associated with CNS disease in humans, but different viruses within the serogroup and different isolates of the same virus show varying extents of neurotropism in laboratory rodents. One isolate of Caraparu virus (CARv) killed 3-day-old or 4-week-old mice in 2–3 days following intraperitoneal or intracranial inoculation, whereas another (CARn) killed such mice more slowly after intracranial inoculation, and 4-week-old mice not at all after intraperitoneal inoculation. Two reassortants were identified by oligonucleotide fingerprinting of the RNA sub-units among ten progeny clones from a mixed infection with these two strains. The genotypes of these reassortants were CARv/CARv/CARn and CARv/CARn/CARn, and both resembled the viscerotropic CARv parent in pathogenic properties. It was concluded, therefore, that the M RNA gene products determined the viscerotropic phenotype, since the parental origin of the L RNA and the S RNA differed in the two reassortants, and only the M RNA had a common origin.

3.2 Mechanisms of Transmission in Invertebrates

Homologous reassortment has been demonstrated in mosquitoes dually infected with group I and group II ts mutants of La Crosse virus, either by simultaneous or interrupted feeding, and by intrathoracic injection (BEATY and BISHOP 1989). Heterologous reassortants have been obtained from mosquitoes dually infected with different California serogroup viruses (BEATY et al. 1981), and putative natural reassortants have been recovered from wild caught mosquitoes as described above. Since transovarial transmission is the normal mechanism of over-wintering of California serogroup viruses, reassortants generated during virus multiplication in the ovaries may be transmitted from insect to insect or directly to the vertebrate host.

Infection of vertebrates is normally acute with a brief period of viraemia, whereas insects are not harmed, and the infection becomes persistent. Nor are cytopathic effects normally observed in insect cell cultures infected with bunyaviruses. Transmission of bunyaviruses from mosquitoes to their vertebrate hosts becomes possible only after virus reaches the salivary glands 7–14 days after ingestion. Infection of the salivary glands depends on dissemination of virus from the mid-gut to other tissues after feeding. BEATY et al. (1982) employed heterologous reassortants to demonstrate that dissemination of California serogroup viruses from the mid-gut was determined by the M RNA. GONZALEZ-SCARANO et al. (1988) have shown further that separate domains of the M RNA determine ability to overcome the mid-gut barrier, on the one hand, and ability to multiply in thoracic organs, on the other. Since the M RNA encodes the viral membrane glycoproteins, it is likely that these tissue tropisms are receptor mediated, however the molecular mechanisms of these phenomena have not yet been explored.

4 Vaccine Development

Experimental inactivated and live virus vaccines have been developed to protect humans and domestic animals from the ravages of Rift Valley fever virus (genus *Phlebovirus*). The live virus vaccines have been developed by empirical routes (SMITHBURN 1949; MOUSSA et al. 1982, 1986) and by mutagen-directed attenuation (CAPLEN et al. 1985; ROSSI and TURELL 1988). The later involved passage of a wild-type strain ZH548 in MRC-5 cells in culture in the continued presence of 200 μg/ml 5-fluorouracil. A single plaque was picked at each cycle of mutagenesis to initiate the next cycle. Virus passage under this regimen became attenuated in terms of LD_{50} in mice, whereas virus passaged similarly in the absence of mutagen did not become attenuated. Reassortment experiments suggested that attenuating mutations were accumulating in all three genome sub-units. TAKEHARA et al. (1989) have sequenced the M RNA of the virulent parent and an attenuated clone from the 12th passage (M12). Coincidentally, there were 12 nucleotide substitutions giving rise to seven inferred amino acid changes, and a new upstream AUG start signal for the G_0 precursor was created. From these results it was estimated that mutations were accumulating at the rate of one substitution per 3885 nucleotides per cycle of mutagenization. The identity of the attenuating mutations will be difficult to determine, particularly as the M RNA of the ZH548 wild type strain, derived originally from a febrile human infection, differed by seven substitutions and three inferred amino acid changes from that of the ZH501 strain, another virulent strain isolated from a fatal haemorrhagic case in the same region of Egypt.

The question of the effect of heterologous reassortment on the efficacy of live virus vaccines has not yet been addressed.

5 Prospects and Priorities

Increasing knowledge of the nature of bunyaviruses indicates clearly that the viruses included in the family have the potential for rapid and explosive variation. Most research to date has concerned a very few viruses presumed to be representative of the individual genera. In future years less narrowly focussed research will be desirable to substantiate these current assumptions and to provide regular and systematic surveillance of those genera which include the more dangerous human and animal pathogens.

The molecular mechanisms underlying the assembly of the tripartite genome, sub-unit reassortment and restriction, tissue tropism and virulence in vertebrates are fundamental problems which should rank high in the order of future research priorities. As with all negative-stranded RNA genome viruses an ability to rescue in vitro generated mutations would greatly facilitate the analysis of gene function and open up new routes to vaccine development. The existence of a discrete group of non-arthropod transmitted viruses within this family of viruses, that are predominantly arthropod vector transmitted, provides an opportunity to define in molecular terms the requirements for vector transmission and alternate replication in vertebrate and invertebrate tissue.

The recent recognition of a bunyavirus (TSWV) replicating both in plants and arthropods re-emphasizes the diversity and evolutionary plasticity of the family Bunyaviridae. Indeed the origin, integrity, and evolution of the family Bunyaviridae now become matters of more than mere academic speculation and concern.

References

Beaty BJ, Bishop DHL (1989) Bunyavirus-vector interaction. Virus Res 10: 289–302

Beaty BJ, Rozhon EJ, Gensemer P, Bishop DHL (1981) Formation of reassortant bunyaviruses in dually infected mosquitoes. Virology 111: 662–665

Beaty BJ, Miller BR, Shope RE, Rozhon EJ, Bishop DHL (1982) Molecular basis of bunyavirus per os infection of mosquitoes: role of the middle-sized RNA segment. Proc Natl Acad Sci USA 79: 1295–1297

Beaty BJ, Sundin DR, Chandler LJ, Bishop DHL (1985) Evolution of bunyaviruses by genome reassortment in dually infected mosquitoes (*Aedes triseriatus*). Science 230: 548–550

Bellocq C, Raju R, Patterson JL, Kolakofsky D (1987) Translational requirement of La Crosse virus S-m RNA synthesis: in vitro studies. J Virol 61: 87–95

Bilsel PA, Tesh RB, Nichol ST (1988) RNA genome stability during serial transovarial transmission in the sandfly *Phlebotomus perniciosus*. Virus Res 11: 87–94

Bishop DHL (1979) Genetic potential of bunyaviruses. In: Arber W et al. (eds) Current topics in microbiology and immunology, vol 86. Springer, Berlin Heidelberg New York, pp 1–33

Bishop DHL (1986) Ambisense RNA genomes of arenaviruses and phleboviruses. Adv Virus Res 31: 1–51

Bishop DHL, Shope RE (1979) Bunyaviridae. In: Fraenkel-Conrat, Wagner RR (eds) Comprehensive virology, vol 14. Plenum, New York, pp 1–156

Bishop DHL, Calisher C, Casals J, Chmako MP, Gaidamovich SY, Hannoun C, Lvov DK, Marshall ID, Oker-Bolm NM, Pettersson RF, Porterfield JS, Russell PK, Shope RE, Westaway EG (1980) Bunyaviridae. Intervirology 18: 125–143

Bishop DHL, Fuller F, Akashi H, Beaty BJ, Shope RE (1984) The use of reassortant bunyaviruses to deduce their coding and pathogenic potentials. In: Kohn A, Fuchs P (eds) Mechanisms of viral pathogenesis: from gene to pathogen. Nijhoff, Boston, pp 49–60

Calisher CH, Karabatsos N (1988) Arbovirus serogroups: definition and geographic distribution. In: Monoth TP (ed) The arboviruses: epidemiology and ecology, vol 1. CRC Press, Boca Raton, pp 19–58

Calisher CH, Gutierrez E, Francy DB, Alava AA, Muth DJ, Laznick JS (1983) Identification of hitherto unrecognised arboviruses from Ecuador: members of serogroups B, C, Bunyamwera, Patois, and Minatitlian. Am J Trop Med Hyg 32: 877–885

Caplen H, Peters CJ, Bishop DHL (1985) Mutagen-directed attenuation of Rift Valley fever virus as a method for vaccine development. J Gen Virol 66: 2271–2277

Chandler LJ, Beaty BJ, Baldridge GD, Bishop DHL, Hewlett MJ (1990) Heterologous reassortment of bunyaviruses in *Aedes triseriatus* mosquitoes and transovarial and oral transmission of newly evolved genotypes. J Gen Virol 71: 1045–1050

Chen D, Burns JW, Estes MK, Ramig RF (1989) Phenotypes of roatavirus reassortants depend upon the recipient genetic background. Proc Natl Acad Sci USA 86: 3743–3747

Cunningham C, Szilagyi JF (1987) Viral RNAs synthesized in cells infected with Germiston bunyavirus. Virology 157: 431–439

David-West TS, Porterfield JS (1974) Dugbe virus: a tick-borne arbovirus from Nigeria. J Gen Virol 23: 297–307

de Haan P, Wagemakers L, Goldbach R, Peters D (1989) Tomato-spotted wilt virus, a new member of the Bunyaviridae? In: Kolakofsky D, Mahy BWJ (eds) Genetics and pathogenicity of negative strand RNA viruses. Elsevier, Amsterdam, pp 287–291

de Haan P. Wagemakers L, Peters D, Goldbach R (1990) The S RNA segment of tomato spotted wilt virus has ambisense character. J Gen Virol 71: 1001–1007

El Said LH, Vorndam V, Gentsch JR, Clewley JP, Calisher CH, Klimas RA, Thompson WH, Grayson M, Trent DW, Bishop DHL (1979) A comparison of La Crosse virus isolates obtained from different ecological niches and an analysis of the structural components of California encephalitis serogroup viruses and other bunyaviruses. Am J Trop Med Hyg 28: 364–386

Elliott RM (1989a) Nucleotide sequence analysis of the small (S) RNA segment of Bunyamwera virus, the prototype of the family Bunyaviridae. J Gen Virol 70: 1281–1285

Elliott RM (1989b) Nucleotide sequence analysis of the large (L) NA segment of Bunyamwera virus, the prototype of the family Bunyaviridae. Virology 173: 426–436

Elliott RM (1990) Molecular biology of the Bunyaviridae. J Gen Virol 71: 501–522

Elliot RM, Wilkie ML (1986) Persistent infection of Aedes albopictus C6/36 cells by Bunyamwera virus, Virology 150: 21–32

Elliott RM, Lees JF, Watret GE, Clark W, Pringle CR (1984) Genetic diversity of bunyaviruses and mechanisms of genetic variation. In: Kohn A, Fuchs P (eds) Mechanisms of viral pathogenesis: from gene to pathogen. Nijhoff, Boston, pp 61–76

Endres MJ, Jacoby DR, Janssen RS, Gonzalez-Scarano F, Nathanson N (1989) The large viral RNA segment of California serogroup bunyaviruses encodes the large viral protein. J Gen Virol 70: 223–228

Gahmberg N (1984) Characterization of two recombination-complementation groups of Uukuniemi virus temperature-sensitive mutants. J Gen Virol 65: 1079–1090

Gentsch JR, Bishop DHL (1976) Recombination and complementation between temperature-sensitive mutants of the bunyavirus snowshoe hare virus. J Virol 20: 351–354

Gentsch JR, Wynne LR, Clewley JP, Shope RE, Bishop DHL (1977) Formation of recombinants between snowshoe hare and La Crosse bunyaviruses. J Virol 24: 893–902

Gentsch JR, Robeson G, Bishop DHL (1979) Recombination between snowshoe hare and La Crosse bunyaviruses. J Virol: 31: 707–711

Gentsch JR, Rozhon EJ, Klimas RA, El Said LH, Shope RE, Bishop DHL (1980) Evidence from recombinant bunyavirus studies that the M RNA gene products elicit neutralising antibodies. Virology 102: 190–204

Gonzalez–Scarano F, Nathanson N (1989) Bunyaviruses. In: Fields BN et al. (eds) Virology. Raven, New York, pp 1195–1228

Gonzalez–Scarano F, Beaty B, Sundin D, Janssen R, Endres MJ, Nathanson N (1988) Genetic determinants of the virulence and infectivity of La Crosse virus. Microbial Pathogen 4: 1–7

Hampson J (1987) A study of the genetics and growth of a tripartite RNA virus using ts mutants. Ph D thesis, University of Warwick

Hunt AR, Calisher CH (1979) Relationships of Bunyamwera serogroup viruses by neutralisation. Am J Trop Med Hyg 28: 740–749

Iapalucci S, Lopez N, Rey O, Zakin MM, Cohen GN, Franze–Fernandez MT (1989) The 5′ region of Tacaribe virus virus L RNA encodes a protein with a potential metal binding domain. Virology 173: 357–361

Iroegbu CU (1981) A genetic study of the Bunyamwera complex of the genus *Bunyavirus* (family Bunyaviridae) Ph D thesis, University of Glasgow

Iroegbu CU, Pringle CR (1981a) Genetic interactions among viruses of the Bunyamwera complex. J Virol 37: 383–394

Iroegbu CU, Pringle CR (1981b) Genetics of the bunyamwera complex. In: Bishop DHL, Compans RW (eds) The replication of negative strand RNA viruses. Elsevier, New York, pp 158–163

Janssen R, Gonzalez–Scarano F, Nathanson N (1984) Mechanisms of bunyavirus virulence. Comparative pathogenesis of a virulent strain of La Crosse and an avirulent strain of Tahyna virus. Lab Invest 50: 447–455

Janssen RS, Nathanson N, Endres MJ, Gonzalez–Scarano F (1986) Virulence of La Crosse virus is under polygenic control. J Virol 59: 1–7

Kacsak RJ, Lyons MJ (1978) Bunyamwera virus. II. The generation and nature of defective interfering particles. Virology 98: 539–546

Karabatsos N (ed) (1985) International catalogue of arboviruses 1985 including certain other viruses of vertebrates. American Society of Tropical Medicine and Hygiene, San Antonio

Klimas RA, Thompson WH, Calisher, CH, Clark GG, Grimstad PR, Bishop DHL (1981) Genotypic varieties of La Crosse virus isolated from different geographic regions of the continental United States and evidence for a naturally occurring intertypic recombinant La Crosse virus. Am J Epidemiol 114: 112–131

Lees JF, Pringle CR, Elliott RM (1986) Nucleotide sequence of the Bunyamwera virus M RNA segment: conservation of structural features in the bunyavirus glycoprotein gene product. Virology 148: 1–14

Lubeck MD, Palese P, Schulman JL (1979) Nonrandom association of parental genes in influenza A virus recombinants. Virology 95: 269–274

Malkova D, Marhoul Z (1976) Influence of temperature corresponding to that of the vector of Tahyna virus. Acta Virol 20: 494–498

Matsuoka, Ihara T, Bishop DHL, Compans RW (1988) Intracellular accumulation of Punta Toro virus glycoprotein expressed from cloned cDNA. Virology 167: 251–260

Millican D, Porterfield JS (1982) Relationship between glycoproteins of the viral envelope of bunyaviruses and antibody-dependent plaque enhancement. J Gen Virol 63: 233–236

Moussa MI, Wood OL, Abdel–Waheb ESE (1982) Reduced pathogenicity associated with a small plaque variant of the Egyptian strain of Rift Valley fever virus (ZH501). Trans R Soc Trop Med Hyg 76: 482–486

Moussa MJ, Abdul–Wahal KSE, Wood OL (1986) Experimental infection and protection of lambs with a minute plague variant of Rift Valley Fever virus. Am J Trop Med Hyg 35: 660–662

Murphy J, Pringle CR (1987) Bunyavirus mutants: reassortment group assignment and G1 protein variants. In: Mahy BWJ, Kolakofsky D (eds) The biology of negative strand viruses. Elsevier, Amsterdam, pp 357–362

Olmsted RA, Baric RS, Sawyer RA, Johnston RE (1984) Sindbis virus mutants selected for rapid growth in cell culture display attenuated virulence in animals. Science 225: 424–442

Ozden S, Hannoun C (1978) Isolation and preliminary characterization of temperature-sensitive mutants of Lumbo virus. Virology 84: 210–212

Ozden S, Hannoun C (1980) Biochemical and genetic characterization of Germiston virus. Virology 103: 232–234

Pensiero MN, Jennings GB, Schmaljohn CS, Hay J (1988) Expression of the Hantaan virus M genome segment by using a vaccinia virus recombinant. J Virol 62: 696–702

Pringle CR (1987) Rhabdovirus genetics. In: Wagner RR (ed) The rhabdoviruses. Plenum, New York, pp 167–244

Pringle CR (1990) The Genetics of paramyxoviruses. In: Kingsbury DW (ed) The paramyxoviruses. Plenum, New York (in press)

Pringle CR, Iroegbu CU (1982) A mutant identifying a third recombination group in a bunyavirus. J Virol 42: 873–879

Pringle CR, Clark W, Lees JF, Elliott RM (1984a) Restriction of sub-unit reassortment in the Bunyaviridae. In: Compans RW, Bishop DHL (eds) Segmented negative strand viruses. Academic, Orlando, pp 45–50

Pringle CR, Lees JF, Clark W, Elliott RM (1984b) Genome subunit reassortment among bunyaviruses analysed by dot hybridization using molecularly cloned complementary DNA probes. Virology 135: 244–256

Raju R, Kolakofsky D (1987) Unusual transcripts in La Crosse virus-infected cells and the site for nucleocapsid assembly. J Virol 61: 667–672

Raju R, Kolakofsky D (1989) The ends of La Crosse virus genome and antigenome RNAs within nucleocapsids are base paired. J Virol 63: 122–128

Raju R, Raju L, Kolakofsky D (1989) The translational requirement for complete La Crosse virus mRNA synthesis is cell-type dependent. J Virol 63: 5159–5163

Rossi CA, Turell MJ (1988) Characterization of attenuated strains of Rift Valley fever virus. J Gen Virol 98: 817–823

Rozhon EJ, Gensemer P, Shope RE, Bishop DHL (1981) Attenuation of virulence of a bunyavirus involving an L RNA defect and isolation of LAC/SSH/LAC and LAC/SSH/SSH reassortants. Virology 111: 125–138

Salvato MS (1989) The ambisense nature of the L RNA of lymphocytic choriomeningitis virus. In: Kolakofsky D, Mahy BWJ (eds) Genetics and pathogenicity of negative strand RNA viruses. Elsevier, Amsterdam, pp 168–173

Salvato MS, Shimomaye EM (1989) The completed sequence of lymphocytic choriomeningitits virus reveals a unique RNA structure and a gene for a zinc finger protein. Virology 173: 1–10

Schmaljohn CS, Dalrymple JH (1983) Analysis of Hantaan virus RNA: evidence for a new genus of Bunyaviridae. Virology 131: 482–491

Shope RE, Rozhon EJ, Bishop DHL (1981) Role of the bunyavirus middle-sized RNA segment in mouse virulence. Virology 114: 273–276

Simons JF, Hellman U, Pettersson RF (1990) Uukuniemi virus S RNA segment: ambisense coding strategy, packaging of complementary strands into virions, and homology to members of the genus *Phlebovirus*. J. Virol 64: 247–255

Smithburn KC (1949) Rift Valley fever: the neurotropic adaptation of the virus and the experimental use of this modified virus as a vaccine. Br J Exp Pathol 30: 1–16

Takehara K, Min MK, Battles JK, Sugiyama K, Emery VC, Dalrymple JM, Bishop DHL (1989) Identification of mutations in the M RNA of a candidate vaccine strain of Rift Valley fever virus. Virology 169: 452–457

Talmon Y, Prased BVV, Clerx JPM, Wang GJ, Chu W, Hewlett MJ (1987) Electron microscopy of vitrified-hydrated La Crosse virus. J Virol 61: 2319–2321

Ulmanen I, Seppala P, Pettersson RF (1981) In vitro translation of Uukuniemi virus-specific RNAs: identification of a non-structural protein and a precursor to the membrane glycoproteins. J Virol 37: 72–79

Ushijima H, Clerx-van Haaster CM, Bishop DHL (1981) Analyses of Patois group bunyaviruses: evidence for naturally occurring recombinant bunyaviruses and existence of immune precipitable and non-precipitable nonvirion proteins induced in bunyavirus-infected cells. Virology 110: 318–332

Verani P, Nicoletti L, Marchi A (1984) Establishment and maintenance of persistent infection by the *Phlebovirus* Toscana in Vero cells. J Gen Virol 65: 367–375

Whitman L, Shope RE (1962) The California complex of arthropod-borne viruses and its relationship to the Bunyamwera serogroup through Guaroa virus. Am J Trop Med Hyg 11: 691–696

Zeller HG, Karabatsos N, Calisher CH, Digoutte JP, Cropp CB, Murphy FA, Shope RE (1989) Electron microscopic and antigenic studies of uncharacterized viruses. II. Evidence suggesting the placement of viruses in the family Bunyaviridae. Arch Virol 108: 211–227

Bunyaviridae—Natural History

B. J. BEATY[1] and C. H. CALISHER[2]

1 Introduction

The purpose of this chapter is to introduce the reader to the remarkable natural histories of viruses in the family Bunyaviridae. This is not a trivial undertaking; there are more than 250 viruses in the family, including such significant human and veterinary pathogens as: Crimean-Congo hemorrhagic fever, Rift Valley fever, La Crosse, Hantaan, Oropouche, Akabane, and Nairobi sheep disease viruses (KARABATSOS 1985). To describe the natural history of each virus in the family is beyond the scope of this chapter. Each has a unique cycle in nature, many have sylvatic and epidemic cycles, and, for most, little is known about

[1] Arthropod-borne and Infectious Diseases Laboratory Colorado State University, Fort Collins, CO 80523, USA
[2] W.H.O. Collaborating Centre for Arbovirus Reference and Research, Division of Vector-borne Infectious Diseases, Centers for Disease Control, US Public Health Service, Department of Health and Human Services, P.O. Box 2087, Fort Collins, CO 80522, USA

Current Topics in Microbiology and Immunology, Vol. 169

their natural history. Even to describe in detail those cycles that are well defined would be impractical in the space allotted. Excellent reviews of the natural histories of viruses in the family Bunyaviridae or specific groups of viruses in the family have been published. Notable are the series of in depth review chapters on specific groups of the family Bunyaviridae in *The Arboviruses: Epidemiology and Ecology* (MONATH 1988) and the books *The Arthropod-borne Viruses of Vertebrates* (THEILER and DOWNS 1973) and *California Serogroup Viruses* (CALISHER and THOMPSON 1983). More general reviews include those by GONZALEZ-SCARANO and NATHANSON (1990), SHOPE (1985), and BISHOP and SHOPE (1979). Many monographs and papers describing in detail the natural history of specific viruses are also referenced in this chapter. Finally, the source for much of the information presented is the *International Catalogue of Arboviruses* (KARABATSOS 1985), which is an invaluable resource for information concerning the arthropod-borne viruses. In many instances, these review articles are cited here for pertinent information. This is in no way meant to diminish the contribution of the many individual scientists who have contributed to our knowledge of the Bunyaviridae; rather, it is simply an attempt to limit the references to a reasonable number.

This chapter has been organized to provide the reader first with an introduction to the family Bunyaviridae, to arthropod-borne viruses, and to Bunyaviridae, cycles for viral maintenance and amplification in nature. Determinants of Bunyaviridae cycle integrity and evolution of Bunyaviridae are then described. Finally, the geographic distribution, known vector and vertebrate hosts and other pertinent information are then presented for most of the viruses in the respective serogroups in the family. More detailed natural histories are presented for selected viruses.

2 Bunyaviridae

The number of viruses in the Bunyaviridae attests to the evolutionary potential of the family; more than 250 viruses are now recognized (Table 1; KARABATSOS 1985). Currently, five genera are recognized in the family; *Bunyavirus*, *Hantavirus*, *Nairovirus*, *Phlebovirus*, and *Uukuvirus*. However, the Bunyaviridae Study Group of the International Committee on the Taxonomy of Viruses (ICTV) has recommended changes in the organization of the family. Because of recently obtained biochemical and molecular biological evidence, the Bunyaviridae Study Group recommended that phleboviruses and uukuviruses be combined into one genus, *Phlebovirus*. In addition, a recommendation has been made to the ICTV to establish a new genus, *Tospovirus*, in the family Bunyaviridae (CALISHER et al. 1990). This genus contains one member, tomato spotted wilt virus, which is transmitted between plants by thrips, plant-feeding arthropods (MILNE and FRANCKI 1984; DE HAAN et al. 1989). In anticipation that the ICTV will

Table 1. Family Bunyaviridae[a]

Genus	Antigenic group	Viruses[b]	Selected significant human/ veterinary pathogens
Bunyavirus	Anopheles A	5	
	Anopheles B	2	
	Bakau	5	
	Bunyamwera	25	
	Bwamba	2	
	Group C	12	
	California	13	La Crosse, Tahyna
	Capim	8	
	Gamboa	5	
	Guama	12	
	Koongol	2	
	Minatitlan	2	
	Nyando	2	
	Olifantsvlei	5	
	Patois	7	
	Simbu	21	Akabane, Oropouche
	Tete	6	
	Turlock	4	
	Ungrouped	3	
Nairovirus	CCHF	4	Crimean–Congo hemorrhagic fever
	Dera Ghazi Khan	5	
	Hughes	4	
	NSD	3	Nairobi sheep disease
	Qalyub	3	
	Sakhalin	5	
Phlebovirus	Phlebotomus fever group	37	Sandfly fever (Naples, Sicilian), Rift Valley fever
	Uukuniemi group	6	
Hantavirus	Hantaan	5	Hantaan, Puumala, Seoul
Tospovirus	Tomato spotted wilt	1	Tomato spotted wilt virus
Unassigned viruses	Bhanja	3	
	Kaisodi	3	
	Mapputta	4	
	Resistencia	3	
	Tanga	2	
	Thiafora	2	
	Upolu	2	
	Yogue	2	
	Ungrouped	21	

[a] Other viruses not listed in the arbovirus catalogue are recognized, however for the sake of simplicity and because many of them are not well-characterized, they are not included in the tables
[b] May include viruses not listed in the catalogue

accept these recommendations, the following five genera are recognized in this chapter: *Bunyavirus*, *Hantavirus*, *Nairovirus*, *Phlebovirus*, and *Tospovirus* (CALISHER et al. 1990).

In this chapter, the term Bunyaviridae is used when referring generically to all of the viruses in the family Bunyaviridae. When referring to members of a particular genus, the terms bunyaviruses, hantaviruses, nairoviruses, phleboviruses, and tospovirus, respectively, are used.

Viruses in the family Bunyaviridae are named typically for the geographic locale of the original isolation. The result is viruses with sometimes exotic and unfamiliar names that do little to reveal relationships between them (CALISHER 1988). However, many of the names are memorable and do provide some information concerning the natural history of the virus.

Originally, the Bunyaviridae were taxonomically categorized on the basis of the results of serologic tests using polyclonal antibody preparations (CALISHER 1988). Historically, the basis for serotype-specific classification is the neutralization (NT) test, and where appropriate the hemagglutination Inhibition (HI) test; both tests assay principally the virion surface glycoprotein G1. The basis for serogroup-specific classification is the complement fixation (CF) test; the N protein is a principal antigen involved in the CF test (CALISHER 1979, 1988). Because they are not exposed on the surface of the virion, N protein epitopes are typically more conserved than G1 epitopes.

Determination of antigenic relationships between viruses resulted in classification of viruses in the family Bunyaviridae into supergroups, serogroups, and ungrouped viruses. The validity of these groupings has subsequently been confirmed by physicochemical, morphological, molecular, and genetic analyses (BISHOP 1990; GONZALEZ-SCARANO and NATHANSON 1990; SCHMALJOHN and PATTERSON 1990; ELLIOTT 1990). The taxonomic relationships of viruses in the family Bunyaviridae are reviewed more or less annually by the Subcommittee on Interrelationships among Catalogued Arboviruses of the American Committee on Arthropod-borne Viruses and periodically by the Bunyaviridae Study Group of the ICTV. The current recommendations of the Bunyaviridae Study Group for classification of the viruses in the family Bunyaviridae are presented in Table 1.

3 Arthropod-borne Viruses

Historically, viruses in the family Bunyaviridae were classified as arthropod-borne (arbo-) viruses. Arboviruses are maintained in nature principally through biological transmission between susceptible vertebrate hosts by hematophagous arthropods. However, in 1987 the genus *Hantavirus*, which contains rodent-borne viruses that are not transmitted by arthropods, was included in the family (SCHMALJOHN and DALRYMPLE 1983). In addition, tospoviruses are not transmitted to vertebrate hosts. Thus, viruses in two genera are not true arboviruses. Nonetheless, the vast majority of viruses in the family are arthropod-borne, and an introduction to the general concepts involving the natural history of arboviruses follows.

Typically an arbovirus is maintained in nature in a distinct cycle involving a limited number of, or a preferred, arthropod vector(s) and vertebrate host(s). Thus, arboviral genes must be expressed in markedly different biological and biochemical systems of the vertebrate and invertebrate hosts. Replication of Bunyaviridae in invertebrate cells is not typically associated with cytopathic effect. Arthropod vectors typically become persistently infected with Bunyaviridae, and no major untoward effects have been noted. Indeed, certain members of the Bunyaviridae have been demonstrated to be very efficiently transmitted venereally and transovarially, denoting a truly remarkable host-parasite relationship (see below).

In contrast, Bunyaviridae replication in vertebrate cells typically causes cytopathic effects (CPE), and infections of tangential vertebrate hosts may be associated with significant morbidity, including fever, encephalitis, hemorrhagic fever, and mortality. Infection of pregnant animals can result in teratogenesis and abortion. Hantaviruses are noteworthy because they do not cause cytopathic effects in vertebrate cells.

More than 530 viruses are now registered in the *International Catalogue of Arboviruses* (KARABATSOS 1985). Most are in the families Togaviridae, Flaviviridae, Reoviridae, Rhabdoviridae, and Bunyaviridae. The only DNA virus unequivocally considered to be an arbovirus is African swine fever virus (KARABATSOS 1985). The genomic plasticity of RNA viruses may permit continuous bridging of the phylogenetic gap separating vertebrate and invertebrate hosts (BEATY et al. 1988). It is remarkable that so many virus families are maintained in such complex, seemingly tenuous cycles. Clearly, factors associated with vector transmission (dissemination in the environment, overwintering, amplification, etc.) compensate for biological and ecological impediments to arbovirus cycle maintenance and amplification.

4 Bunyaviridae Cycles

Bunyaviruses, nairoviruses, and phleboviruses are transmitted between susceptible vertebrate hosts by a wide variety of hematophagous vectors, including culicine and anopheline mosquitoes, argasid and ixodid ticks, biting midges, and sandflies. It is noteworthy that the virus genera are typically associated with a given vector taxon: bunyaviruses are transmitted principally by mosquitoes, nairoviruses by ticks, and phleboviruses by sandflies (with the exception of the uukuviruses, which are transmitted by ticks). Arboviruses in the family Bunyaviridae are associated with a wide variety of reservoir hosts, principally mammals (large and small) and birds. Taxonomic restrictions in vertebrate reservoir host utilization occur, especially for certain viral serogroups, but generally not to the extent noted for vector hosts.

As noted previously, hantaviruses and tospoviruses are not maintained in classical arbovirus cycles. Hantaviruses are maintained in rodent populations and have no known vectors. Tospoviruses are transmitted between plant hosts by thrips.

Arbovirus cycles can be viewed as dynamic interactions between vector, vertebrate host, and virus that serve to maintain and amplify the virus in nature (Fig. 1). Typically, these cycles occur in rural or suburban areas. Humans and domesticated animals are typically tangential or dead-end hosts, which become infected when they impinge upon the natural cycle. Only in rare instances do humans serve as amplifying hosts for the Bunyaviridae and then typically only during human epidemics. Thus, diseases caused by viruses in the family Bunyaviridae can be considered zoonoses.

Horizontal transmission is classically considered the principal mechanism for arbovirus amplification (Fig. 1). Conventionally, this is referred to as "transmission between susceptible vertebrate hosts by the vector." The newly infected vertebrate host develops a viremia and serves to infect subsequently feeding vectors, thereby amplifying the virus. Ingestion of virus by a nonsusceptible arthropod or transmission of virus to nonsusceptible vertebrate host is a dead-end for the virus. California group viruses can also be transmitted horizontally from male to female mosquitoes venereally (THOMPSON and BEATY 1977).

Vertical transmission is efficiently exploited for viral maintenance and amplification by viruses in the family Bunyaviridae. Vertical transmission can contribute to viral maintenance during adverse climatic conditions; virus can be

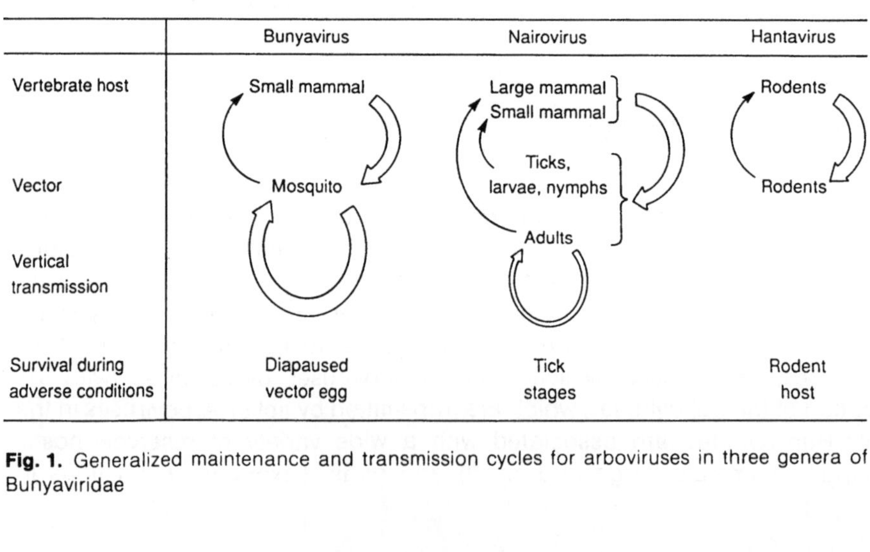

Fig. 1. Generalized maintenance and transmission cycles for arboviruses in three genera of the Bunyaviridae

maintained in life stages of vectors that persist through conditions such as winter and dry season. For example, most California group viruses overwinter in diapaused eggs of the vector (WATTS et al. 1974; BEATY and THOMPSON 1975), and infected mosquitoes emerge in the spring to initiate transmission cycles in forested areas. Vertical transmission can also contribute to amplication of virus; infected females that successfully complete gonadotrophic cycles can produce many infected progeny. This type of amplification may be important in areas where high antibody prevalence rates exist in preferred vertebrate hosts. Transmission of virus to a vertebrate host with antibody is a dead-end for the virus, thereby interrupt the amplification cycle.

The specificity or restricted host involvement of arbovirus cycles is remarkable; typically only one or a few species of vectors and vertebrate hosts are associated with amplification and transmission of the virus. Even in areas where multiple virus cycles intersect (Fig. 2), the integrity of the arbovirus cycle is retained. The specificity of these cycles is undoubtedly determined in part by biological and behavioral attributes of the vector (e.g., host feeding preference, diet and seasonal activity, etc.) and of the vertebrate host (e.g., geographic distribution, defensive behavior, etc.) Molecular and anatomic determinants may also function to maintain arbovirus cycle integrity (see below).

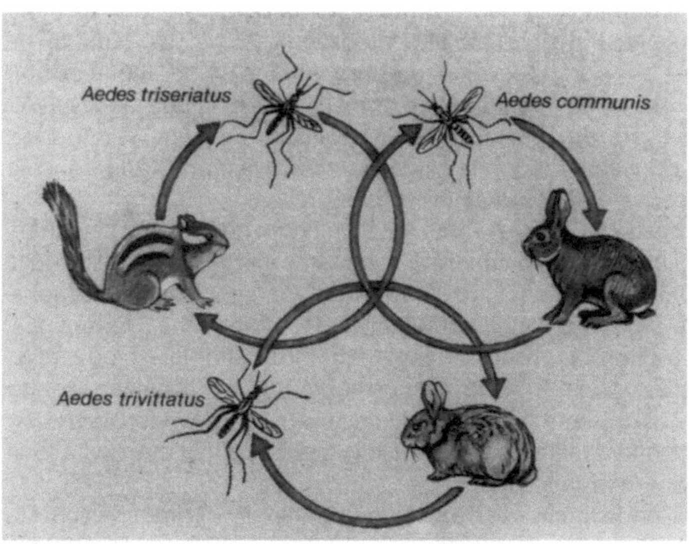

Fig. 2. Sympatric nature of certain California group viruses in nature. These three arbovirus cycles (La Crosse virus: *Aedes triseriatus* and chipmunks; snowshoe hare virus: *Aedes communis* and showshoe hares; trivittatus virus: *Aedes trivittatus* and cottontail rabbits) overlap in portions of their respective ranges. Determinants that maintain the integrity of each cycle in such circumstance are poorly understood

4.1 Vertebrate Host–Virus Interactions

The pathogenesis of viruses of the family Bunyaviridae in vertebrates is discussed elsewhere in this volume. Frequently, discussion of pathogenesis in vertebrate hosts focuses on the pathologic consequences of viral infection of tangential rather than infection of reservoir hosts (PARSONSON and McPHEE 1985). Reservoir hosts are typically not untowardly affected by arboviral infection, presumably because they have evolved a relative benign host–parasite relationship over time. In contrast, tangential hosts may experience considerable morbidity and mortality when infected with the virus. Such infections frequently occur in the context of epidemics in human or large animal populations.

Reservoir hosts must develop a viremia of sufficient amplitude to infect vectors. The potential for vertebrates to serve as reservoirs has frequently been assessed in laboratory studies; the respective arbovirus is inoculated into the putative host, and blood is sampled periodically for virus. If viremia is not detected, the animal is considered to be an improbable host for the arbovirus. However, the recent demonstration of "non-viremic" transmission of an arbovirus by ticks (JONES et al. 1987) suggests that, at least for infection of tick vectors, detectable viremia may not be necessary.

It is noteworthy that small mammals and birds frequently serve as reservoir hosts for viruses that are transmitted by dipteran and other short-lived vectors (KARABATSOS 1985). The fecundity of small mammals and rodents ensures that a new generation of susceptible vertebrate hosts will be available one or more times a year. Thus, viral transmission and amplification are not likely to be interrupted by high antibody prevalence (herd immunity) in the reservoir population. In contrast, large mammals are sometimes implicated as reservoir hosts for viruses that are transmitted by ticks (KARABATSOS 1985). Long-lived argasid and ixodid ticks could theoretically outlast herd immunity and transmit the virus when susceptible vertebrates are again present.

Many viruses of the family Bunyaviridae are significant human and veterinary pathogens and may cause a wide variety of symptoms in infected hosts (SHOPE 1985). Clinical manifestations include: inapparent or subclinical infections; febrile, self-limiting illnesses (e.g., Bunyamwera, Oropouche, Group C); encephalitis (e.g., California group viruses); abortion and teratology (e.g., Akabane, Rift Valley fever—RVF); hemorrhagic fever (Crimean-Congo hemorragic fever—CCHF), and hemorrhagic fever with renal syndrome (HFRS, Hantaan). The pathogenic mechanisms causing these disease manifestations are discussed elsewhere in this volume.

As noted, viruses in the family are transmitted principally between vertebrate hosts by arthropod vectors; hantaviruses are notable exceptions. Human-to-human or vertebrate reservoir host-to-human transmission is rare for most Bunyaviridae, again with the exception of the hantaviruses. However, nosocomial and iatrogenic infections are relatively common with three important pathogens: CCHF, RVF, and Hantaan viruses. Many cases of nosocomial infections have been documented in medical personnel providing care to index CCHF cases, and these secondary cases are frequently severe (WATTS et al. 1988; GONZALEZ-SCARANO and NATHANSON 1990). Infections with RVF virus occur in persons in direct contact with animals infected with the virus, which is clearly an important mode of transmission during epidemics (MEEGAN and BAILEY 1988). In both CCHF and RVF, human infections occur because of direct contact with blood and other tissues of the infected animals. Hantavirus infection of laboratory workers has been documented, and in some instances the number of cases can be substantial. In one outbreak in the Soviet Union, over 100 cases of HFRS were attributed to exposure to aerosolized animal excreta (CALISHER and SHOPE 1988).

4.2 Vector–Virus Interactions

Arboviruses are transmitted either biologically or mechanically by vectors.

Biological Versus Mechanical Transmission. For biological transmission, the vector ingests the virus when feeding upon a viremic vertebrate host. The vector then becomes infected with the virus, which replicates in vector tissues and organs. After the extrinsic incubation period (see below), the virus can be transmitted when the vector seeks another blood meal. In terms of the natural history of these viruses, the vector cannot biologically transmit the virus immediately; an extrinsic incubation period is required. However, once infected, the vector is infected for life and typically is capable of transmitting the virus for life. Arboviruses are maintained and amplified in nature by biological transmission (TURELL 1988).

In mechanical transmission, vectors need not become infected with the virus. Vector mouth parts become contaminated while feeding upon a viremic host, and if the vector feeds upon another host relatively promptly, it can transmit the virus. Mechanical transmission can occur only for a short period of time, i.e., the amount of time required before virus is inactivated on the mouth parts. Obviously, the more stable the virus, the longer the potential period of mechanical transmission. The potential for mechanical transmission is also determined in large part by the titer of viremia in the vertebrate host; those with high-titer plasma viremias are more likely to serve as hosts for mechanical transmission. In terms of the natural history of the Bunyaviridae, the vectors are capable of mechanical transmission of viruses immediately; an extrinsic incubation period is not required (TURELL 1988). Thus, virus can be rapidly amplified in nature. In major epizoodemics of RVF, virus is undoubtedly amplified by mechanical transmission because of the high-titer viremia in vertebrate hosts (MEEGAN and BAILEY 1988).

Arbovirus–Vector Interactions. The "black box" understanding of vector and arbovirus interactions, in which virus goes into the vector and subsequently comes back out, has been replaced by the realization that vector–virus interactions are quite complex and involve multiple infection and dissemination barriers (HARDY 1988; KRAMER et al. 1981; BEATY and BISHOP 1988). The normal sequence of events in the extrinsic incubation period is: (1) infection of vector midgut cells, (2) replication in these cells, (3) escape into the hemolymph, and (4) dissemination to an infection of epidemiologically significant target organs. Infection of salivary gland cells permits subsequent transmission to vertebrates, and infection of ovaries permits transovarial transmission to progeny. The shorter this extrinsic incubation period, the greater is the likelihood of viral transmission. With most arthropod-borne viruses, the duration of the extrinsic incubation period is a function of temperature (TURELL 1989). Increased temperature reduces the extrinsic incubation period. Vectors with short extrinsic incubation periods are typically considered to be vector competent.

The accepted view has been that arbovirus infection is not detrimental to natural vectors, in contrast to the sometimes dramatic effects on vertebrate hosts. However, recent evidence has demonstrated that there may be some effect associated with arboviral infection of vectors (BEATY et al. 1980; GRIMSTAD et al. 1980; WEAVER et al. 1988).

As noted previously, arboviruses are typically maintained in specific cycles involving a restricted number of vector(s) and vertebrate host(s). The permissiveness of vector hosts for viruses may be a major determinant of this specificity. One of the fundamental observations in arbovirology was the differential susceptibility of potential vectors to different viruses (MITCHELL 1983; HARDY 1988). Both intra- and interspecific differences in vector competence occur. There have been many reports of a midgut barrier in mosquitoes and ticks. The arthropod ingests the virus in the blood meal, but midgut cells do not become infected (HARDY et al. 1983). In many instances, these refractory arthropods can be infected by intrathoracic inoculation, which bypasses the midgut. The midgut barrier can be breached by mechanical means; piercing the midgut with a needle permits efficient infection of previously resistant mosquitoes (HARDY et al. 1983). The midgut barrier may also be breached by biological means; ingestion of filariae, which penetrate midguts, increases the susceptibility of mosquitoes for RVF virus (TURELL et al. 1984). Typically there is a threshold for midgut infection; when the amount of virus in an ingested blood meal is of sufficient titer, most of the arthropods become infected. The more susceptible the vector, the greater its competence to transmit virus (MITCHELL 1983). The mechanisms involved in the midgut barrier are not known but may involve such phenomena as: immobilization of virus in the peritrophic membrane, lack of midgut receptors, improper proteolytic processing of viral proteins, proteolytic inactivation of virus in the nonpreferred arthropod host, inability to fuse with midgut cells, etc. (HARDY et al. 1983; TURELL 1988; BEATY and BISHOP 1988).

The virogenesis or mode of development of some of the Bunyaviridae in their respective vector species has been determined. Two such systems are discussed below as generalized examples of virus–vector interactions: La Crosse (LAC) virus in *Aedes triseriatus* mosquitoes and Dugbe (DUG) virus in *Amblyomma variegatum* ticks.

LAC in Aedes triseriatus. This system has been extensively studied in both natural and laboratory settings. This mosquito is an efficient oral (WATTS et al. 1972), transovarial (PANTUWANTANA et al. 1974; WATTS et al. 1973), and venereal transmitter of LAC virus (THOMPSON and BEATY 1977, 1978). In temperate regions of the United States, the virus overwinters in diapaused eggs of the vector (WATTS et al. 1974; BEATY and THOMPSON 1975). Vertical transmission of the virus does not result in reduced fecundity of females or teratology in the offspring (MILLER et al. 1977), and rates of transovarial transmission (proportion of females transmitting to progeny) and filial infection (proportion of progeny infected) can approach 80%. Arboviruses in the families Flaviviridae and Rhabdoviridae have also been demonstrated to be vertically transmitted, but not with the efficiency of viruses in the family Bunyaviridae (TURELL and LEDUC 1983; TURELL 1988).

Immunofluorescence techniques were used to delineate anatomic bases of LAC virus development in and transmission by *Aedes triseriatus* (BEATY and THOMPSON 1978). Subsequent to oral infection, viral antigen was first detected in the cuboidal epithelial cells in the pyloric portion of the midgut (6 days postinfection). By 10 days postinfection, virus had disseminated from the midgut, and antigen was detected in most secondary organ systems, including fat body, pericardial cells, and ovarian sheath at approximately the same time, suggesting that the virus was disseminated via the hemolymph. Antigen was subsequently detected in the salivary glands. Viral infection was virtually pantropic in metabolically active organs of the arthropod, with the notable exception of thoracic muscle, and most organ systems contained large quantities of virus antigen.

Detection of viral antigen in ovarian follicles and isolation of virus from surface-sterilized eggs demonstrated that vertical transmission was indeed transovarial and not transovum (viral contamination of the surface of the eggs). Viral antigen was detected in most organ systems of first instar larvae, and viral titer increased with development through the instars (BEATY and THOMPSON 1976). Viral titers decreased during metamorphosis, when many tissues are histolyzed, and increased in adults. Viral antigen was detected in the salivary glands of newly emerged adults, indicating that the females were capable of transmission of virus upon emergence.

During studies of transovarially infected mosquitoes, substantial viral antigen was detected in the accessory sex glands of transovarially infected male mosquitoes. Immunofluorescent techniques revealed viral antigen in many portions of the male reproductive tract, including fat body surrounding the testes, vas, seminal vesicles, and accessory sex glands. Indeed, substantial amounts of viral antigen were detected in fluids extruded from the accessory sex glands, but viral antigen was not detected in or attached to sperm. In addition, substantial amounts of viral antigen were detected in the bursa copulatrix of females immediately after mating (THOMPSON and BEATY 1977). Venereal transmission rates were subsequently demonstrated to be dependent upon the physiological history of the female mosquito. Rates were higher in those females that had previously ingested a blood meal than in those that had not. The physiological or anatomical basis for this phenomenon remains to be determined (PATRICAN and DEFOLIART 1987).

Aedes triseriatus mosquitoes from different geographic origins differ in their vector competence for LAC virus (GRIMSTAD et al. 1977). Frequently, the difference in vector permissiveness to a productive viral infection is associated with a midgut dissemination barrier (BEATY et al. 1982; BEATY and BISHOP 1988; PAULSON et al. 1989; PAULSON and GRIMSTAD 1989). Virus infects mosquito midgut cells, and transcription and translation of at least certain genes occur. However, dissemination of virus from midgut cells to infect secondary target organs is restricted. Salivary gland and ovarian transmission barriers also have been reported (BEATY et al. 1981; GRIMSTAD 1988). There is little information concerning the anatomical and molecular basis of these events.

LAC virus overwinters efficiently in the diapaused eggs of the vector mosquito (WATTS et al. 1973; BEATY and THOMPSON 1975). In an intensive study of LAC virus overwintering near La Crosse, Wisconsin, virus was isolated from larvae collected from tree holes in each of four study areas before the seasonal emergence of adults (BEATY and THOMPSON 1975). Furthermore, infected larvae could be detected in tree holes that had been screened shut 2 years previously. Thus, installment hatching of the vector eggs could serve as a mechanism to maintain viral persistence in natural areas without transmission for several years. Many other California group viruses are transovarially transmitted efficiently, and presumably these viruses overwinter in a similar manner. Vertical transmission of arboviruses has been the subject of several excellent reviews (TURELL 1988; GRIMSTAD 1988).

DUG in Amblyomma variegatum *Ticks.* DUG virus is common in parts of Africa and has been isolated from ticks and cattle. DUG is a nairovirus related to CCHF virus. Because CCHF virus is a significant pathogen of humans (HOOGSTRAAL 1979; WATTS et al. 1988), and DUG virus is not (SWEET and COATES 1990). DUG virus has been used to investigate nairovirus–tick interactions.

Tissue tropisms of DUG virus in *Amblyomma variegatum* have been determined histochemically and by viral titration in both inoculated and orally infected ticks (BOOTH et al. 1990). DUG antigen is detected primarily in the epidermis, hemocytes, and digestive cells in the gut lumen. Virus disseminates to infect target organs via hemocytes and is not detected free in the hemolymph. Hemocytes also are the principal site of transstadial maintenance of DUG virus. Infection of salivary gland cells is enhanced by ingestion of a blood meal.

Comparative studies have demonstrated that *Amblyomma variegatum* is a competent vector of DUG virus, while a tick with a similar African distribution, *Rhipicephalus appendiculatus*, is not (STEELE and NUTTALL 1989). Both ticks become infected after ingestion of a blood meal; however, in *Rhipicephalus appendiculatus* ticks the viral infection is apparently restricted to the midgut.

Virus could not be detected after moulting. Apparently there is a midgut dissemination barrier for nairoviruses in ticks, as was described for bunyaviruses in mosquitoes.

DUG virus has been isolated from eggs of *Amblyomma variegatum* (CORNET et al. 1987; GONZALEZ et al. 1989). Thus, nairoviruses can be vertically transmitted in their tick vectors.

4.3 Viral Genetic Determinants of Virus–Vector Interactions

A genetic approach, using parent and progeny reassortant viruses, was used to elucidate the relationships between gene structure and biological function of bunyaviruses. LAC, snowshoe hare (SSH), and LAC-SSH reassortant viruses were used to investigate the molecular basis of productive infections of *Aedes triseriatus* (BEATY et al. 1981, 1982; BEATY and BISHOP 1988). These two viruses are maintained in distinct cycles in nature, involving different vectors and vertebrate hosts, even in areas where the two viruses are sympatric (Fig. 2). Thus, reassortant viruses were used to determine viral genetic determinants of vector infection and transmission in preferred and nonpreferred vertebrate hosts.

Aedes triseriatus mosquitoes ingested blood meals containing equivalent titers of the respective parent or reassortant virus. After 14 days of extrinsic incubation, mosquitoes were assayed for viral antigen by immunofluorescence. Detection of viral antigen in the midgut revealed that the respective virus had infected midgut cells. Detection of viral antigen in head tissues revealed that virus had disseminated from infected midgut cells and had infected secondary target organs. All the viruses, irrespective of genotype, infected the mosquito midgut cells (Table 2). However, viruses containing the middle-sized (M) RNA segment from the LAC parent efficiently escaped from midgut cells and infected secondary target organs. Similarly, efficient transmission of bunyaviruses was determined principally by the parental origin of the M RNA segment. After 14 days of extrinsic incubation, 93% of the mosquitoes infected with viruses containing the M RNA from the LAC parent transmitted virus, whereas only 35% of those infected with viruses containing the M RNA from the SSH parent transmitted (Table 2).

Thus, the M RNA segment is a major determinant of dissemination of LAC virus from infected *Aedes triseriatus* mosquito midgut cells to infect secondary target organs and of oral transmission of the virus. Studies using the natural vector of SSH virus are yet to be conducted. Such studies would reveal whether these functions resulted from an evolved vector–virus relationship or are due to enhanced infectivity of the gene products of the LAC M RNA or reduced infectivity of the gene products of the SSH RNA.

Because the M RNA codes for three gene products, an alternative approach was used to determine which protein is responsible for midgut infection. A monoclonal antibody escape variant of LAC virus that was markedly reduced in fusion function and was not neuroinvasive in mice (GONZALEZ-SCARANO et al.

Table 2. Infection, dissemination, and transmission of La Crosse, snowshoe hare, and reassortant viruses in *Aedes triseriatus*[a]

Virus genotype[b]	Percentage infected	Percentage mosquitoes disseminated	Percentage transmitted
LAC/LAC/LAC SSH/LAC/LAC SSH/LAC/SSH LAC/LAC/SSH	98 (115/117)[c]	98 (113/115)[d]	93 (126/136)[e]
SSH/SSH/LAC LAC/SSH/LAC LAC/SSH/SSH SSH/SSH/SSH	92 (92/100)	26 (24/92)	35 (36/104)

[a] Results adapted from Beaty et al. 1981, 1982; reproduced from Beaty and Bishop 1988, with permission
[b] Genotype of parent and reassortant viruses is presented according to the parental origin of the L/M/S RNA segments
[c] Number of mosquitoes with detectable viral antigen in midgut cells/number examined
[d] Number of mosquitoes with detectable viral antigen in headsquash tissues/number examined
[e] Number of mosquitoes that transmitted virus to mice/number with detectable viral antigen in salivary glands

1985) was investigated for its ability to infect mosquitoes. The variant (V22) was also restricted in its ability to infect mosquitoes (Sundin et al. 1987). The *Aedes triseriatus* infection rate was 74% for the parent virus but only 5% for the variant virus (Table 3). Interestingly, the viruses were equally infectious for mosquitoes when intrathoracically inoculated. V22 virus was passaged in mosquitoes and then compared with the parent virus in its ability to infect mosquitoes. It was as infectious as the parent virus, was neutralized by the monoclonal antibody used to derive variant V22, and exhibited restored fusion function (Table 3). This evidence strongly suggests that the variant V22 had reverted at the specific epitope. This epitope is apparently crucial for mosquito infection.

A specific epitope has been associated with mosquito cell infectivity (Grady and Kinch 1985), but the specific role of the epitope remains to be determined. However, an attractive hypothesis is that since LAC virus is an enteric virus of the

Table 3. Phenotypic analyses of parent La Crosse virus, a monoclonal escape variant (V22), and a revertant[a]

Virus	*Aedes triseriatus* infection rate	NT titer	Fusion index (pH 6)
Parent LAC	74% (14/19)[b]	100[c]	>0.8
Variant V22	5% (4/47)	<10	0.2
Revertant V22	85% (50/59)	320	>0.6

[a] Adapted from Sundin et al. 1987; reproduced from Beaty and Bishop 1988, with permission
[b] Number infected/number examined
[c] Reciprocal NT titer using monoclonal antibody V22

vector, LAC glycoproteins may require proteolytic processing before full expression of infectivity in the midgut. The G1 glycoprotein is cleaved by proteolytic enzymes (KINGSFORD and HILL 1983; GONZALEZ-SCARANO 1985), and it is likely that viral proteins could be altered in the mosquito midgut. Hydrophobic amino acid sequences, which could mediate fusion with midgut cells, could be exposed by cleavage of G1. Alternatively, cleavage of G1 could expose G2 sequences (LUDWIG et al. 1989). The latter investigators have demonstrated that proteolytic processing of LAC virions with trypsin yields virions that bind to vector cells but not to vertebrate cells. They hypothesize that the proteolytic cleavage exposes G2 epitopes that can bind to receptors present on arthropod cells but missing on vertebrate cells. Definitive answers concerning the molecular basis of midgut infection should be forthcoming in the near future.

4.4 Evolution of *Bunyaviridae* in Arbovirus Cycles

Elucidation of the evolutionary potential of arboviruses is of more than academic significance. In addition to resulting in viruses with altered virulence and other characteristics, new viral phenotypes could be capable of infecting new vector species. Because vectors conduct natural and ongoing experiments in testing viruses in new systems (MATTINGLY 1969), new viruses could be introduced into new arbovirus cycles with potentially significant epidemiologic consequences (BEATY et al. 1988).

The plethora of viruses and the multiplicity of serogroups that comprise the Bunyaviridae suggest that viruses in the family possess considerable evolutionary potential. Arbovirus cycles provide multiple opportunities for evolution of the Bunyaviridae, both by genetic drift and genetic shift (RNA segment reassortment). For example, arboviruses can replicate in the vector throughout its life span, providing ample opportunity for intramolecular changes in the viral genome, such as point mutations, sequence deletions, inversions, etc. (genetic drift). Adult dipteran vectors may feed several times during their life span and may also be infected transovarially and venereally, allowing ample opportunity for dual infection and subsequent segment reassortment.

Chances of dual infection and segment reassortment or viral evolution via intramolecular changes may be enhanced in ticks. Ixodid (hard) ticks can live for many years and exhibit different host feeding preferences at different life stages. Therefore, the ticks may intersect different arbovirus cycles at the larval, nymph, and adult stages. Soft ticks, which are extremely long-lived, feed many times during their lifetimes, providing multiple opportunities for viral evolution through genetic shift. Because many arboviruses are sympatric in nature, dual infection with different members of the Bunyaviridae is likely. Indeed, several reassortant members of the Bunyaviridae have been isolated from nature (see below).

Genetic Drift. Long-term, persistent infection of vectors, and in some instances vertebrate hosts, might facilitate virus evolution (BEATY et al. 1988). Most of the life cycle of bunyaviruses, nairoviruses, and phleboviruses occurs in the vector host. In the case of transovarial transmission, the viruses may persist through generations of infected arthropods without being transmitted to a vertebrate host. Such circumstances should promote nucleotide sequence deletions, inversions, and point mutations, as has been documented to occur with other negative-strand RNA viruses (HOLLAND et al. 1982). The high base substitution error frequency (10^{-3}) exhibited by these viruses is presumably due to the accumulation of defective interfering RNA species, the poor fidelity of RNA replicative enzymes, and the lack of proofreading enzymes (HOLLAND et al. 1982). In this regard, no two isolates of LAC virus recovered from nature have identical genome sequences, as evidenced by difference in oligonucleotide fingerprints (BISHOP and BEATY 1988). This applies to viruses isolated from the same place at the same time, the same place at different times, and different places at different times. Low-passage SSH virus isolates from mosquitoes exhibit dramatically different fingerprints (BEATY, unpublished data). In contrast, SSH and LAC viruses derived from the original prototype virus and maintained by lytic passage in different laboratories for many years have identical oligonucleotide fingerprints.

Clearly, intramolecular changes occur in viruses in nature. However, laboratory studies involving long-term persistent infections of vectors with selected bunyaviruses suggest that changes do not occur (BILSEL et al. 1988; BALDRIDGE et al. 1989). In the latter studies, plaque-purified LAC virus was passaged transovarially for two generations and horizontally to mice during each generation. Virus was isolated at each stage and examined by oligonucleotide fingerprinting. The virus genome remained stable during these studies; the oligonuclotide fingerprints were identical (BALDRIDGE et al. 1989). However, the virus used was plaque purified and adapted to the laboratory systems, and the viruses were amplified in BHK-21 cells to prepare the viral RNA for fingerprinting. These are standard techniques for analysis of the arboviral genomes, but direct amplification of viral sequences by polymerase chain reaction could now be used to determine unequivocally the extent of intramolecular changes in portions of the viral genomes during laboratory passage or in natural circumstances.

Genetic Shift. The tripartite genome of the Bunyaviridae provides the opportunity for virus evolution via RNA segment reassortment during dual infection of cells. High-frequency reassortment of viruses in the family Bunyaviridae has been documented to occur in vitro, in vivo, and in nature (BISHOP and SHOPE 1979; BISHOP 1985a, 1990; BISHOP and BEATY 1988; BEATY and BISHOP 1988; PRINGLE et al. 1984).

Bunyavirus reassortment does occur in nature. Genotypic analyses of natural isolates of LAC virus suggest that RNA segment reassortment occurs between topotypes of LAC virus (KLIMAS et al. 1981), and Shark River and Pahayokee viruses are apparently reassortant viruses (BISHOP and SHOPE 1979).

The Group C viruses in the Utinga Forest near Belem, Brazil, are probably the best example of the potential for RNA segment reassortment in natural circumstances (SHOPE 1985). Seven Group C viruses have been isolated from arthropods and vertebrates in the forest (Fig. 3). These viruses segregate into three groups by HI and NT tests, which assay gene products of the M RNA segment (Fig. 3A). However, the viruses segregate into three alternate groups by CF tests, which assays principally the gene product of the S RNA segment. Thus, reassortment of RNA segments is apparently occurring between the Group C viruses in the forest (Fig. 3B). The natural history of these viruses in this restricted ecosystem has been summarized by SHOPE (1985). Caraparu and Itaqui viruses are transmitted by *Culex* (Melanoconion spp.) mosquitoes between spiny rats and rice rats at the forest floor. Apeu, Marituba, and Murutucu viruses infect canopy marsupials and primates. Apeu is transmitted by *Aedes* mosquitoes, and Marituba and Murutucu are transmitted by *Culex* mosquitoes. Although these viruses share the same confined tropical Utinga forest ecosystem, their preferred vertebrate and vector hosts serve to preserve arbovirus cycle integrity. Caraparu and Itaqui viruses share vector and vertebrate hosts and are identical by CF test. However, their M RNA segment gene products differ, and thus infection with one does not induce NT or HI antibodies to the other in the vertebrate host.

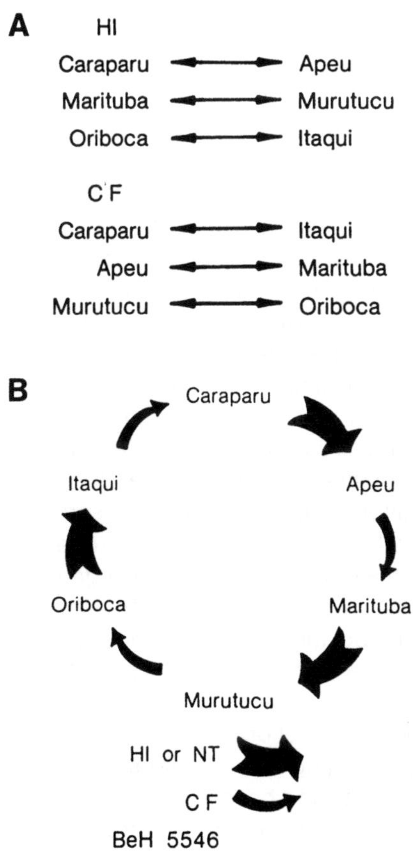

Fig. 3 A, B. Group C viruses in the Utinga Forest, Brazil. **A** Serologic relationships between group C viruses as determined by hemagglutination (*HI*) test or by complement fixation (*CF*) test. **B** Potential explanation for serologic relationships: RNA segment reassortment (see text)

Apeu and Caraparu viruses share NT or HI antigenic determinants. However, their vertebrate hosts differ and live in different zones of the forest. Thus, it is not likely that transmission of one virus is impeded by antibodies to the other. Natural reassortants of these viruses have been isolated; thus, either a vector or vertebrate host is occasionally dually infected with two Group C viruses, and a new viral genotype results. Survival of new genotypes depends upon whether or not they have a selective advantage in one or more components of the arbovirus cycle.

Reassortment is typically restricted to viruses from a given serogroup, which suggests that serogroups constitute gene pools that are evolving divergently (BISHOP 1985b; PRINGLE et al. 1984). Serologic relationships between groups—e.g., Guaroa virus, which is serologically related by NT test to California group viruses and by CF test to Bunyamwera group viruses—suggest that under certain circumstances reassortment can occur between groups (WHITMAN and SHOPE 1962).

In arbovirus cycles, either vertebrates or vectors could serve as sites for reassortment. However, infection of either host with two or more viruses is a prerequisite for segment reassortment. Because many Bunyaviridae cycles are sympatric, vectors that are catholic in feeding preferences could intersect multiple cycles and become infected with different viruses. Such vectors could also serve to infect vertebrate hosts with multiple viruses.

High-frequency reassortment of Bunyaviridae occurs in vitro in vertebrate cell cultures but has not been detected in vivo in vertebrates (BISHOP and BEATY 1988). Lack of reassortment may be due to the ephemeral nature of infection in most reservoir hosts, the relatively low-level viremia, and the production of antibody, all of which limit the probability that two viruses infect the same cell.

In contrast, high-frequency reassortment does occur in vectors, following intrathoracic or oral infection. Both homologous reassortment between temperature-sensitive mutants of LAC virus and heterologous reassortment between LAC and SSH viruses have been demonstrated (BEATY and BISHOP 1988). Interestingly, there is only a small window of time for dual oral infection to occur. Within 2–3 days after ingestion of one virus, mosquitoes become resistant to superinfection with a second, closely related virus (BEATY et al. 1985; SUNDIN and BEATY 1988). Interference obviously constrains the potential for dual infection of vectors. However, many mosquitoes, including *Aedes triseriatus*, exhibit a behavior called interrupted feeding, which would preclude interference. When defensive behavior of a host interrupts the mosquito during engorgement, the vector frequently engorges to repletion on an alternate vector (BEATY et al. 1988). Thus, the mosquito would ingest a blood meal from two hosts, each potentially viremic with a different virus, in a time frame that could permit dual infection and subsequent segment reassortment.

An alternate mechanism exists for dual infection: superinfection of transovarially infected mosquitoes. Females emerging from diapaused eggs in the spring could feed upon viremic vertebrate hosts and become dually infected. Because many members of the Bunyaviridae are efficiently transmitted transovarially, dual infection via this route may be more likely than dual oral infection. However, one would expect that interference to superinfection would occur in transovarially infected mosquitoes, which would reduce the probability of dual infection via this route. Interestingly, the nature of viral infection in transovarially infected mosquitoes differs from that in orally infected mosquitoes; titers are generally lower, and the two routes of infection result in differential CO_2 sensitivity (TURELL 1988). Concomitantly, transovarially infected vectors seem to be less resistant to superinfection; approximately 30% of transovarially infected mosquitoes can be superinfected (BEATY, unpublished data).

Reassortment of RNA segments would be epidemiologically significant only if the newly envolved genotypes were expressed in the arbovirus cycle. Transmission of newly evolved genotypes to susceptible vertebrate hosts has also been demonstrated (BEATY et al. 1985). In addition, heterologous reassortment of LAC and SSH bunyaviruses and subsequent transovarial transmission of the newly evolved genotypes was recently demonstrated in *Aedes triseriatus* mosquitoes (CHANDLER et al. 1990). Reassortment occurred in the ovaries of dually infected mosquitoes, and the relative frequency of reassortment was correlated with the number of blood meals ingested. Apparently, dual infection was enhanced by gonadotropic activity stimulated by the ingestion of the blood meal. Metabolically active and dividing cells in the oocyte might have been permissive for dual infection. Newly evolved viruses can be expressed in the arbovirus cycle both by horizontal and vertical transmission.

The actual potential for evolution of viruses in the family Bunyaviridae by segment reassortment and the phylogenetic constraints on reassortment in nature remain to be determined (BEATY and BISHOP 1988). Determination of the evolutionary potential of the Bunyaviridae via antigenic shift and drift may permit prediction of the epidemiologic consequences of these events.

5 Natural History of Bunyaviridae Viruses

This section presents the information that is known concerning the natural history of each virus in the family *Bunyaviridae*. The geographic distribution and the presumed vector(s) and vertebrate host(s) obtained from the *International Catalogue of Arboviruses* (KARABATSOS 1985) are presented in the tables for each genus and serogroup. More detailed information is presented for selected viruses that are of significant medical or veterinary importance.

The following abbreviations are used in the tables. Vectors: M mosquito, a anopheline, c culicine; T tick, i ixodid (hard ticks), a argasid (soft ticks); P phlebotomine (sandfly); C Culicoid (biting midge). Vertebrate hosts: Ba Bat, B bird, H human, M marsupial, P primate, R rodent, S sentinel animal, O other (e.g., bovine, goat, sheep, shrew, etc.). Detailed information about the vector and vertebrate host associations is presented in the International Catalogue of

Arthropod-borne Viruses (KARABATSOS (1985) and BISHOP and SHOPE (1979) and many of the other review articles noted previously. A few viruses listed in the tables have not been registered in the *Catalogue* (denoted by an asterisk in the tables). Listing of these viruses here is not intended to constitute priority of publication.

5.1 Genus *Bunyavirus*

Viruses belonging to the Bunyamwera and certain other serogroups were originally classified in a single supergroup because of antigenic interrelationships not only among viruses in each serogroup but also between one or more viruses in one group and one or more viruses in another group (Fig. 4). The former Bunyamwera supergroup now comprises the genus *Bunyavirus* (CALISHER 1988). More than 155 viruses are now recognized in the genus.

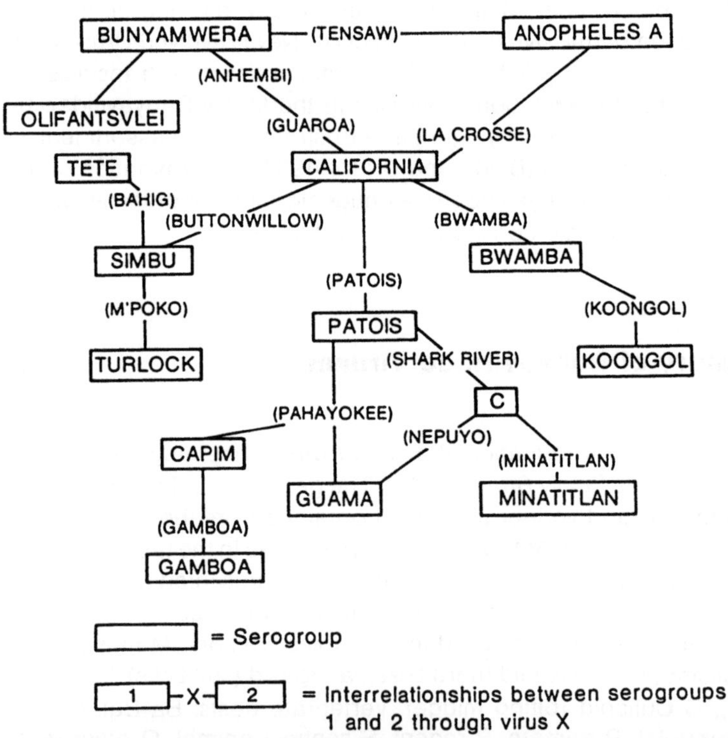

Fig. 4. Schematic representation of the antigenic relationships among viruses and serogroups within the family Bunyaviridae, genus Bunyavirus. (From CALISHER 1988, with permission)

Table 4. Anopheles A group viruses

Virus	Geographic distribution	Vector	Vertebrate
Anopheles A	South America	M a	
Las Maloyas	South America	M a	
Lukuni	South America	M a + c	
Tacaiuma	South America	Ma + c	H,P,S
Virgin River	North America	M c	

1. Members of the Anopheles A group viruses (Table 4) have been isolated from anopheline and culicine mosquitoes (KARABATSOS 1985). Tacaiuma virus has been isolated from a patient with a febrile illness and from a sentinel monkey (CAUSEY et al. 1961). Serologic surveys have detected viral antibodies in humans, domestic animals, birds, and rodents. Distribution of the viruses is in the New World. Serologic cross-reactions between Anopheles A and California group viruses have been detected (CALISHER et al. 1973).

Table 5. Anopheles B group viruses

Virus	Geographic distribution	Vector	Vertebrate
Anopheles B	South America	M a	
Boraceia	South America	M a + c	

2. Anopheles B viruses have been isolated only from mosquitoes collected in South America (ROCA-GARCIA 1944; Table 5). Little is known about the vertebrate hosts, and human infections apparently do not occur (KARABATSOS 1985).

Table 6. Bakau viruses

Virus	Geographic distribution	Vector	Vertebrate
Bakau	Asia	M c, T a	P
Ketapang	Asia	M c	
Nola	Africa	M c	
Tanjong Rabok	Asia		S
Telok Forest	Asia		P

3. Members of the recently recognized group of Bakau viruses have been isolated from mosquitoes, ticks, and monkeys in Asia and Africa (ZELLER et al. 1989; Table 6). Little is known of their natural histories.

Table 7. Bunyamwera group viruses

Virus	Geographic distribution	Vector	Vertebrate
Anhembi	South America	M c	R
Batai	Asia, Europe	M c + a	
Birao	Africa	M a	
Bozo	Africa	M c + a	
Bunyamwera	Africa	M c	H
Cache Valley	North America	M c + a	M
Fort Sherman	Central America		H
Germiston	Africa	M c	H, R, S
Iaco	South America	M c	
Ilesha	Africa	M a	H
Kairi	South America	M c	P, R, S
Lokern	North America	M c + a, C	O
Macaua	South America	M c	R, B
Maguari	South America	M c + a	S, O
Main Drain	North America	M c + C	O
Mboke	Africa	M c	
Ngari	Africa	M c + a	
Northway	North America	M c	R, S
Playas	South America	M c	
Santa Rosa	North America	M c	
Shokwe	Africa	M c + a	H, R
Sororoca	South America	M c	
Tensaw	North America	M c + a	R, S, O
Tlacotalpan	North America	M c + a	
Wyeomyia	North, South America	M c + a	

4. The Bunyamwera group of viruses was proposed in 1960 (CASALS 1963; Table 7). These viruses are transmitted principally by mosquitoes, although Lokern and Main Drain viruses have been isolated from *Culicoides* as well. Vertebrate hosts for these viruses range from rodents and birds to large animals, including domesticated animals and humans. The viruses are also widely distributed geographically, with the notable exception of Australia. Many of the viruses (e.g., Bunyamwera, Batai, Germiston, Guaroa, Ilesha, Tensaw, and Wyeomyia) have been associated with naturally occurring human infections, and five have been isolated from humans (KARABATSOS 1985; GONZALEZ and GEORGES 1988). Main Drain virus has been isolated from a horse with encephalitis (EMMONS et al. 1983). These viruses are not associated with major epizootics or epidemics.

Bunyamwera virus, the prototype virus of the genus, was isolated from *Aedes* mosquitoes in 1943 in Uganda (SMITHBURN et al. 1946). Subsequently, it was isolated from other *Aedes*, *Mansonia*, and *Culex* mosquitoes, and transmission of the virus by *Aedes aegypti*, *Culex fatigans*, and *Anopheles quadrimaculatus* has been demonstrated in the laboratory. Bunyamwera virus has also been isolated from viremic humans in various countries in Africa, and antibodies to the virus have been detected in sera from humans and primates and from domestic animals (KARABATSOS 1985). Antibody prevalence rates in birds and rodents are exceedingly low (BISHOP and SHOPE 1979). Human infection can result in fever, rash, nucchal rigidity, and other signs and symptoms (GONZALEZ and GEORGES 1988). Germiston, Ilesha, Shokwe, Wyeomyia, and Tensaw viruses have all been associated with human infections (KARABATSOS 1985).

Cache Valley virus is widely distributed in North America (CALISHER et al. 1986). It has been isolated from a variety of mosquitoes, but *Culiseta inornata* is probably the principal vector (KARABATSOS 1985). High antibody prevalence rates to Cache Valley virus have been detected in large domestic animals and wild ungulates (CALISHER et al. 1986). The virus has been implicated in abortion and teratogenesis, including arthrogryposis and hydranencephaly, in sheep (EDWARDS et al. 1989; CHUNG et al. 1990).

Table 8. Bwamba group viruses

Virus	Geographic distribution	Vector	Vertebrate
Bwamba	Africa	Ma	H
Pongola	Africa	Ma+c	H

5. Bwamba and Pongola viruses are geographically restricted to Africa (KARABATSOS 1985), and both have been isolated from humans with febrile illnesses (Table 8). Antibodies to them have been detected in human, domestic animal, and avian sera; seroprevalence rates of 97% have been detected in regions of Africa (BISHOP and SHOPE 1979). These viruses may be serologically related to viruses of the California group (CASALS 1963).

Table 9. Group C viruses

Virus	Geographic distribution	Vector	Vertebrate
Apeu	South America	Mc	H, M, S
Caraparu	North, South America	Mc	H, R, S
Gumbo Limbo	North America	Mc	R
Itaqui	South America	Mc	H, R, M, S
Madrid	North America	Mc	H, R, S
Marituba	South America	Mc	H, M, S
Murutucu	South America	Mc	H, R, M, S, O
Nepuyo	North, South America	Mc	H, R, Ba, S
Oriboca	South America	Mc	H, R, M, S
Ossa	North America	Mc	H, R, S
Restan	South America	Mc	H
Vinces	South America	Mc	S

6. Group C viruses (Table 9) have been isolated from mosquitoes and from rodents or marsupials in South, Central, and/or North America (KARABATSOS 1985). Excellent reviews of their vectors and vertebrate hosts are presented

elsewhere (SHOPE 1985, 1988). Ten of the viruses also have been isolated from humans with febrile illnesses (SHOPE 1988). The viruses do not cause epidemics or fatal disease, but they commonly cause a severe, self-limited, disabling febrile illness (SHOPE 1988). These viruses cause infections in people whose occupations require them to work in forested areas. A brief description of the natural cycles of the Group C viruses in the Utinga Forest near Belem, Brazil, was presented previously.

Table 10. California group viruses

Virus	Geographic distribution	Vector	Vertebrate
California encephalitis	North America	M c	
Guaroa	North, South America	M a	H
Inkoo	Europe	M c	
Jamestown Canyon	North America	M c+O	S
Jerry Slough	North America	M c	
Keystone	North America	M c+a	R,S
La Crosse	North America	M c	H,R,S
Melao	South America	M c	
San Angelo	North America	M c+a	
Serra do Navio	South America	M c	
Snowshoe hare	North America	M c	R,S,O
Tahyna	Africa, Asia, Europe	M C+a	H,S
Trivitattus	North America	M c	R,S

7. California encephalitis virus was isolated from mosquitoes in 1943 and was subsequently implicated serologically in three cases of encephalitis (HAMMON and REEVES 1952). California group viruses (Table 10) have a widespread distribution, with either virus isolations or serologic demonstration of their presence from North and South America, Eurasia, and Africa (CALISHER 1983). These viruses are associated with mosquito vectors, and most are associated with rodent or lagomorph hosts (KARABATSOS 1985). Many of the viruses infect humans (e.g., California encephalitis, Jamestown Canyon, LAC, Inkoo, snowshoe hare, Tahyna, and trivittatus), and antibody prevalence rates can be high (GRIMSTAD 1988). Minor cross-reactivities have been detected with members of other serogroups of the *Bunyavirus* genus; Bunyamwera, Group C, Bwamba, Simbu, Anopheles A, and Patois group viruses (BISHOP and SHOPE 1979). Guaroa virus is related to California group viruses by HI and NT but to Bunyamwera group viruses by CF (WHITMAN and SHOPE 1962). Guaroa may be an example of reassortment between viruses of two serogroups of the genus *Bunyavirus*. Particularly noteworthy is that many of these viruses have been demonstrated to the efficiently transovarially transmitted by their respective vectors (TURELL and LEDUC 1983; GRIMSTAD 1988).

Prototype LAC virus was isolated at autopsy from the brain of a child who was hospitalized with encephalitis in La Crosse, Wisconsin (THOMPSON et al. 1965). LAC virus is the major cause of pediatric arboviral encephalitis in the

United States (KAPPUS et al. 1983; GRIMSTAD 1988). The principal vector of LAC virus is *Aedes triseriatus*, a mosquito found east of the Rocky Mountains in the United States and southern Canada (PAULSON et al. 1989). A detailed discussion of interactions between LAC virus and *Aedes triseriatus* is provided above. The mosquito oviposits in tree holes and other water-containing vessels. Oak trees form particularly good tree holes, and in the climax oak-hickory forests of the upper midwest, a plethora of potential breeding sites exist. In these areas, chipmunks and tree squirrels are the preferred vertebrate hosts for viral amplification (YUILL 1983). These rodents do not develop apparent illness as a consequence of LAC virus infection but do develop viremia of sufficient titer to infect feeding mosquito vectors. Viremia titers of 3.0–4.0 $\log_{10}$ $TCID_{50}$ per milliliter are frequently detected in these species, and viremia occurs 2–5 days after infection (PANTUWATANA et al. 1972; PATRICAN et al. 1985a, b).

Oak-hickory forests and wood lots provide excellent habitat not only for mosquito vectors but also for their rodent hosts. *Aedes triseriatus* does not travel far from sheltered areas and breeding sites (GRIMSTAD 1988). Infections with LAC virus occur when humans impinge upon the natural cycle in the forested areas. Indeed, many of the encephalitis cases in La Crosse, Wisconsin, in the 1960s occurred when suburban homes were built in forested areas (THOMPSON and GUNDERSEN 1983).

Recognition that *Aedes triseriatus* would also oviposit in discarded tires, which are frequently found near homes and garages, revealed a major risk factor for infection (WATTS et al. 1974). Water-retaining tires provide excellent habitat for larval development, and when larvae are transovarially infected, the tires serve as a source of adult mosquitoes, which greatly enhances chances of human infection. Tires can also serve to transport infected mosquitoes to new areas. In one well-studied outbreak of LAC encephalitis in eastern Wisconsin, a previously nonendemic area, the source of infected mosquitoes was traced to tires that had been transported from the LAC endemic zone in western Wisconsin.

Realization that tires and tree holes are significant risk factors for LAC virus infections led to a major source reduction campaign in La Crosse County, Wisconsin. Tree holes were closed, and tires were removed and discarded (PARRY 1983). These efforts were remarkably effective, and few if any cases of LAC encephalitis have resulted from infections in La Crosse County in recent years (PARRY 1983; GUNDERSEN 1989, personal communication).

The natural cycle of LAC virus is well defined in the upper midwest. During the summer months, the virus is transmitted by *Aedes triseriatus* mosquitoes between susceptible amplifying hosts, such as chipmunks and tree squirrels. Alternate vertebrate hosts, e.g., foxes and domestic dogs (AMUNDSON and YUILL 1981; GODSEY et al. 1988), may in some circumstances serve to amplify the virus. Clearly, however, chipmunks and tree squirrels are the principal vertebrate hosts (YUILL 1983). In these animals, antibody prevalence rates decline each year with the introduction of new susceptibles and increase to virtually 100% in many forested areas by the end of the transmission season (BEATY and THOMPSON, unpublished data). The virus overwinters in the diapaused eggs of the vector. New susceptible vertebrate hosts are available each spring and early summer at the time that transovarially infected mosquitoes emerge.

Transovarial transmission may serve as an important amplifying mechanism for the virus in areas where antibody prevalence rates are significant. Attempted horizontal transmission of virus to seropositive vertebrate hosts would be unsuccessful. However, an immune vertebrate host could still serve as a blood meal source for infected mosquitoes, which could then oviposit many infected eggs (WATTS et al. 1973).

Transovarial transmission also results in infection of male mosquitoes, which can subsequently transmit virus venereally to uninfected female mosquitoes (THOMPSON and BEATY 1977). The importance of venereal transmission in terms of amplifying LAC virus in nature remains to be determined (DEFOLIART 1983).

LAC virus infections have now been documented in most states abutting and east of the Mississippi River (KAPPUS et al. 1983). Most cases occur in upper midwestern states, including Minnesota, Wisconsin, Iowa, Illinois, Indiana, and Ohio. Ecological conditions are apparently maximal in these areas for transmission of the virus.

Three apparent subgroups of LAC virus, as determined by oligonucleotide fingerprinting, occur in the United States (KLIMAS et al. 1981). Two types, A and B, occur in the upper midwest and are sympatric in certain areas. An intertypic reassortant between types A and B was detected in an area of overlap (KLIMAS et al. 1981). The distribution of these two genotypes coincides with the areas with the greatest incidence of LAC encephalitis. Interestingly, the oligonucleotide fingerprints of the two isolates from human fatalities in the La Crosse area are remarkably similar even though they were obtained 20 years apart (BISHOP 1985a). Whether these viruses are more neurovirulent than other LAC isolates, or whether a more virulent genotype is selected for by human passage remains to be determined.

The third type of LAC virus, C, is found in viruses isolated from the eastern and southern United States. The epidemiology of LAC virus may differ in these latter areas. In New York and the eastern states, LAC seroprevalence rates in chipmunks and tree squirrels are not as high as in the upper midwest, virus isolations are not as restricted to *Aedes triseriatus*, and the mosquito is not as efficient in transovarial transmission as its midwestern counterpart (MILLER et al. 1982; GRAYSON et al. 1983). In the southern United States, ecological conditions differ from those in the upper midwest. High temperatures may affect LAC virus persistence and transmission by *Aedes triseriatus*.

Approximately 75 cases of LAC encephalitis are reported annually in the United States (GRIMSTAD 1988). This is surely an underrepresentation of the ture number. Cases occur as early as June, because of probably infection by emerging transovarially infected females, and increase in number through the summer with the amplification of virus in the natural cycle. Cases cease with the first severe frosts, which kill the adult vectors. LAC virus infects all age groups, but most clinical cases occur in children under 15 years of age, and a male: female ratio of 2:1 or 3:2 is frequently reported (KAPPUS et al. 1983; GRIMSTAD 1988). The sex differences are presumably due to boys being more active in forested areas where the vectors are prevalent. Humans are dead-end hosts for the virus; person-to-person transmission does not occur. Except in years of epidemic Saint Louis encephalitis activity, LAC is the most commonly reported arboviral encephalitis.

Snowshoe hare (SSH) and Jamestown Canyon viruses are California group viruses that also cause human infections and disease in North America (GRIMSTAD 1988). SSH virus has frequently been isolated in Canada from a variety of *Aedes* species mosquitoes (TURELL and LEDUC 1983; GRIMSTAD 1988). Human encephalitis resulting from SSH virus infection occurs principally in Ontario and Quebec (ARTSOB 1983) Jamestown Canyon virus has also been documented to cause encephalitis (SRIHONGSE et al. 1984; GRIMSTAD 1988). Unlike LAC encephalitis, cases predominate in adults rather than in children. Jamestown Canyon virus frequently infects humans; in Michigan 25% of young adults are seropositive (GRIMSTAD et al. 1986), and 5%–10% of sera from other states contain antibodies to the virus. Deer serve as the vertebrate host for Jamestown Canyon virus, and the great increase in deer populations in midwestern and eastern States in recent decades is thought to be a contributing factor in human infections with this virus.

Tahyna virus was originally isolated from *Aedes caspius* mosquitoes collected near the village of Tahyna in Czechoslovakia and has subsequently been isolated from a variety of mosquitoes in many parts of Europe. The virus causes influenza-like illness and has been isolated from human sera in Europe (KARABATSOS 1985; GRIMSTAD 1988). Many isolates have been obtained from *Aedes vexans* mosquitoes, and the virus has been isolated from *Culiseta annulata* larvae. Lumbo virus, isolated from mosquitoes in Africa, is considered to be essentially identical to Tahyna virus (KARABATSOS 1985).

Antibodies to Tahyna virus have been detected in a variety of mammals, including rodents, lagomorphs, and domestic animals. Seroprevalence rates in humans approach 100% in certain areas of Czechoslovakia (GRIMSTAD 1988).

Table 11. Capim group viruses

Virus	Geographic distribution	Vector	Vertebrate
Acara	North, South America	M c	R,S
Benevides	South America	M c	S
Benfica	South America	M c	R,S
Bushbush	South America	M c	S
Capim	South America	M c	R,M,S
Guajara	North, South America	M c	R,S
Juan Diaz	North America		S
Moriche	South America	M c	

8. Capim group viruses (Table 11) are vectored principally by *Culex* mosquitoes (KARABATSOS 1985) and are associated with rodent hosts. None has been associated with human disease. These viruses have been detected only in the Americas.

Table 12. Gamboa group viruses

Virus	Geographic distribution	Vector	Vertebrate
Gamboa	Central America	M c	
Pueblo Viejo	South America	M c	
San Juan	South America	M c	
Alajuela*	Central America	M c	
Brus Laguna*	Central America	M c	

9. Gamboa group viruses have been isolated only in Central and South America (Table 12). All isolates have been obtained from *Aedeomyia squamipennis* mosquitoes, including larvae and adult male mosquitoes (KARABATSOS 1985). The virus is efficiently transmitted transovarially by *Aedeomyia squamipennis* (DUTARY et al. 1989). None of the viruses has been implicated in human infections.

Table 13. Guama group viruses

Virus	Geographic distribution	Vector	Vertebrate
Ananindeua	South America	M c	R,B,M,S
Bertioga	South America		S
Bimiti	South America	M c	R,S
Cananeia	South America	M c	S
Catu	South America	M c+a	H,R,M,S,B
Guama	North, South America	M c	H,R,B,M,S
Guaratuba	South America	M c	B,S
Itimirim	South America		R
Mahogany Hammock	North America	M c	R
Mirim	South America	M c	S
Moju	South America	M c	R,M,S
Timboteua	South America		R,S

10. Guama group viruses (Table 13) are restricted to tropical regions in the Western Hemisphere (KARABATSOS 1975; SHOPE 1988). They are vectored by mosquitoes from many different genera and are associated with rodent and marsupial hosts. The viruses have also been isolated from birds and bats as well as a variety of sentinel animals. Catu and Guama viruses have been associated with disease in humans. Infections typically occur in forest workers (SHOPE 1988).

Table 14. Koongol group viruses

Virus	Geographic distribution	Vector	Vertebrate
Koongol	Australia	M c+a (?)	
Wongal	Australia	M c	

11. Both Koongal and Wongal viruses were isolated from *Culex annulirostris* mosquitoes in Australia (DOHERTY et al. 1963; Table 14). Little is known of their respective vertebrate hosts; isolates have not been made from vertebrates. However, serologic surveys have detected antibodies in cattle and other vertebrates (BISHOP and SHOPE 1979). These viruses exhibit weak cross-reactivities with Bwamba, California, and Bunyamwera groups by HI test (CASALS 1963) and with other bunyaviruses by immunofluorescence (ZELLER et al. 1989).

Table 15. Minatitlan group viruses

Virus	Geographic distribution	Vector	Vertebrate
Minatitlan	North America		S
Palestina	South America	M c	S

12. Minatitlan viruses have been isolated from *Culex* mosquitoes and sentinel hamsters in Mexico and Ecuador, respectively (KARABATSOS 1985; Table 15). Little is known of their natural vertebrate hosts. They are not known to cause disease in humans.

Table 16. Nyando group viruses

Virus	Geographic distribution	Vector	Vertebrate
Nyando	Africa	M a+c	H
Eretmapodites 147*	Africa	M	

13. Both viruses of the Nyando group have been isolated from mosquitoes in Africa (Table 16). Nyando is a human pathogen and has been isolated from a patient with a febrile illness (KARABATSOS 1985).

Table 17. Olifantsvlei group viruses

Virus	Geographic distribution	Vector	Vertebrate
Bobia	Africa	M c	
Botambi	Africa	M c	
Dabakala	Africa	M c	
Olifantsvlei	Africa	M c	
Oubi	Africa	M c	

14. Olifantsvlei viruses are geographically restricted to Africa (KARABATSOS 1985; Table 17). All were isolated from *Culex* species mosquitoes; little is known about their vertebrate hosts. The viruses have not been associated with human disease.

Table 18. Patois group viruses

Virus	Geographic distribution	Vector	Vertebrate
Abras	South America	M c	
Babahoyo	South America	M c	S
Estero Real*	North America	T o	
Pahayokee	North America	M c	
Patois	Central America	M c	R,S
Shark River	North America	M c+a	R,S
Zegla	Central America		R,S

15. Patois group viruses are geographically restricted to the Western Hemisphere (Table 18). They are vectored by mosquitoes and associated with rodent hosts (KARABATSOS 1985). None of the viruses has been implicated in human disease.

16. Simbu group viruses are widely distributed in nature (Table 19). Most are

Table 19. Simbu group viruses

Virus	Geographic distribution	Vector	Vertebrate
Aino	Asia, Australia	M c+C	
Akabane	Africa, Australasia	M c+C	O
Buttonwillow	North America	C	O,S
Douglas	Australia	C	O
Ingwavuma	Africa, Asia	M c	B,O
Inini	South America		B
Kaikalur	Asia	M c	
Manzanilla	South America		P
Mermet	North America		B
Oropouche	South America	M c+C	H,O
Para	South America	M c	S
Peaton	Australia	C	O
Sabo	Africa	C	O
Sango	Africa	M c+C	O
Sathuperi	Africa, Asia	M c+C	O
Shamonda	Africa	C	O
Shuni	Africa	M c+C	H,O
Simbu	Africa	M c	
Thimiri	Africa, Asia		B
Tinaroo	Australia	C	
Utinga	South America		O

found in Africa and Australasia; some are found in the Western Hemisphere (BISHOP and SHOPE 1979; KARABATSOS 1985). These viruses are vectored by either mosquitoes or culicoids, and the viruses have been most frequently isolated from livestock and birds. Akabane virus is of significant veterinary importance, and Oropouche and Shuni viruses have been isolated from humans (KARABATSOS 1985; ST. GEORGE and STANDFAST 1988).

Akabane virus was isolated in Japan from mosquitoes in 1959 (OYA et al. 1961); however, association of the virus with an epidemic involving premature, stillborn, and deformed (arthrogryposis and hydranencephaly) calves occurred in the 1970s. Major epizootics of Akabane disease have been documented in many countries, including Israel, Japan, and Australia (ST. GEORGE and STANDFAST 1988). The economic impact of these outbreaks can be substantial. In one epizootic in Japan, 42000 calves were born dead or aborted (KUROGI et al. 1975). Human illnesses apparently do not occur.

Akabane virus has been isolated from a variety of mosquitoes in Japan, but in Australia most virus isolations have been made from *Culicoides brevitarsis* (ST. GEORGE and STANDFAST 1988). This midge is active throughout the year in northern Australia, thereby precluding the need for an overwintering mechanism. The midge oviposits in fresh cow dung, and adult flies emerge 8–10 days later. Infected midges can be transported long distances by the wind, thereby introducing virus into new areas (SELLERS and PEDGLEY 1985).

Cows and sheep are vertebrate hosts. Both develop viremia, but it is only of 3–4 days duration (ST. GEORGE and STANDFAST 1988). Antibodies have been found in deer, but not in marsupials in Australia, and a wide range of wild ruminants in Africa have antibodies to Akabane.

In Australia, five Simbu group viruses are considered to have teratogenic potential; Akabane, Aino, Douglas, Tinaroo, and Peaton (MCPHEE and DELLA-PORTA 1988; ST. GEORGE and STANDFAST 1988). None cause apparent effects other than low-grade fever in nonpregnant cattle or sheep. It is noteworthy that Cache Valley virus has recently been implicated in cases of arthrogryposis and hydranencephaly in sheep in the western United States (CHUNG et al. 1990).

Aino virus infection of pregnant cows or sheep has been associated with teratology, but the evidence is less strong than that for Akabane (PARSONSON and MCPHEE 1985).

Oropouche virus was first isolated from the blood of a febrile forest worker in Trinidad in 1955 (ANDERSON et al. 1961). The epidemic potential of the virus was not recognized until 1961, when the first of many major epidemics of Oropouche fever occurred in Belem, Brazil (LEDUC and PINHEIRO 1988). Epidemics involving thousands of cases have subsequently been recorded in other parts of Brazil, and all of the epidemics have occurred during rainy seasons. Incidence rates range from 30% to 60%, and nearly two-thirds are clinically apparent infections (LEDUC and PINHEIRO 1988). The disease episode lasts 2–7 days. Thus, epidemics of Oropouche virus clearly exert considerable social and economic impact. Interestingly, an Oropouche epidemic has recently occurred in Panama.

Two distinct cycles have been proposed for Oropouche virus: a sylvatic maintenance cycle and an epidemic cycle. In the sylvatic cycle, primates, sloths, and birds have been proposed as reservoir vertebrate hosts. However, despite processing thousands of tissues from wild birds and mammals. Oropouche virus has been isolated only four times; each isolate was from sloths (PINHEIRO et al. 1981). Nearly 12% of Brazilian primate sera contained antibodies to Oropouche. Elucidation of the sylvatic vector has been equally difficult. More than a million arthropods collected between Oropouche epidemics have been processed; virus has been isolated only from a single pool of *Aedes serratus* in Brazil and once from *Coquillettidia venezuelensis* in Trinidad (PINHEIRO et al. 1981). Thus, the sylvatic cycle of Oropouche remains unknown.

In contrast, the epidemic cycle is fairly well understood (LEDUC and PINHEIRO 1988). The virus is transmitted between humans by the biting midge, *Culicoides paraensis*. However, the midge does not appear to be a very competent vector, as determined by laboratory studies (PINHEIRO et al. 1982). *Culex pipiens quinquefasciatus*, a peridomestic mosquito, has yielded many virus isolates (LEDUC and PINHEIRO 1988). Nonetheless, the consensus is that *Culicoides paraensis* is the principal vector in the epidemic cycle, and its relative incompetence is compensated for by the sheer abundance of the vector during epidemics. Extensive searches for nonhuman amplifying hosts during Oropouche epidemics have not been successful. Humans apparently serve as amplifying hosts in the epidemic cycle, which is unusual for viruses in the family Bunyaviridae.

Table 20. Tete group viruses

Virus	Geographic distribution	Vector	Vertebrate
Bahig	Africa, Europe	T i	B
Batama	Africa		B
Matruh	Africa, Europe	T i	B
Tete	Africa		B
Tsuruse	Asia		B
Weldona	North America	C	

17. Most Tete group virus isolations have been from birds; Bahig and Matruh viruses have been isolated from ixodid ticks (KARABATSOS 1985; Table 20). The viruses have been isolated in Europe, Africa, Asia, and recently in North America in Colorado. None of these viruses has been associated with human infections.

Table 21. Turlock group viruses

Virus	Geographic distribution	Vector	Vertebrate
Lednice	Europe	M c	
M' Poko	Africa	M c	
Turlock	North, South America	M c	B,S,O
Umbre	Asia	M c	B

18. Turlock group viruses are vectored by mosquitoes and have been isolated in North and South America, Africa, Asia, and Europe (KARABATSOS 1985; Table 21).

Table 22. Unassigned Bunyaviruses

Virus	Geographic distribution	Vector	Vertebrate
Kaeng Khoi	Asia	O	B,S
Mojui dos Campos*	South America		B
Termeil*	Australia	M c	

19. Mojui dos Campos virus (Table 22) is serologically related to members of the California and Nyando serogroups (ZELLER et al. 1989).

5.2 Genus *Nairovirus*

Six antigenic groups comprise the genus *Nairovirus* (KARABATSOS 1985). Antigenic relationships have been demonstrated between members of the groups (CASALS and TIGNOR 1980). Significant human and veterinary pathogens are found in the CCHF and Nairobi sheep disease (NSD) groups.

Table 23. Crimean-Congo hemorrhagic fever (CCHF) viruses

Virus	Geographic distribution	Vector	Vertebrate
Congo	Africa, Asia, Europe	T i+C	H,R,O
CCHF	Africa, Asia, Europe	T i+a	H
Hazara	Asia	T i	
Khasan	Asia	T i	

1. CCHF virus is the type virus of the genus *Nairovirus* (KARABATSOS 1985; Table 23). Both Congo and Crimean hemorrhagic fever viruses are listed in the *International Catalogue of Arboviruses*, although they are serologically indistinguishable. Hazara and Khasan viruses apparently do not cause human disease and little is known of their natural histories (KARABATSOS 1985). Viruses in the group are vectored principally by ixodid ticks and are restricted to Africa and Eurasia. Excellent in-depth reviews of the epidemiology and ecology of CCHF are available (HOOGSTRAAL 1979; WATTS et al. 1988). CCHF was described clinically in 1944 during an epidemic in the Crimea in the Soviet Union. The virus was isolated in mice in 1967 from blood obtained from clinically diagnosed CCHF cases. CCHF virus has been associated with thousands of cases of hemorrhagic fever in the Soviet Union and in other Asian countries (WATTS et al. 1988). Viremia titer can be substantial, accounting for many of the reported nosocomial infections (BUTENKO and CHUMAKOV 1990).

Congo virus was isolated in 1956 from a patient in the former Belgian Congo, now Zaire (SIMPSON et al. 1967). Although Congo virus has also been associated with human illness in Africa, epidemics there have not typically involved the numbers of patients as in Eurasia. However, early speculation that the disease in Africa is less severe than that seen in Asia has not been confirmed; fatal and severe disease clearly can result from Congo virus infection (WATTS et al. 1988).

Human infections are generally "agriculture associated," and index cases are typically associated with tick bite or contact with infected domesticated animals (HOOGSTRAAL 1979). Nosocomial infections are common; physicians and nurses frequently become infected while tending patients, especially during the hemorrhagic phase of the disease. The disease is apparently more severe in nosocomial infections than in tick bite infections (WATTS et al. 1988), perhaps due to the dose of the inoculum. The ratio of inapparent to apparent infections is estimated to be 5:1, and mortality rates of approximately 20% or greater are reported (WATTS et al. 1988). CCHF virus apparently does not cause disease in livestock.

CCHF virus has been isolated from at least 29 species/subspecies of ixodid and argasid ticks, including one-, two-, and three-host ticks (WATTS et al. 1988). In sub-Saharan Africa, 17 tick species have been implicated as vectors (CAMICAS et al. 1990). The biological significance of these isolates remains to be determined. The dangers associated with working with CCHF virus have limited the number of studies to investigate potential vector and vertebrate hosts. Nonetheless, it is clear that the greatest risk factor for CCHF is being in areas where ticks of the genus *Hyalomma* are present, and where the virus has been documented to occur.

CCHF virus has been isolated from eggs of certain ticks, from field-collected unfed nymphs, and from vertebrate hosts of nymphs (HOOGSTRAAL 1979), all of which suggest vertical transmission of the virus. Virus survives transstadially in ticks, and the virus can overwinter in the respective tick stage. Little is known of amplifying or reservoir vertebrate hosts. Antibodies have been detected in a variety of vertebrates, and virus has been isolated from man, cattle, goats, sheep, hares, hedgehogs, and mice (WATTS et al. 1988). However, the biological significance of these observations also remains to be determined.

The geographic distribution of CCHF is extensive, ranging from South Africa (SWANEPOEL et al. 1983), through southern Europe, to western China. Indeed, this is probably the greatest range of any of the nonavian, tick-borne viruses.

Table 24. Dera Ghazi Khan viruses

Virus	Geographic distribution	Vector	Vertebrate
Abu Hammad	Africa, Asia	T a	
Dera Ghazi Khan	Asia	T i	
Kao Shuan	Asia, Australia	T a	
Pathum Thani	Asia	T a	
Pretoria	Africa	T a	

2. Dera Ghazi Khan group viruses (Table 24) apparently do not cause human infections, nor have they been isolated from vertebrates (KARABATSOS 1985). Most of the isolates have been made from argasid ticks collected in African and Australasia.

Table 25. Hughes group viruses

Virus	Geographic distribution	Vector	Vertebrate
Hughes	North, South America	T a	B
Punta Salinas	South America	T a	
Soldado	Africa, Europe, South America	T a	
Zirqa	Asia	T a	

3. Hughes group viruses (Table 25) apparently do not infect humans. Most isolates have been from argasid ticks, and Hughes virus has been isolated from birds. The viruses are widely distributed in nature.

Table 26. Nairobi sheep disease (NSD) viruses

Virus	Geographic distribution	Vector	Vertebrate
Dugbe	Africa	M c+C+T+ i	H,R,B,O
Ganjam	Asia	M c+T i	H
NSD	Africa	T i+C	H,O

4. NSD virus (Table 26) is a significant veterinary pathogen. Dugbe and Ganjam viruses have also been isolated repeatedly from ticks obtained from domesticated animals (KARABATSOS 1985). All three viruses can cause febrile illnesses in humans. Ganjam virus may be an antigenic variety of NSD virus. The viruses have been isolated from ixodid ticks, mosquitoes, and culicoids, but ticks are considered to be the vectors. The viruses have been isolated in Africa and in Asia. The infectious nature of NSD was demonstrated in 1910 (MONTGOMERY 1917); the disease could be transmitted to sheep by inoculation of blood from an infected sheep. Infection of domestic sheep can result in exceptional morbidity, and mortality rates of up to 75% can occur (WEINBREN et al. 1958). Many isolates of NSD have been obtained from *Ripicephalus appendiculatus* and domesticated animals in many parts of Africa (BISHOP and SHOPE 1979).

Table 27. Qalyub group viruses

Virus	Geographic distribution	Vector	Vertebrate
Bandia	Africa	T a	R
Omo	Africa		R
Qalyub	Africa	T a	

5. Qalyub group viruses (Table 27) have been isolated from argasid ticks in Africa and from rodents (KARABATSOS 1985). The viruses apparently do not infect humans.

Table 28. Sakhalin group viruses

Virus	Geographic distribution	Vector	Vertebrate
Avalon	North America	T i	B
Clo Mor	Europe	T i	
Paramushir	Asia	T i	
Sakhalin	Asia	T i	
Taggert	Australia	T i	

6. Sakhalin group viruses (Table 28), apparently do not infect humans. The viruses are widely distributed geographically and have been isolated from ixodid ticks and birds (KARABATSOS 1985).

5.3 Genus *Phlebovirus*

Currently, the genus *Phlebovirus* is comprised of the 37 viruses of the phlebotomus fever (sandfly fever, SF) group viruses (KARABATSOS 1985; TESH 1988). Recent biochemical and molecular biological studies have revealed similar coding strategies and common 5′ and 3′ sequences for phleboviruses and uukuviruses (SCHMALJOHN and PATTERSON 1990). This information prompted the Bunyaviridae Study Group to propose to the ICTV the incorporation of the uukuviruses into the *Phlebovirus* genus. In anticipation of the acceptance of the recommendation, the two genera are combined into the genus *Phlebovirus*, which is comprised of the phlebotomus fever group and the uukuniemi group viruses.

1. Phlebotomus fever group viruses (Table 29) are widely distributed (KARABATSOS 1985). The group is comprised of 37 viruses, and most have been associated with plebotomine files. Indeed, transovarial transmission of phleboviruses in sandflies probably serves to maintain the viruses during adverse climatic conditions (TESH 1988). However, several members of the group (e.g., Arumowot, Itaporanga, and RVF) have been isolated repeatedly from mosquitoes. Nine members of the group have been isolated from humans, and several are significant human or veterinary pathogens. Little is known concerning reservoir vertebrate hosts for the viruses. Virus isolations and serologic surveys have incriminated a variety of animals (KARABATSOS 1985), but disease has not been observed in any of these presumed vertebrate hosts.

Table 29. Phlebotomus fever group viruses

Virus	Geographic distribution	Vector	Vertebrate
Aguacate	North America	P	
Alenquer	South America		H
Anhanga	South America		O
Arbia	Europe	P	
Arboledas	South America	P	
Arumowot	Africa	M c	R,B,O
Buenaventura	North, South America	P	
Bujaru	South America		R
Cacao	North America	P	
Caimito	North America	P	
Candiru	South America		H
Chagres	North America	M c+P	H
Chilibre	North America	P	
Corfou	Europe	P	
Frijoles	North America	P	
Gabek Forest	Africa		R,O
Gordil	Africa		R
Icoaraci	South America	M c+a+P	R,B,S
Itaituba	South America		M
Itaporanga	South America	M c	B,M,S
Karimabad	Asia	P	
Munguba	South America	P	
Nique	North America	P	
Odrenisrou	Africa	M c	
Oriximina	South America	P	
Pacui	South America	P	R
Punta Toro	North, South America	P	H
Rift Valley fever	Africa	M c	H,O
Rio Grande	North America		R
Sandfly fever–Naples	Africa, Asia, Europe	P	H
Sandfly fever–Sicilian	Africa, Asia, Europe	P	H
Saint-Floris	Africa		R
Salehabad	Asia	P	
Tehran	Asia	P	
Toscana	Europe	P	H
Turuna	South America	P	
Urucuri	South America		R

Because of the frequent isolations of viruses from male sandflies, demonstration of efficient transovarial transmission of many members of the group and establishment of lines of *Phlebotomus perniciosus* capable of persistently transovarially transmitting Toscana virus, some investigators suggest that the role of the vertebrate may be solely to provide a blood meal for egg development and not to serve as a source for horizontal amplification of the virus (TESH 1988). Phlebotomus fever group viruses have been isolated in Europe, Africa, central Asia, and the Americas (KARABATSOS 1985). Interestingly, none has been isolated in Southeast Asia or Australia, presumably due not only to the absence of virus but also to the sparse sandfly fauna in these areas (TESH 1988).

The prototype virus of the group is SF-Sicilian virus. This virus and the related SF-Naples virus were originally isolated from human sera collected during epidemics of the disease occurring in American troops in Italy in 1943 and 1944 (SABIN 1955). However, the disease was described clinically in 1886, and a filterable agent and a vector, *Phlebotomus papatasi*, were implicated in the disease in 1909 (TESH 1988). This was the second demonstration of an arthropod vector for an arboviral disease. The disease is characterized by high fever, severe headache, retroorbital pain, conjunctivitis, and general malaise (GAIDAMOVICH et al. 1990).

Historically, phlebotomus fever has been of significant military importance, but epidemics and endemic disease have occurred in civilian population as well (TESH 1988; GAIDAMOVICH et al. 1990). One outbreak in Serbia resulted in over one million cases (GUELMINO and JEVTIC 1955). *Phlebotomus* spp. are the principal vectors in the Old World and are associated with dry environments, agricultural and rural niches. *Lutzomyia* spp. are the principal vectors in the New World and are associated with tropical niches. Sandflies are poor fliers and actually travel in short hops (TESH 1988). This significantly limits their range, and as a consequence they typically feed on virtually any vertebrate host that they encounter. Sandfly fever virus isolates have been obtained from humans from Italy to Bangaladesh and from a variety of *Phlebotomus* spp. (BISHOP and SHOPE 1979; TESH 1988). Antibodies to SF-Sicilian virus have been detected in a variety of species, including cattle and sheep, throughout the Middle East and central Europe.

RVF virus was first isolated in 1930 in Kenya from the blood of a lamb (DAUBNEY et al. 1931) and has been repeatedly isolated since that time from sheep, cattle, humans, particularly veterinarians, abattoir workers, and others who handle sick animals. The virus causes severe disease in human and animals, and in the 1977 epidemic in Egypt attack rates, morbidity, and mortality in human and domestic animals were extensive (MEEGAN 1979a; MEEGAN and BAILEY 1988; SWARTZ et al. 1981).

Historically, RVF virus has been associated with sporadic epidemics involving domesticated animals and humans in sub-Saharan Africa. These epidemics were of considerable economic importance (MEEGAN and BAILEY 1988). High attack rates occur in domestic animals. Depending upon the species, mortality rates in adults approach 30%, and abortion rates in pregnant animals approach 100% (MEEGAN and BAILEY 1988). RVF outbreaks have been documented in 24 African nations; however, the extension of RVF from sub-Saharan Africa to Egypt in a massive epizoodemic in 1977 was unprecedented (MEEGAN et al. 1979). RVF virus in sub-Saharan Africa is not characterized by endemicity; rather, outbreaks typically follow periods of excessive rain (MCINTOSH and JUPP 1981; SWANEPOEL 1981).

RVF virus has been isolated from mosquitoes of 23 species as well as from a few other hematophagus arthropods (MEEGAN and BAILEY 1988). In addition, a variety of arthropods have been demonstrated to be capable of transmitting RVF virus in the laboratory (MEEGAN and BAILEY 1988; TURELL and PERKINS 1990; LINTHICUM et al. 1989). Incrimination of such an extensive number of arthropods as vectors did little to reveal species involved in viral maintenance. Isolation of RVF virus from *Aedes lineatopennis* larvae, a floodwater mosquito, collected from dambos (water-retaining depressions) in Africa suggests that transovarial transmission of the virus is a mechanism for interepidemic maintenance of RVF virus (LINTHICUM et al. 1985). When excessive rainfall floods the dambos, massive numbers of mosquito eggs are induced to hatch. A small dambo can serve as a source for literally millions of *Aedes lineatopennis* (LINTHICUM and BAILEY 1985). Emerging, transovarially infected mosquitoes can then feed upon susceptible vertebrate hosts that frequent the water-retaining dambos and initiate an epidemic. These potential breeding sites can be detected by satellite-based remote sensing (LINTHICUM et al. 1987), providing epidemiologic information concerning potential RVF virus activity.

Many vertebrate hosts develop very high-titer plasma viremias (MCINTOSH and JUPP 1981), which can then serve to infect a variety of vectors and humans in contact with infected animals. Because of the high-titer viremias, mechanical transmission undoubtedly contributes to amplify transmission during the epidemics. This and similar maintenance and amplification cycles are thought to characterize the epidemiology of RVF in sub-Saharan Africa.

In Egypt, where the epidemic continued for 2 years, *Culex pipiens* was incriminated as a vector. Because floodwater mosquito populations analogous to those in sub-Saharan Africa are uncommon in Egypt, the virus must have been maintained in a different fashion, perhaps by continual horizontal transmission in a highly susceptible vertebrate population (MEEGAN and BAILEY 1988).

Domestic animals, e.g., cattle, sheep, and goats, are considered the principal amplifying hosts of the virus. Death and abortion in these animals foreshadow human cases. Domestic animals serve directly as a source of infection for humans. Many human cases occur due to contact with tissues or blood of infected animals or animal tissues (GONZALEZ-SCARANO and NATHANSON 1990). During an epizootic, when there are many dead and dying animals, this type of transmission may become very important.

In 1987 and 1988, an extensive RVF epidemic occurred in Senegal and Mauritania (JOUAN et al. 1988). Human febrile illnesses and fatal hemorrhagic disease were noted. Domestic animals were involved, with widespread abortion and high mortality in young sheep. Little is known of vectors involved in transmission or maintenance of virus in the area. Recent construction of a large dam on the Senegal River may have contributed to the epidemic by altering the ecological conditions in the area. This epidemic reemphasizes the pressing need to understand better the natural history of the virus (SHOPE et al. 1982).

2. Viruses in the Uukuniemi group (Table 30) have been obtained from ticks in Eurasia and Australia (KARABATSOS 1985). The prototype virus Uukuniemi was isolated from an engorged *Ixodes ricinus* tick collected in 1960 from a cow in Uukuniemi, Finland (OKER-BLOM et al. 1964). Grand Arbaud, Ponteves, and

Table 30. Uukuniemi group viruses

Virus	Geographic distribution	Vector	Vertebrate
Grand Arbaud	Europe	T a	
Manawa	Asia	T i+a	
Ponteves	Europe	T a	
Precarious Point	Australia	T i	
Uukuniemi	Europe	T i	R,B
Zaliv Terpeniya	Asia	T i	

Manawa viruses have been isolated from argasid ticks (BISHOP and SHOPE 1979). Many isolates of these viruses have been obtained from ticks, birds, and rodents. Antibodies to Uukuniemi have been detected in humans (SEKEYOVA et al. 1970) and other vertebrates.

5.4 Genus *Hantavirus*

Table 31. Hantaan group viruses

Virus	Geographic distribution	Vector	Vertebrate
Hantaan	Asia		H,R
Prospect Hill	North America		R
Puumala	Europe		R
Seoul	North South America, Asia		R
Thottapalyam	India		O

Based on compelling physicochemical and biochemical data (SCHMALJOHN and DALRYMPLE 1983), in 1987 the ICTV approved the incorporation of the *Hantavirus* genus in the family Bunyaviridae (KARABATSOS 1985; Table 31). This was unprecedented because hantaviruses are not transmitted by arthropod vectors; they are rodent-borne. They cause persistent infections in their natural hosts and are spread by aerosolized excreta and fomites (LEE 1982; LEDUC 1989; GAVRILOVSKAYA et al. 1990). Human infections occur under conditions of close contact with infected rodents. Five hantaviruses are currently listed in the *International Catalogue of Arthropod-borne Viruses* (KARABATSOS 1985); however, other hantaviruses are recognized, and six or seven serotypes have been proposed (LEE et al. 1990). Hantaviruses are almost worldwide in distribution (CHUMAKOV et al. 1990; LEE et al. 1990). Four of the hantaviruses have well-defined rodent reservoir associations: Hantaan virus with *Apodemus agrarius*, Puumala virus with *Clethrionomys glareolus*, Seoul virus (and other rat-borne viruses) with *Rattus* spp., and Prospect Hill virus with *Microtus pennsylvanicus* (CHUMAKOV et al. 1990; GLASS et al. 1990). The prototype virus of the genus, Hantaan virus, was first isolated from a field mouse, *Apodemus agrarius* (LEE et al. 1978). Viral antigen was detected in lung tissue of the mouse using sera from patients suffering from hemorrhagic fever with renal syndrome (HFRS). Isolation of the agent quickly led to identification of other hantaviruses (LEDUC 1987). The zoonotic reservoir of Puumala virus is *Clethrionomys glareolus*, a vole common in areas of Scandinavia where the disease is endemic (BRUMMER-KORVENKONTIO et al. 1980). Seoul virus is maintained in both *Rattus rattus* and *Rattus norvegicus* populations (LEE et al. 1982), and persistent virus infections occur. Prospect Hill virus was isolated from *Microtus pennsylvanicus* in the United States (LEE et al. 1985). However, either Hantavirus antigen or antibodies have been detected in a large number of alternate rodent species (CHUMAKOV et al. 1990). Thottapalayam was isolated in India from a healthy shrew (KARABATSOS 1985).

Hantaviruses include the etiologic agents of HFRS. More than 200000 cases of hospitalized HFRS occur annually, with an estimated mortality rate of 2%–10% (YAN ET AL. 1986; GLASS et al. 1990; LEE et al. 1990). Korean hemorrhagic fever was recognized among United Nations soldiers in Korea (SMADEL 1951; GONZALEZ-SCARANO and NATHANSON 1990). Three clinical manifestations of HFRS are recognized: severe HFRS associated with Hantaan virus infections, nephropathia epidemica (MYHRMAN 1951) associated with Puumala virus infections, and mild epidemic hemorrhagic fever associated with Seoul virus infections (GLASS et al. 1990). Antibodies to Prospect Hill virus have been detected in humans, but illness has not been documented (LEDUC 1989).

In much of Asia, human Hantaan virus infection is associated with close contact with the reservoir, *Apodemus agrarius*. The mouse develops an asymptomatic infection, but considerable amounts of viral antigen can be detected in organs, particularly lungs and kidneys. The virus can be secreted in saliva, feces, and urine for as long as 1 year or more after infection (LEE et al. 1982; TSAI 1987; GAVRILOVSKAYA et al. 1990). Despite considerable effort, an arthropod vector for the virus has not been demonstrated (LEE 1982). Human infections occur when individuals come into contact with mouse excreta, and most cases occur in agricultural workers or military personnel. In Korea there are two seasonal peaks of infections, in spring and in fall, presumably corresponding to the planting and harvest seasons (LEE 1982). HFRS also occurs in Europe, where *Apodemus flavicollis* or alternate rodents may serve as a reservoir (ANTONIADIS et al. 1987; GLIGIC et al. 1989; AVSIC-ZUPANC et al. 1990; LEDUC et al. 1986a). Interestingly, 50% of *Mus musculus* collected in a study in Bulgaria were infected (VASILENKO et al. 1990). The significance of such a high infection rate in a peridomestic rodent remains to be determined.

Clearly, the genus *Hantavirus* includes significant human pathogens. Remarkable progress has been made in understanding the pathology, molecular biology, and epidemiology of these viruses in a relatively short period of time. The detection of Seoul-like viruses in rat populations in major seaports throughout the world is remarkable; infection prevalence rates can exceed 50%, and these rats can serve as a source of human infection (LEDUC et al. 1986b). Antibodies have also been found in local human populations, and statistically significant correlations with antibodies to Seoul virus and chronic renal problems and other clinical manifestations (GLASS et al. 1990) suggest that the hantavirus story will have many more chapters.

5.5 Genus *Tospovirus*

Table 32. Genus *Tospovirus*

Virus	Geographic distribution	Vector	Plants
Tomato spotted wilt virus	Widespread	Thrips	370 species

Tomato spotted wilt virus (TSWV; Table 32) has many properties in common with viruses in the family Bunyaviridae (MILNE and FRANCKI 1984; DE HAAN et al. 1989). Based upon this information, the Bunyaviridae Study Group has recommended that a new genus *Tospovirus* be established in the family. TSWV, the only member of the genus, is an important plant pathogen. It infects a wide range of flowering plants (370 species in 50 families, comprising both monocotyledons and dicotyledons) in tropical, subtropical, and temperate regions (DE HAAN et al. 1989). The virus is horizontally transmitted between plant hosts in a circulative fashion by nine species of thrips (Thysanoptera), but the virus can also be vertically transmitted through the seeds of certain plant hosts (MANDAHAR 1981). In certain *Senecio* spp., seed transmission rates of more than 70% have been reported, and of 96% in *Senecio cruentus*. Vertical transmission of tospovirus is particularly interesting in light of efficient vertical transmission of certain other members of the Bunyaviridae in their respective vectors.

5.6 Bunyaviridae-Like Viruses

Most of the viruses in the remaining minor groups are considered to be members of the family Bunyaviridae on the basis of morphologic (electron microscopy) or other evidence. The Bhanja, Kaisodi, and Upolu group viruses have been isolated from ticks; most of the rest of the groups have been associated with mosquitoes (KARABATSOS 1985).

Table 33. Bhanja group viruses

Virus	Geographic distribution	Vector	Vertebrate
Bhanja	Africa, Asia, Europe	T i	H,R,O
Forecariah	Africa	T i	
Kismayo*	Africa	T i	

1. Bhanja group viruses have been isolated only from ixodid ticks (KARABATSOS 1985; Table 33). The virus has been isolated in 15 countries in Africa and Eurasia, and antibodies to the virus have been detected in 15 additional countries (HUBALEK 1987). Ruminants are the apparent vertebrate hosts. Bhanja virus has been known to infect humans.

Table 34. Kaisodi group viruses

Virus	Geographic distribution	Vector	Vertebrate
Kaisodi	Asia	T i	B
Lanjan	Asia	T i	P
Silverwater	North America	T i	S

2. Kaisodi group viruses (Table 34) have been isolated from ticks collected in Asia and in North America (KARABATSOS 1985). None of these has been found to infect humans.

Table 35. Mapputta group viruses

Virus	Geographic distribution	Vector	Vertebrate
Gan Gan	Australia	M c	
Mapputta	Australia	M a	
Maprik	Australia	M c+a	
Trubanaman	Australia	M a	

3. Mapputta viruses (Table 35) are restricted to Australia. Little is known concerning their vertebrate hosts. They are not known infect humans (KARABATSOS 1985).

Table 36. Resistencia group viruses

Virus	Geographic distribution	Vector	Vertebrate
Antequera	South America	M c	
Barranqueras	South America	M c	
Resistencia	South America	M c	

4. Resistencia group viruses (Table 36) have been isolated from only culicine mosquitoes in Argentina (KARABATSOS 1985). They are known to infect humans.

Table 37. Tanga group viruses

Virus	Geographic distribution	Vector	Vertebrate
Okola	Africa	M c	
Tanga	Africa	M a	

5. Tanga group viruses (Table 37) were isolated from mosquitoes collected in Tanzania and Cameroon. Little is known about their natural histories. Tanga virus has been isolated from a febrile human (ZELLER et al. 1989).

Table 38. Thiafora group viruses

Virus	Geographic distribution	Vector	Vertebrate
Erve	Europe		O
Thiafora	Africa		O

6. Thiafora group viruses were isolated from shrews collected in France and Cameroon (Table 38). Little is known of their natural history. A minor serologic cross-reaction with polyvalent sera prepared to nairoviruses has been demonstrated (ZELLER et al. 1989). If these are nairoviruses, ticks vectors are expected.

Table 39. Upolu group viruses

Virus	Geographic distribution	Vector	Vertebrate
Aransas Bay	North America	T a	
Upolu	Australia	T a	

7. Upolu viruses have been isolated only from argasid ticks (Table 39). None of these viruses are known to infect humans (KARABATSOS 1985).

Table 40. Yogue group viruses

Virus	Geographic distribution	Vector	Vertebrate
Kasokero*	Africa		Ba
Yogue	Africa		Ba

8. Yogue virus were isolated from the same species of bat in Uganda and Senegal (ZELLER et al. 1989; Table 40). Kasokero virus readily infects and causes disease in laboratory workers (KARABATSOS 1985).

Table 41. Ungrouped viruses

Virus	Geographic distribution	Vector	Vertebrate
Belmont	Australia	M c	
Enseada	South America	M c	
Issyk-Kul	Asia	M c+a, T a	H,B
Kowanyama	Australia	M a	
Pacora	North America	M c	
Tai	Africa	M c	
Tataguine	Africa	M c+a	H
Witwatersrand	Africa	M c	R,S
Yacaaba	Australia	M c	
Batken	Asia	M c, T i	
Chim	Asia	T i	
Keterah	Asia	T a	B
Lone Star	North America	T i	
Razdan	Asia	T i	
Sunday Canyon	North America	T a	
Tamdy	Asia	T a	H
Wanowrie	Africa, Asia	M c, T i	H
Bangui	Africa		H
Belem	South America		B
Bobaya	Africa		B
Santarem	South America		R

9. Twenty-one ungrouped viruses have been listed according to their vector associations (Table 41). Some apparently are mosquito-borne, some tick-borne, and the others have no known vector associations.

6 Summary

Obviously, the family Bunyaviridae is comprised of a large number of epidemiologically diverse viruses. They vary dramatically in their vector and vertebrate host relationships, geographic distributions, and epidemic potential in humans and animals. Public health practitioners, veterinarians, virologists, entomologists, biologists, ecologists, molecular biologists, and other scientists will all benefit from increased study and knowledge of this fascinating group of viruses.

References

Amundsen TE, Yuill TM (1981) Natural La Crosse virus infection in the red fox (*Vulpes fulva*), gray fox (*Urocyon cinereargenteus*), raccoon (*Procyon lotor*), and oppossum (Didelphis virginiana). Am J Trop Med Hyg 30: 706–714

Anderson CR, Spence L, Downs WG, Aitken THG (1961) Oropouche virus: a new human disease from Trinidad, West Indies. Am J Trop Med Hyg 10: 574–578

Antoniadis A, LeDuc JW, Daniel-Alekiou S (1987) Clinical and epidemiological aspects of hemorrhagic fever with renal syndrome (HFRS) in Greece. Eur J Epidemiol 3: 295–301

Artsob H (1983) Distribution of California serogroup viruses and virus infections in Canada. In: Calisher CH, Thompson WH (eds) California serogroup viruses. Liss, New York, pp 277–290

Avsic-Zupanc T, Likar M, Novakovic S, Cizman B, Kraigher A, van der Groen G, Stojanovic R, Obradovic M, Gligic A, LeDuc JW (1990) Evidence of the presence of two hantaviruses in Slovenia, Yugoslavia. Arch Virol, pp 87–94

Baldridge GD, Beaty BJ, Hewlett MJ (1989) Genomic stability of La Crosse virus during vertical and horizontal transmission. Arch Virol 108: 89–99

Bardos V, Pesko J (1981) Biological and antigenic variants among Tahyna virus strains isolated in Czechoslovakia . Arch Virol 68: 65

Beaty BJ, Bishop DHL (1988) Bunyavirus-vector interactions. Virus Res 10: 289–302

Beaty BJ, Thompson WH (1975) Emergence of La Crosse virus from endemic foci. Fluorescent antibody studies of overwintered *Aedes triseriatus*. Am J Trop Med Hyg 24: 685–691

Beaty BJ, Thompson WH (1976) Delineation of La Crosse virus in developmental stages of transovarially infected *Aedes triseriatus*. Am J Trop Med Hyg 25: 505–512

Beaty BJ, Thompson WH (1978) Tropsims of La Crosse virus in *Aedes triseriatus* (Diptera:Culicidae) following infective blood meals. J Med Entomol 14: 499–503

Beaty BJ, Tesh RB, Aitken THG (1980) Transovarial transmission of yellow fever virus in *Stegomyia* mosquitoes. Am J Trop Med Hyg 29: 125–132

Beaty BJ, Holterman M, Tabachnick W, Shope RE, Rozhon EJ, Bishop DHL (1981) Molecular basis of bunyavirus transmission by mosquitoes: role of the middle-sized RNA segment. Science 211: 1433–1435

Beaty BJ, Miller BR, Shope RE, Rozhon EJ, Bishop DHL (1982) Molecular basis of bunyavirus per os infection of mosquitoes: role of the middle-sized RNA segment. Proc Natl Acad Sci USA 79: 1295–1297

Beaty BJ, Sundin DR, Chandler LJ, Bishop DHL (1985) Evolution of bunyaviruses by genome reassortment in dually infected mosquitoes (*Aedes triseriatus*). Science 230: 548–550

Beaty BJ, Trent DW, Roehrig JT (1988) Virus variation and evolution: mechanisms and epidemiological significance. In: Monath TP (ed) The arboviruses: epidemiology and ecology, vol I. CRC Press, Boca Raton, pp 59–85

Bilsel P, Tesh RB, Nichol ST (1988) RNA genome stability of Toscana virus during serial transovarial transmission in the sandfly *Phlebotomus perniciosus*. Virus Res 11: 87–94

Bishop DHL (1985a) Replication of arenaviruses and bunyaviruses. In: Fields BN (ed) Virology, 1st edn. Raven, New York, pp 1083–1119

Bishop DHL (1985b) The genetic basis for describing viruses as species. Intervirology 24: 79–93

Bishop DHL (1990) Bunyaviridae and their replication. I. Bunyaviridae. In: Fields BN (ed) Virology, 2nd edn. Raven, New York, pp. 1155–1173

Bishop DHL, Beaty BJ (1988). Molecular and biochemical studies of the evolution, infection and transmission of insect bunyaviruses. Philos Trans R Soc Lond [Bisl] B321: 463–483

Bishop DHL, Shope RE (1979) Bunyaviridae. In: Frankel-Conrat H, Wagner RR (eds) Comprehensive virology, vol 14. Plenum, New York, pp 1–156

Booth TF, Marriott AC, Steele GM, Nuttall PA (1990) Dugbe virus in ticks: histological localization studies using light and electron microscopy. Arch Virol, pp 207–218

Brummer-Korvenkontio M, Vaheri A, Hovi T (1980) Nephropathia epidemica: detection of antigen in bank voles and serologic diagnosis of human infection. J Infect Dis 141: 131–134

Butenko AM, Chumakov MP (1990) Isolation of Crimean-Congo hemorrhagic fever virus from patients and from autopsy specimens. Arch Virol, pp 295–301

Calisher CH (1979) Antigenic relationships of the arboviruses: an ecological and evolutionary approach. In: (Institute of Virology, Slovak Academy of Sciences) New aspects in ecology of arboviruses, Prague, pp 117–152

Calisher CH (1983) Taxonomy, classification, and geographic distribution of California serogroup bunyaviruses. In: Calisher CH, Thompson WH (eds) California serogroup viruses. Liss, New York, pp 1–16

Calisher CH (1988) Evolutionary significance of the taxonomic data regarding bunyaviruses of the family Bunyaviridae. Intervirology 29: 268–276

Calisher CH, Sasso DR, Maness KSC, Gheorghiu VN, Shope RE (1973) Relationships of Anopheles A group arboviruses. Proc Soc Exp Biol Med 143: 465–468

Calisher CH, Thompson WH (1983) Progress in clinical and biological research, vol 123. California serogroup viruses. Liss, New York

Calisher CH, Francy DB, Smith GC, Muth DJ, Lazuick JS, Karabatsos N, Jakob WL, McLean RG (1986)

Distribution of Bunyamwera serogroup viruses in North America, 1956–1984. Am J Trop Med Hyg 35: 429–443
Calisher CH, Shope RE (1988) Bunyaviridae: the bunyaviruses. In: Lennette EH, Halonen P, Murphy FA (eds) Viral, rickettsial, and chlamydial diseases. Springer, Berlin Heidelberg New York, pp 626–646 (Lasboratory diagnosis of infections diseases—principles and practice, vol 2)
Calisher, CH (ed) Hemorrhagic fever with renal syndrome, tick- and mosquito-borne viruses. Arch Virol (suppl. 1), Springer, Vienna, 1990
Calisher CH, Beaty BJ, Dalrymple JM, Elliott RM, Karabatsos N, Lee HW, Lvov DK, Nuttall PA, Peters D, Petersson R, Schmaljohn CS, Shope RE (1990) Report of the Bunyaviridae Study Group. Arch Virol [Suppl]
Camicas JL, Wilson ML, Cornet JP, Digoutte JP, Calvo MA, Adam F, Gonzalez JP (1990) Ecology of ticks as potential vectors of Crimean-Congo hemorrhagic fever virus in Senegal: epidemiological implications. Arch Virol, pp 303–322
Casals J (1963) New developments in the classification of arthropod-borne animal viruses. Ann Microbiol 11: 13–34
Casals J, Tignor GH (1980). The *Nairovirus* genus: serological relationships. Intervirology 14: 144–147
Causey OR, Causey CE, Maroja OM, Macedo DG (1961) The isolation of arthropod-borne viruses, including members of two hitherto undescribed serological groups, in the Amazon region of Brazil. Am J Trop Med Hyg 10: 227–249
Chandler LJ, Beaty BJ, Baldridge GD, Bishop DHL, Hewlett MJ (1990) Heterologous reassortment of bunyaviruses in *Aedes triseriatus* mosquitoes and transovarial and oral transmission of newly evolved genotypes. J Gen Virol 71: 1045–1050
Chumakov MP, Gavrilovskaya IN, Myasnikov YA (1990) Reservoirs and modes of spread of hemorrhagic fever with renal syndrome, a zoonotic nontransmissible human disease. Arch Virol, pp 49–56
Chung S-I, Livingston CW, Edwards JF, Crandell RW, Shope RE, Shelton MJ, Collisson EW (1990) Evidence that Cache Valley virus induces congenital malformations in sheep. Vet Microbiol 21: 297–307
Cornet JP, Georges AJ, Gonzalez JP (1987) Passage transovarien in natura du virus Dugbe chez la tique *Amblyomma variegatum*. Ann Inst Pasteur Virol 138: 269–271
Daubney R, Hudson JR, Garnham PC (1931) Enzootic hepatitis or Rift Valley fever. An undescribed virus disease of sheep, cattle and man from East Africa. J Pathol Bacteriol 34: 545
DeFoliart GR (1983) *Aedes triseriatus*: vector biology in relationship to the persistence of La Crosse virus in endemic foci. In: Calisher CH, Thompson WH (eds) California serogroup viruses. Liss, New York, pp 89–104
DeHaan P, Wagemakers L, Peters D, Goldbach R (1989) Molecular cloning and terminal sequence determination of the S and M RNAs of tomato spotted wilt virus. J Gen Virol 70: 3469–3473
Doherty RL, Carley JG, Mackerras MJ, Marks EN (1963) Studies of arthropod-borne virus infections in Queensland. III. Isolation and characterization of virus strains from wild-caught mosquitoes in north Queensland. Aust J Exp Biol Med Sci 41: 17–40
Dutary BE, Peterson JL, Peralta PH, Tesh RB (1989) Transovarial transmission of Gamboa virus in a tropical mosquito, *Aedeomyia squamipennis*. Am J Trop Med Hyg 40: 108–113
Edwards JF, Livingston CW, Chung SI, Collisson EW, (1989) Ovine arthrogryposis and central nervous system malformations associated with in utero Cache Valley virus infection: spontaneous disease. Vet Pathol 26: 33–39
Elliott RM (1990) Molecular biology of the Bunyaviridae. J Gen Virol 71: 501–522
Emmons RW Woodie JD, Laub RL Oshiro LS (1983) Main Drain virus as a cause of equine encephalomyelitis. J Am Vet Med Assoc 183: 555–558
Gaidamovich SY, Khutoretskaya NV, Asyamov YV, Tsyupa VI, Melnikova EE (1990) Sandfly fever in central Asia and Afghanistan. Arch Virol, pp 287–293
Gavrilovskaya IN, Apekina NS, Bernshtein AD, Demina VT, Okulova NM, Myasnikov YA, Chumakov MP (1990) Pathogenesis of hemorrhagic fever with renal syndrome virus infection and mode of horizontal transmission of hantavirus in bank voles. Arch Virol, pp 57–62
Glass GE, Childs JE, Watson AJ, LeDuc JW (1990) Association of chronic renal disease, hypertension and infection with a rat-borne hantavirus. Arch Virol, pp 69–80
Gligic A, Frusic M, Obradovic M, Stojanovic R, Hlaca D, Gibbs CJ Jr, Yanagihara R, Calisher CH, Gadjusek DC (1989) Hemorrhagic fever with renal syndrome in Yugoslavia: antigenic characterization of hantaviruses isolated from *Apodemus flavicollis* and *Clethrionomys glareolus*. Am J Trop Med Hyg 41: 109–115

Godsey MS Jr, Amoo F, Yuill TM, DeFoliart GR (1988) California serogroup virus infections in Wisconsin domestic animals. Am J Trop Med Hyg 39: 409–416

Gonzalez JP, Georges AJ (1988) Other bunyaviral fevers: Bunyamwera, Ilesha, Germiston, Bwamba, and Tataguine. In: Monath TP (ed) The arboviruses: epidemiology and ecology, CRC Press, Boca Raton, FL, pp 87–98

Gonzalez JP, Cornet JP Camicas JL (1989) Dugbe virus replication in nymph and adult *Amblyomma variegatum*. Res Virol 140: 333–336

Gonzalez-Scarano F (1985) La Crosse virus G1 glycoprotein undergoes a conformational change at the pH of fusion. Virology 140: 209–216

Gonzalez-Scarano F, Nathanson N (1990) Bunyaviruses. In: Fields BN (ed) Virology, 2nd edn. Raven, New York, pp 1195–1228

Gonzalez-Scarano F, Janssen RS, Najjar JA, Pobjecky N, Nathanson N (1985) An avirulent G1 glycoprotein variant of La Crosse bunyavirus with defective fusion function. J Virol 54: 757–763

Grady LJ, Kinch W (1985) Two monoclonal antibodies against La Crosse virus show host-dependent neutralizing activity. J Gen Virol 66: 2773–2776

Grayson MA, Srihongse S, Deibel R, Calisher CH, (1983) California serogroup viruses in New York State: a retrospective analysis of subtype distribution patterns and their epidemiologic significance, 1965–1981. In: Calisher CH, Thompson WH (eds) California serogroup viruses. Liss, New York, pp 257–267

Grimstad PR (1988) California group virus disease. In: Monath TP (ed) The arboviruses: epidemiology and ecology, vol II. CRC Press, Boca Raton, FL, pp 99–136

Grimstad PR, Craig GB, Ross QE, Yuill TM (1977) *Aedes triseriatus* and La Crosse virus: geographic variation in vector susceptibility and ability to transmit. Am J Trop Med Hyg 26: 990–996

Grimstad PR, Ross QE, Craig GB Jr (1980) *Aedes triseriatus* (Diptera: Culicidae) and La Crosse virus. II. Modification of mosquito feeding behavior by virus infection. J Med Entomol 17: 1–7

Grimstad PR, Calisher CH, Harroff RN, Wentworth BB (1986) Jamestown Canyon virus (California serogroup) is the etiologic agent of widespread infection in Michigan humans. Am J Trop Med Hyg 35: 376–386

Guelmino DJ, Jevtic M (1955) An epidemiological and hematological study of sandfly fever in Serbia. Acta Trop (Basel) 12, 179

Hammon WM, Reeves WC (1952) California encephalitis virus, a newly described agent. I. Evidence of natural infection in man and other animals. Calif Med 77: 303–309

Hardy JL (1988) Susceptibility and resistance of vector mosquitoes. In: Monath TP (ed) The arboviruses: epidemiology and ecology, vol I. CRC Press, Boca Raton, FL, pp 87–126

Hardy JL, Houk EJ, Kramer LD, Reeves WC (1983) Intrinsic factors affecting vector competence of mosquitoes for arboviruses. Ann Rev Entomol 28: 229–262

Holland J, Spindler K, Horodyski F, Grabau E, Nichol S, VandePol S (1982) Rapid evolution of RNA genomes. Science 215: 1577–1585

Hoogstraal H (1979) The epidemiology of tick-borne Crimean-Congo hemorrhagic fever in Asia, Europe, and Africa. J Med Entomol 15: 307–417

Hubalek Z (1987) Geographic distribution of Bhanja virus. Folia Parasitol 34: 77–86

Jones LD, Davies CR, Steele GM, Nuttall PA (1987) A novel mode of arbovirus transmission involving a nonviremic host. Science 237: 775–777

Jouan A, LeGuenno B, Digoutte JP, Philippe B, Riou O, Adam F (1988) A RVF epidemic in southern Mauritania. Ann Inst Pasteur Virol 139: 307–308

Kappus KD, Monath TP, Kaminski RM, Calisher CH (1983) Reported encephalitis associated with California serogroup virus infections in the United States, 1963–1981. In: Calisher CH, Thompson WH (eds) California serogroup viruses. Liss, New York, pp 31–41

Karabatsos N (1985) International catalogue of arboviruses including certain other viruses of vertebrates. American Society of Tropical Medicine and Hygiene, San Antonio

Kingsford L, Hill DW (1983) The effect of proteolytic cleavage of La Crosse virus G1 glycoprotein on antibody neutralization. J Gen Virol 64: 2147–2156

Klimas R, Thompson WH, Calisher CH, Clark GG, Grimstad PR, Bishop DHL (1981) Genotypic varieties of La Crosse virus isolated from different geographic regions of the continental United States and evidence for a naturally occurring intertypic recombinant La Crosse virus. Am J Epidemiol 114: 112–131

Kramer LD, Hardy JL, Presser SB, Houk EJ (1981) Dissemination barriers for western equine encephalomyelitis virus in *Culex tarsalis* infected after ingestion of low viral doses. Am J Trop Med Hyg 30: 190–197

Kurogi H, Inaba Y, Goto Y, Miura Y, Takahashi H, Sato K, Omori T, Matumoto M (1975) Serological evidence for etiologic role of Akabane virus in epizootic abortion—arthrogryposis-hydranencephaly in cattle in Japan, 1972–1974. Arch Virol 47: 71–83

LeDuc JW (1987) Epidemiology of Hantaan and related viruses. Lab Anim Sci 37: 413–418

LeDuc JW (1989) Epidemiology of hemorrhagic fever viruses. Rev Infect Dis 11: S730–S735

LeDuc JW, Antoniades A, Siamopoulos K (1986a) Epidemiological investigations following an outbreak of hemorrhagic fever with renal syndrome in Greece. Am J Trop Med Hyg 35: 654–659

LeDuc JW, Smith GA, Childs JE (1986b) Global survey of antibody to Hantaan-related viruses among peridomestic rodents. Bull WHO 64: 139–144

LeDuc JW, Pinheiro FP (1988) Oropouche fever. In: Monath TP (ed) *The arboviruses: epidemiology and ecology*, vol IV. CRC Press, Boca Raton, FL, pp 1–14

Lee HW (1982) Korean hemorrhagic fever. Prog Med Virol 28: 96–113

Lee HW, Lee PW, Johnson KJ (1978) Isolation of the etiologic agent of Korean hemorrhagic fever. J Infect Dis 137: 298–308

Lee HW, Baek LJ, Johnson KM (1982) Isolation of Hantaan virus, the etiologic agent of Korean hemorrhagic fever, from wild urban rats. Infect Dis 146: 638–644

Lee PW, Amyx HL, Yanagihara R, Gajdusek DC, Goldgaber D, Gibbs CJ Jr (1985) Partial characterization of Prospect Hill virus isolated from meadow voles in the United States. J Infect Dis 152: 826–829

Lee HW, Lee PW, Baek LJ, Chu YK (1990) Geographical distribution of hemorrhagic fever with renal syndrome and hantaviruses. Arch Virol, pp 5–18

Linthicum KJ, Bailey CL (1985) Observations on the dispersal and survival of a population of *Aedes lineatopennis* (Ludlow) (Diptera: Culicidae) in Kenya. Bull Entomol Res 75: 661

Linthicum KJ, Davies FG, Kairo A, Bailey CL (1985) Rift Valley fever virus collected during an interepizootic period in Kenya. J Hyg (Cambridge) 95: 197–209

Linthicum KJ, Bailey CL, Davies FG, Tucker CJ (1987) Detection of Rift Valley fever viral activity in Kenya by satellite remote sensing imagery. Science 235: 1656–1659

Linthicum KJ, Logan TM, Bailey CL, Hohm DJ, Moulton JR (1989) Transtadial and horizontal transmission of Rift Valley fever virus in *Hyalomma truncatum*. Am J Trop Med Hyg 41: 491–496

Ludwig GV, Christensen BM, Yuill TM, Schultz KT (1989) Enzyme processing of La Crosse virus glycoprotein G1: a bunyavirus-vector infection model. Virology 171: 108–113

Mandahar CL (1981) Virus transmission through seed and pollen. In: Maramorosch K, Harris KF (eds) Plant diseases and vectors, ecology and epidemiology. Academic, New York, pp 241–292

Mattingly PF (1969) *The biology of mosquito-borne diseases*. George, Allen and Unwin, London

McIntosh BM, Jupp PG (1981) Epidemiological aspects of Rift Valley fever in South Africa with references to vectors. Contrib Epid Biostat 3: 92–99

McPhee DA, Della-Porta AJ (1988) Biochemical and serological comparisons of Australian bunyaviruses belonging to the Simbu serogroup. J Gen Virol 69: 1007–1017

Meegan JM (1979) The Rift Valley fever epizootic in Egypt 1977–78. I. Description of the epizootic and virological studies. Trans R Soc Trop Med Hyg 73: 618–723

Meegan JM, Bailey CL (1988) Rift Valley fever. In: Monath TP (ed) The arboviruses: *epidemiology and ecology*, vol IV. CRC Press, Boca Raton, FL, pp 51–76

Meegan JM, Hoogstraal H, Moussa MI (1979). An epizootic of Rift Valley fever in Egypt in 1977. Vet Rec 105: 124–125

Miller BR, DeFoliart GR, Yuill TM (1977) Vertical transmission of La Crosse virus (California encephalitis group): transovarial and filial infection rates in *Aedes triseriatus* (Diptera: Culicidae). J Med Entomol 14: 437–440

Miller BR, Beaty BJ, Lorenz LH (1982) Variation of La Crosse virus filial infection rates in geographic strains of *Aedes triseriatus* (Diptera: Culicidae). J Med Entomol 19: 213–214

Milne RG, Francki RIB (1984) Should tomato spotted wilt virus be considered as a possible member of the family Bunyaviridae? Intervirology 22: 72–76

Mitchell CJ (1983) Mosquito vector competence and arboviruses. In: Harris KF (ed) Current topics in vector research, vol 1. Praeger, New York, pp 63–92

Monath TP (1988) The arboviruses: epidemiology and ecology. CRC Press, Boca Raton

Montgomery RE (1917) On a tick-borne gastro-enteritis of sheep and goats occurring in British East Africa. J Comp Pathol Ther 30: 28–57

Myhrman G (1951). Nephropathia epidemica: a new infectious disease in northern Scandinavia. Acta Med Scand 140: 52–56

Oker-Blom N, Salminin A, Brummer-Korvenkontio M (1964) Isolation of some viruses other than typical tick-borne encephalitis viruses from *Ixodes ricinus* ticks in Finland. Ann Med Exp Fenn 42: 100–112
Oya A, Ogata T, Kobayashi I, Matsuyama T (1961) Akabane, a new arbovirus isolated in Japan. Jpn J Med Sci Biol 14: 101–108
Pantuwatana S, Thompson WH, Watts DM, Hanson RP (1972) Experimental infection of chipmunks and squirrels with La Crosse and trivittatus virus by *Aedes triseriatus*. Am J Trop Med Hyg 21: 476–481
Pantuwatana S, Thompson WH, Watts DM, Yuill TM, Hanson RP (1974) Isolation of La Crosse virus from field collected *Aedes triseriatus*. Am J Trop Med Hyg 21: 476–481
Parry JE (1983) Control of *Aedes triseriatus* in La Crosse, Wisconsin. In: Calisher CH, Thompson WH (eds) California serogroup viruses. AR Liss, New York, pp 355–363
Parsonson IM, McPhee DA (1985) Bunyavirus pathogenesis. Adv Virus Res 30: 279–316
Patrican LA, DeFoliart GR (1987) *Aedes triseriatus* and La Crosse virus—similar venereal infection rates in females given the first bloodmeal immediately before mating or several days after mating. Am J Trop Med Hyg 36: 648–652
Patrican LA, DeFoliart GR, Yuill TM (1985a) La Crosse virus viremias in juvenile, subadult, and adult chipmunks (*Tamias striatus*) following feeding by transovarially infected *Aedes triseriatus*. Am J Trop Med Hyg 34: 596–602
Patrican LA, DeFoliart GR, Yuill TM (1985b) Oral infection and transmission of La Crosse virus by an enzootic strain of *Aedes triseriatus* feeding on chipmunks with a range of viremia levels. Am J Trop Med Hyg 34: 992–998
Paulson SL, Grimstad PR (1989) Replication and dissemination of La Crosse virus in the competent vector *Aedes triseriatus* and the incompetent vector *Aedes hendersoni* and evidence for transovarial transmission by *Aedes hendersoni* (Diptera:Culicidae). J Med Entomol 26: 602–609
Paulson SL, Grimstad PR, Craig GB Jr (1989) Midgut and salivary gland barriers to La Crosse virus dissemination in mosquitoes of the *Aedes triseriatus* group. Med Vet Entomol 3: 113–123
Pinheiro FP, Travassos da Rosa AP, Travassos da Rosa JFS, Ishak R, Freitas RB, Gomes ML, LeDuc JW, Oliva OFP (1981) Oropouche virus. I. A review of clinical, epidemiological and ecological findings. Am J Trop Med Hyg 30: 149–160
Pinheiro FP, Travassos da Rosa APA, Gomes MLC, LeDuc JW, Hoch AL (1982) Transmission of Oropouche virus from man to hamster by the midge, *Culicoides paraensis*. Science 215: 1251–1252
Pringle CR, Lees JF, Clark W, Elliott RM (1984) Genome subunit reassortment among bunyaviruses analysed by dot hybridization using molecularly cloned complementary DNA probes. Virology 135: 244–256
Roca-Garcia MJ (1944) The isolation of three neurotropic viruses from forest mosquitoes in eastern Colombia. J Infect Dis 75: 160–169
Sabin AB (1955) Recent advances in our knowledge of dengue and sandfly fever. Am J Trop Med Hyg 4: 198–207
Schmaljohn CS, Dalrymple JM (1983) Analysis of Hantaan virus RNA: evidence for a new genus of Bunyaviridae. Virology 131: 482–491
Schmaljohn CS, Patterson JL (1990) Bunyaviridae and their replication part II. Replication of Bunyaviridae. In: Fields BN (ed) Virology, 2nd edn. Raven, New York, pp 1175–1194
Sekoyova M, Gresikova M, Stupalova S (1970) Serological study on distribution of Uukuniemi virus in man. Folia Parasitol 17: 341
Sellers RF, Pedgley DE (1985). Possible windborne spread to western Turkey of blue tongue virus in 1977 and of Akabane virus in 1979. J Hyg (Lond) 95: 149–158
Shope RE (1985) Bunyaviruses. In: Fields BN (ed) Virology, 1st edn. Raven, New York, pp 1055–1082
Shope RE (1988) Group C viruses. In: Monath TP (ed) *The arboviruses: epidemiology and ecology*, vol III. CRC Press, Boca Raton, FL, pp 37–52
Shope RE, Peters CJ, Davies FG (1982) The spread of Rift Valley fever and approaches to its control. Bull WHO 60: 299–304
Simpson DIH, Knight EM, Courtois G, Williams MC, Weinbren MP, Kibukamusoke JW (1967) Congo virus: a hitherto undescribed virus occurring in Africa. I. Human isolations—clinical notes. East Afr Med J 44: 87–92
Smadel JE (1951) Epidemic hemorrhagic fever. Am J Public Health 43: 1327–1330
Smithburn KC, Haddow AJ, Mahaffy AF (1946) A neurotropic virus isolated from *Aedes* mosquitoes caught in the Semliki forest. Am J Trop Med Hyg 26: 189–208

Srihongse S, Grayson MA, Diebel R (1984) California serogroup viruses in New York State: the role of subtypes in human infections. Am J Trop Med Hyg 33: 1218–1227

St George TD, Standfast HA (1988) Simbu group viruses with teratogenic potential. In: Monath TP (ed) The arboviruses: epidemiology and ecology, vol IV. CRC Press, Boca Raton, pp 145–166

Steele GM, Nuttall PA (1989) Difference in vector competence of two species of sympatric ticks, *Amblyomma variegatum* and *Rhipicephalus appendiculatus*, for Dugbe virus (*Nairovirus*, Bunyaviridae). Virus Res 14: 73–84

Sundin DR, Beaty BJ (1988) Interference to oral superinfection of *Aedes triseriatus* infected with La Crosse virus. Am J Trop Med Hyg 38: 428–432

Sundin DR, Beaty BJ, Nathanson N, Gonzalez-Scarano F (1987) A G1 glycoprotein epitope of La Crosse virus: a determinant of infection of *Aedes triseriatus*. Science 235: 591–593

Swanepoel R (1981) Rift Valley fever in Zimbabwe. Contrib Epid Biostat 3: 83–91

Swanepoel R, Struthers JK, Shepherd AJ, McGillivray GM, Nel MJ, Jupp PG (1983) Crimean-Congo haemorrhagic fever in South Africa. Am J Trop Med Hyg 32: 1407–1415

Swartz TA, Klingberg MA, Goldblum N (1981) Rift Valley fever: contributions to epidemiology and biostatistics. Karger, Basel

Sweet C, Coates DM (1990) Studies of the pathogenesis of Dugbe virus in normal and in immunosuppressed mice. Arch Virol, pp 181–195

Tesh RB (1988) Phlebotomus fevers. In: Monath TP (ed) The arboviruses: epidemiology and ecology, vol IV. CRC Press, Boca Raton, pp 15–27

Theiler M, Downs WG (1973) *The arthropod-borne viruses of vertebrates.* Yale University Press, New Haven

Thompson WH, Beaty BJ (1977) Venereal transmission of La Crosse (California encephalitis) arbovirus in *Aedes triseriatus* mosquitoes. Science 196: 530–531

Thompson WH, Beaty BJ (1978) Venereal transmission of La Crosse virus from male to female *Aedes triseriatus*. Am J Trop Med Hyg 27: 187–196

Thompson WH, Gundersen CB (1983) La Crosse encephalitis: occurrence of disease and control in a suburban area. In: Calisher CH, Thompson WH (eds) California serogroup viruses, Liss, New York, pp 225–236

Thompson WH, Kalfayan B, Anslow RO (1965) Isolation of California encephalitis group virus from a fatal human illness. Am J Epidemiol 81: 245–253

Tsai TF (1987) Hemorrhagic fever with renal syndrome: mode of transmission to humans. Lab Anim Sci 37: 428–430

Turell MJ (1988) Horizontal and vertical transmission of viruses by insect and tick vectors. In: Monath TP (ed) *The arboviruses: epidemiology and ecology*, vol I. CRC Press, Boca Raton, pp 127–152

Turell MJ (1989) Effect of environmental temperature on the vector competence of *Aedes fowleri* for Rift Valley fever virus. Res Virol 140: 147–154

Turell MJ, LeDuc JW (1983) The role of mosquitoes in the natural history of California serogroup viruses. In: Calisher CH, Thompson WH (eds) California serogroup viruses, Liss, New York, pp 43–55

Turell MJ, Perkins PV (1990) Transmission of Rift Valley fever virus by the sand fly, *Phlebotomus duboscqi* (Diptera: Psychodidae). Am J Trop Med Hyg 42: 185–188

Turell MJ, Rossignol PA, Spielman A, Rossi CA, Bailey CL (1984) Enhanced arboviral transmission by mosquitoes that concurrently ingested microfilariae. Science 225: 1039–1041

Vasilenko S, Brodvarova I, Topov Y, Shindarov L, Kebedzhiev G (1990) Hemorrhagic fever with renal syndrome in Bulgaria: isolation of hantaviruses and epidemiologic considerations. Arch Virol, pp 63–67

Watts DM, Morris CD, Wright RE, DeFoliart GR, Hanson RP (1972) Transmission of La Crosse virus (California encephalitis group) by the mosquito *Aedes triseriatus*. J Med Entomol 9: 125–127

Watts DM, Pantuwatana S, DeFoliart GR, Yuill TM, Thompson WH (1973) Transovarial transmission of La Crosse virus (California encephalitis group) in the mosquito, *Aedes triseriatus*. Science 182: 1140–1141

Watts DM, Thompson WH, Yuill TM, DeFoliart GR, Hanson RP (1974) Overwintering of La Crosse virus in *Aedes triseriatus*. Am J Trop Med Hyg 23: 694–700

Watts DM, Ksiazek TG, Linthicum KJ, Hoogstraal H (1988) Crimean-Congo hemorrhagic fever. In: Monath TP (ed) *The arboviruses: epidemiology and ecology*, vol II. CRC Press, Boca Raton, pp 177–222

Weaver SC, Scott TW, Lorenz LH, Lerdthusnee K, Romoser WS (1988) Togavirus-associated pathologic changes in the midgut of a natural mosquito vector. J Virol 62: 2083–2090

Weinbren MP, Gourlay RN, Lumsden WHR, Weinbren BM (1958) An epizootic of Nairobi sheep disease in Uganda. J Comp Pathol Ther 68: 174–187
Whitman L, Shope RE (1962) The California complex of arthropod-borne viruses and its relationship to the Bunyamwera group through Guaroa virus. Am J Trop Med Hyg 11: 691–696
Yan DY, Xie YJ, Zhang CA (1986) New isolates of HFRS virus in Sichuan, China and characterization of antigenic differences by monoclonal antibodies. Lancet i: 1328
Yuill TM (1983) The role of mammals in the maintenance and dissemination of La Crosse virus. In: Calisher CH Thompson WH (eds) California serogroup viruses. Liss, New York, pp 77–87
Zeller HG, Karabatsos N, Calisher CH, Digoutte J-P, Cropp CB, Murphy FA, Shope RE (1989) Electron microscopic and antigenic studies of uncharacterized viruses. II. Evidence suggesting the placement of viruses in the family Bunyaviridae. Arch Virol 108: 211–227

Virion Structure*

M. J. HEWLETT[1] and W. CHIU[2]

1 Introduction

A detailed three-dimensional image of a virus particle is essential for understanding its mechanism of assembly, its antigenic properties, its interaction with receptors (ROSSMANN and REUCKERT 1987; HARRISON 1990) and for the formation of a biophysical basis for rational design of antiviral agents (ROSSMANN 1989). X-ray crystallographic analysis has resulted in the structural solution, at the atomic level, of several nonenveloped animal viruses (HOGLE et al. 1985; ROSSMANN et al. 1985; ACHRYA et al. 1989). The diameters of these viruses are smaller than 400 Å. So far, its has been difficult to grow crystals of larger viruses with sufficient diffraction quality. Electron cryomicroscopy coupled with image analysis has emerged as a useful technique for providing three-dimensional structures of icosahedral virus particles which cannot readily form large crystalline arrays

* This research was supported by NIH grant number NS25877.
[1] Department of Molecular and Cellular Biology, University of Arizona, Tucson, Arizona, 85721, USA
[2] Verna and Marrs McLean Department of Biochemistry, Baylor College of Medicine, One Baylor, Plaza, Houston, Texas 77030, USA

Current Topics in Microbiology and Immunology, Vol. 169

(ADRIAN et al. 1984; VOGEL et al. 1986; PRASAD et al. 1988; BAKER et al. 1988; SCHRAG et al. 1989). The availability of monospecific antisera directed against viral capsid proteins has permitted the use of these reagents to locate the domains of antigenic sites on the surface of viral capsid proteins (PRASAD et al. 1990).

Bunyavirus structure has been discussed in several recent reviews (PETTERSON and VON BONSDORFF 1987; BISHOP 1990; ELLIOTT 1990; GONZALEZ-SCARANO and NATHANSON 1990; SCHMALJOHN and PETTERSSON 1990). In this chapter we focus our attention on structural features which have been revealed primarily by electron microscopic investigations. We also pose questions which still need to be answered with respect to the structures of these viruses.

2 Methods of Analysis

A detailed description of the methods used for structural analysis is beyond the scope of this chapter. Rather, we have chosen to focus on a few biochemical and electron microscopic methods which have been useful for obtaining structural information about the virus particles and their nucleocapsids, especially with regard to their biological activities.

2.1 Virus Purification

The examination of features of virion structure depends to some extent on the ability to purify reasonable quantities of the virus particles. In the case of bunyaviruses, growth and purification have been carried out by relatively standard procedures. Host cells (e.g., BHK21, chicken embryo cells) are infected, and released virus is purified from the culture medium after an appropriate time. Virus particles are isolated by differential and gradient centrifugation steps. Centrifugation through sucrose gradients (GOLDMAN et al. 1977) or glycerol/tartrate gradients (OBIJESKI et al. 1976b) has been employed. These steps are quite effective for the production of large quantities of bunyavirions at mature but not at intermediate states. Since these treatments subject the virus to nonphysiological conditions (high *g* forces, high osmolality), they could result in the alteration of some structural features. Therefore a caveat must be given in the interpretation of data resulting from such preparations.

2.2 Electron Microscopy

Electron microscopy has been a commonly used technique to study the morphologies of virus-infected cells, purified viruses, and nucleocapsids. The information content of the electron micrographs depends on the way in which the samples are prepared. In the case of virus-infected cells the samples are usually prepared in a sequence of steps including fixation, dehydration, embedding, sectioning, and staining. This kind of study would reveal the pathway of viruses entering the cells in an infection process and the morphologies of the viruses and subcellular components at different stages of infection (NERMUTT et al. 1987).

For purified viruses and nucleocapsids, one would use one of the following procedures: metal shadowing, negative staining, freeze-facture, or frozen hydration (NERMUT et al. 1987). Metal shadowing would give the surface topology of the virus particle whereas freeze-fracture would show the contour of the fracture face along the viral membrane. Negative stain would give a rough estimate of the particle size and some of the structural details in the stain-excluding regions of the viral surface proteins.

Frozen hydration is a more recent cryopreservative technique by which the virus particles are embedded in a thin layer of vitreous ice by rapid freezing. The frozen, hydrated samples are imaged under low temperature and low electron dose conditions. (ADRIAN et al. 1984; CHIU 1986; DUBOCHET et al. 1988; JENG et al. 1988). This technique can preserve the structural details of viral proteins to near atomic resolution (JENG and CHIU 1987). Furthermore, computer analysis of the electron images can provide information as to whether the particle possesses any icosahedral symmetry. If so, the three-dimensional structure can be determined by Fourier reconstruction (CROWTHER 1971). Because of the difference in scattering density between the protein and ice, the structures details of not only the outer shell but also the inner shell viral proteins can be retrieved (SCHRAG et al. 1989; PRASAD et al. 1990).

3 Composition

The chemical composition has not been determined for members of the Bunyaviridae family, except in the case of Uukuniemi virus. In this particular case, the particles were found to consist of 7% carbohydrate, 33% lipid, 58% protein, and 2% RNA by weight (R.F. PETTERSSON, personal communication).

The virion particles of this family are composed of nucleocapsids containing three different RNA species (L, M, and S) complexed with the nucleocapsid protein (N) and the virion transcriptase/polymerase (L). The nucleocapsids are packaged inside a lipid envelope during budding at internal cellular membranes, appearing in Golgi vesicles (LYONS and HEYDUK 1973; MURPHY et al. 1973; KUISMANEN et al. 1982; SMITH and PIFAT 1982). The lipid envelope contains two viral glycoproteins, G1 and G2. Molecular properties of these viral structural proteins are discussed in detail in other chapters of this volume. G1 and G2 are transmembrane proteins, and their presence dominates the surface features of the viral particles. The nucleocapsids are ribonucleoprotein complexes of viral RNA and N protein (with the L protein as a minor constituent). We discuss aspects of these structural features below with an emphasis on their known and presumed functions in the biology of the bunyaviruses.

4 Particle Morphology and Function

4.1 Size and Shape

Particles of highly purified La Crosse virus have been observed by electron cryomicroscopy. Their images appear spherical, as shown in Fig. 1a. This is in contrast to the common notion that bunyaviruses are pleomorphic, based on images of negative-stained particles (TALMON et al. 1987; JENG et al. 1988). A number of enveloped viruses, including Sendai virus and Pichinde virus (W. CHIU, unpublished observations) have been found to be spherical under neutral pH and 4 °C conditions, using electron cryomicroscopy. We interpret these findings to mean that the irregular shapes often observed by negative staining are a function of the desiccation which occurs during specimen preparation in this method.

We have reported that the size of La Crosse virus in our preparations varied from 750 to 1150 Å in diameter (TALMON et al. 1987). This size heterogeneity has also been observed for preparations of vitrified-hydrated Sendai virus and Pichinde virus (W. CHIU, unpublished observations). In contrast, cryopreserved Sindbis virus appears to be quite homogeneous in size (FULLER 1987). It is not known how the size of the virion might influence its infectivity.

The lower limit to the size of a viable virion should be determined by the necessity to assemble a biological active particle. For the case of segmented-genome, enveloped viruses (bunyaviruses, influenza viruses, and arenaviruses), this requirement implies some mechanism to ensure that the proper number of nucleocapsids are present in a particle. This may involve interactions between nucleocapsids and membrane sites containing viral glycoproteins. While most enveloped RNA viruses have a matrix protein, the bunyaviruses (and the unrelated arenaviruses) have no matrix protein. Therefore, assembly of nucleocapsids and envelope must take place due to interactions between the nucleocapsids and the internal domains of the glycoproteins or with the membrane itself. The assembly mechanisms by which the correct number of genomes are packaged for these enveloped viruses are not known.

The upper limit to virion size may be a function of several factors. The stability of the particles formed by the budding process may determine their maximum diameter. Additionally, associations between nucleocapsids and membrane glycoproteins may be limited by the number of G1 and/or G2 molecules embedded in the bilayer. Figure 1a is an electron image of the La Crosse virions

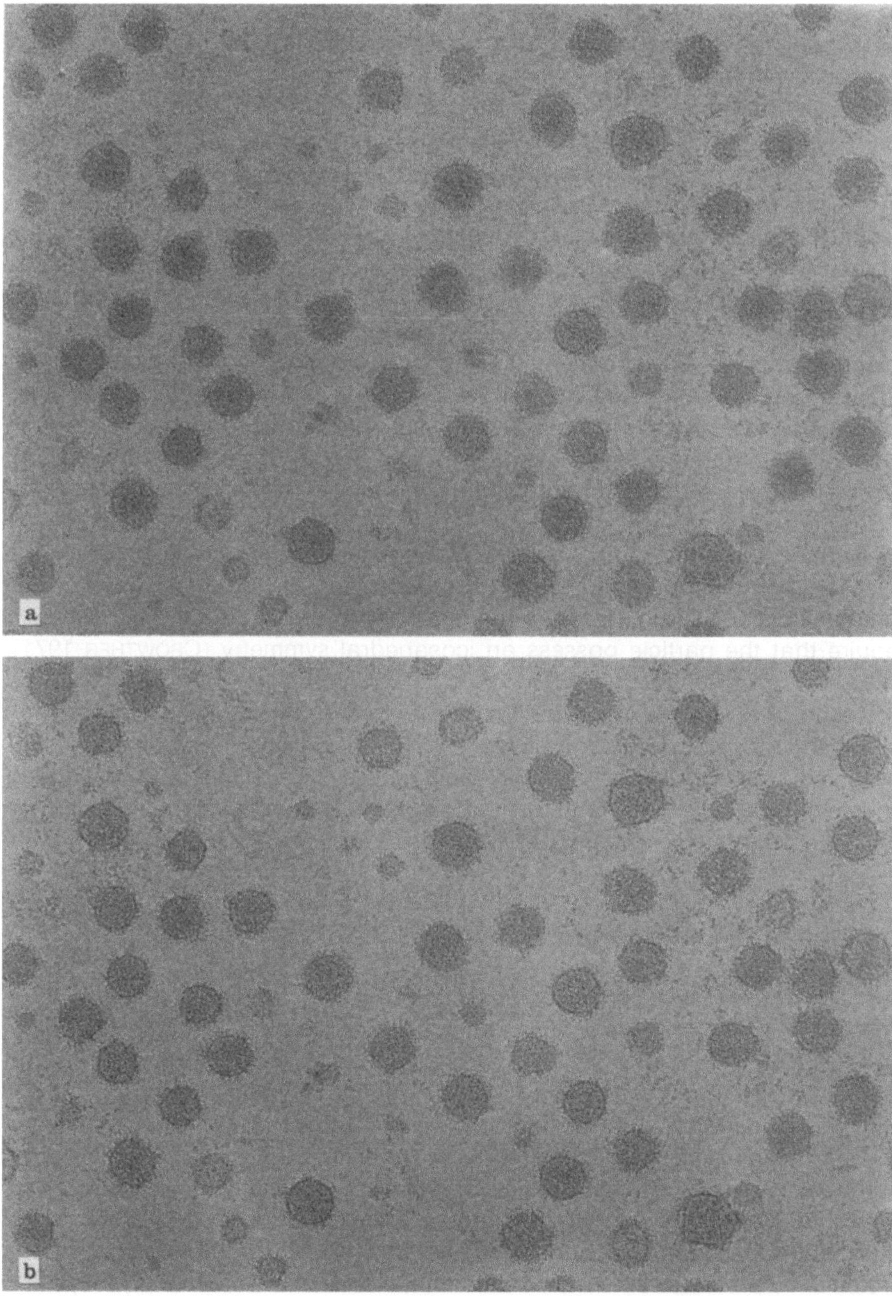

Fig. 1 a, b. Electron images of frozen, hydrated La Crosse viruses. **a** Taken close to optimal defocus. **b** Taken further away from optimal defocus. Different defocuses can enhance the visibility of certain structural features, such as the membrane bilayer or spike proteins at different spatial resolutions. Particle sizes range from 750 to 1150 Å (TALMON et al. 1987)

taken at close to optimal defocus value where the bilayer is visible. Because of the overlap of protein masses in a projection image, we cannot derive any reliable information regarding the molecular interactions around the membrane region. OBIJESKI et al. (1976b) have determined for La Crosse virus that there are approximately 640 G1 and G2 proteins per virion. Using the particle dimensions determined from frozen, hydrated specimens (TALMON et al. 1987), we have calculated that the available surface area on a La Crosse virion is 164–415 $\times 10^4$ Å^2. If we assume a spike diameter of 20 Å, then closest packing of the surface with such complexes would result in 4700–12000 spikes per particle. These data suggest that the glycoproteins in the La Crosse virus membrane are not organized in a tightly packed array but rather are quite widely distributed. Based on a direct measurement of the distance between two adjacent spikes, shown in Fig. 1b (taken with coarser defocus value to enhance the contrast), we have found them to be around 60–80 Å apart. Using the same calculation, we estimated the number of spikes to range from 270 to 1400 per virion. These later calculations are within a range comparable with the estimate of OBIJESKI et al. (1976b).

Based on the projection images, we cannot delineate the sites of interactions between the neighboring spike proteins. We have attempted to determine the three-dimensional structures from images shown in Fig. 1. This procedure would require that the particle possess an icosahedral symmetry (CROWTHER 1971). Based on an extended analysis, we have not yet been able to confirm the existence of the symmetry in the images of frozen, hydrated La Crosse virus (B.V.V. PRASAD, personal communication). However, the images of chemically fixed and stained samples of Uukuniemi viruses appear to have icosahedral symmetry (VON BONSDORFF and PETTERSSON 1975). However, this suggestion has not been quantitatively verified.

4.2 Membrane Glycoproteins

Particles of the family Bunyaviridae contain two glycoproteins, designated G1 and G2. Features of these proteins have been reviewed recently (BISHOP 1990; ELLIOTT 1990; GONZALEZ-SCARANO and NATHANSON 1990; SCHMALJOHN and PATTERSON 1990) and are covered in more detail elsewhere in this volume. The coding sequence for both glycoproteins is contained within the large, single open reading frame of the M RNA segment (see ELLIOT et al., this volume). Computer analysis of the predicted sequence of the G1 glycoprotein of snowshoe hare virus (BUNYAVIRUS genus; FAZAKERLEY et al. 1988) reveals the expected transmembrane anchor region near the carboxy terminus of the protein. The predicted external domain contains several potential glycosylation sites and a trypsin cleavage site, determined by gel electrophoresis. The predicted sequence for the G2 protein contains a hydrophobic region near residues 187–245 at the middle of the protein. This region has been interpreted as the transmembrane anchor (FAZAKERLEY et al. 1988) or as a potential sequence involved in fusion (POBJECKY et al. 1989; see Sect. 4.3). Since nothing is known about the orientation of this protein in the membrane, no conclusions can be drawn at this time.

The glycoproteins on the surface of the bunyavirus virion function as a site for interaction between the particle and the host cell. Neutralizing monoclonal antibodies for La Crosse virus have been prepared against G1 but not G2 (KINGSFORD and HILL 1983; KINGSFORD et al. 1983; KINGSFORD 1984; GONZALEZ-SCARANO et al. 1982, 1983; GRADY et al. 1983; GRADY and KINCH 1985; NIJJAR et al. 1985). Several of these neutralizing monoclonals also inhibit hemagglutination. These data suggest that the G1 protein of La Crosse is involved in attachment to a receptor on the cell surface, although the nature of this receptor is unknown. However, other members of the Bunyaviridae family exhibit neutralization activity with monoclonals directed against one or both of the membrane glycoproteins. For instance, MAbs against both G1 and G2 from Hantaan virus (ARIHAWA et al. 1989) show neutralizng properties. The G2 of Rift Valley fever virus induces neutralizing MAbs (KEEGAN and COLLETT 1986).

4.3 Fusion

Many enveloped viruses, such as Semliki Forest virus, enter the host cell by attachment to surface receptors, entry into coated pits, and penetration in endocytic vesicles (MARSH and HELENIUS 1980). These vesicles become acidified after fusion with lysosomes, as described originally for macroglobulins (TYCKO and MAXFIELD 1982). In this low-pH environment, the viral membrane fuses with the membrane of the lysosome, releasing the viral nucleocapsids into the cell cytoplasm. At present there is no direct evidence that bunyaviruses follow this pathway of entry. Electron micrographs of cells infected with Rift Valley fever virus show virus particles entering the cell in phagocytic vacuoles, consistent with this mode of entry (ELLIS et al. 1988).

When cells infected with La Crosse virus are subjected to low pH environments ($pH < 6.0$), the cells undergo fusion into heterocaryons (GONZALEZ-SCARANO et al 1984). In addition, the virus can mediate fusion of uninfected cells when it is present in the medium at the lower pH (GONZALEZ-SCARANO et al. 1984). It has been shown that G1 undergoes a conformational change at low pH (GONZALEZ-SCARANO 1985). In addition, monoclonal antibody resistant variants have been isolated for G1 (GONZALEZ-SCARANO 1985). One of these, V22, also exhibits an altered pH dependency for the fusion event (GONZALEZ-SCARANO 1985) and is unable to cause a disseminated infection in mosquitoes when introduced orally (GONZALEZ-SCARANO et al. 1988). These data initially pointed to a direct role for G1 in the fusion event.

More recently it has been argued that G2 may be the fusion protein of La Crosse virus. Liposomes containing only G1 are unable to induce fusion at low pH, whereas virus particles with most of G1, but not G2, removed by protease could still cause cells to fuse (POBJECKY et al. 1989). In addition, protease removal of G1 enhanced the ability of La Crosse to bind to mosquito midgut cells (LUDWIG et al. 1989). It is therefore possible that G1 functions to interact with a receptor on the cell surface, and that G2 is in fact the fusion protein. In this model, G1, at low pH, would undergo a conformational change which would allow efficient interaction of G2 with the membrane bilayer.

One of the advantages in electron cryomicroscopy is the capability of observing the bilayer features of the viral membrane in the images, as shown in Fig. 1a. These images were taken with an appropriate defocus to reveal this 40 to 50 Å detail. Fusion is a phenomenon of merging the viral membrane to (for example) the lysosomal membrane. Cryomicroscopy may offer an opportunity to visualize this process. Preliminary images have been recorded which show that La Crosse virus particles, when incubated at optimal fusion pH and temperature, form large aggregates. In some cases, virus particles can be seen fused together. Low pH dependent aggregation of particles is enhanced by incubation at 37 °C, compared with 4 °C. These data suggest that the surface of the virus particle is altered by exposure to the low pH environment. This change in the virion does not require interaction with a membrane surface. We have not yet determined whether particles without G1 undergo this aggregation.

5 Nucleocapsid Morphology and Function

5.1 Viral RNA

RNA is purified from bunyavirus particles by phenol extraction (HEWLETT et al. 1977) or by lysis with ionic detergent coupled with purification through CsCl gradients (RAJU and KOLAKOFSKY 1989). Purified viral RNA can be observed as circular structures when prepared for electron microscopy under mildly denaturing conditions (SAMSO et al. 1976; HEWLETT et al. 1977). However there is no direct evidence that the viral RNA is in circular form inside the virion. The sequence analsis of bunyaviral RNAs reveals the presence of inverted complimentary sequences at the 3′ and 5′ ends (see ELLIOT et al., this volume). Recently, evidence has been obtained (RAJU and KOLAKOFSKY 1989) indicating that the viral RNA within nucleocapsids maintains this base-paired, panhandle organization of the terminal regions, since these paired RNA sequences could be cross-linked by treatment with psoralen.

5.2 Viral Nucleocapsids

Nucleocapsids (ribonucleoprotein complexes) are made up of N protein complexed with RNA and small (probably stoichiometric) amounts of L protein.

They can be purified from intact bunyavirus particles by nonionic detergent lysis and equilibrium centrifugation in CsCl (RARU and KOLAKOFSKY 1989) or by sedimentation through sucrose density gradients (PETTERSSON and VON BONSDORF 1975). Bunyavirus nucleocapsids have been observed by electron microscopy and appear as circular structures (PETTERSSON and VON BONSDORF 1975; SAMSO et al. 1975; OBIJESKI et al. 1976a). The morphology of bunyavirus nucleocapsids is generally described as helical, based on electron images. However, there is not yet any experimental evidence to show quantitatively the presence of helical symmetry in this protein RNA complex. Furthermore, its mode of packing inside the virion is not yet determined.

So far, the best characterized interaction of viral proteins and RNA is the tobacco mosaic virus, the X-ray structure of which was determined by fiber diffraction analysis (NAMBA and STUBBS 1986). In this case the RNA is protected by the viral coat protein. In the case of the nucleosome, the nucleic acid (DNA) is exposed and wraps around the histone proteins (RICHMOND et al. 1984). These two systems have been the intuitive physical models for the bunyaviral nucleocapsids. However, based on electron images of these nucleocapsids, they appear to be more like other non-sequence specific binding proteins, such as those found in the case of bacteriophage or *Escherichia coli* (LOHMAN and BUJALOWSKI 1990). The resolution of these pictures is not sufficient to see whether the proteins wrap around the RNA or the RNA around the proteins. However, it is not unreasonable to speculate that the bunyavirus nucleocapsids bear resemblance in the mode of interaction to some of these other non-sequence specific binding proteins.

6 Conclusion

Details of the molecular biology of the bunyaviruses have been elucidated at a rapid pace during the past few years. Precise structural information has awaited the development of techniques suitable for enveloped viruses. The most ideal solution is to crystallize the whole virion. Because of the heterogeneity in sizes among the purified virions and the relatively large size of the virion, it is unlikely that one can attain this goal in the near future. The more feasible approach to tackle the structural problem of this class of viruses is with electron cryomicroscopy coupled with image reconstruction. Because of the good contrast found in images of the ice-embedded specimens, we anticipate the application of this technique for answering a variety of functional questions relating to viral assembly, receptor interaction, and neutralization.

Acknowledgements. We thank G.-J. Wang for providing the figures, and R. Collins and M.J. Perez for preparing this manuscript.

References

Achrya R, Fry E, Stuart D, Fox G, Rowlands D, Brown F (1989) The three-dimensional structure of foot and mouth disease virus at 2.9 Å resolution. Nature (London) 337: 709–716

Adrian M, Dubochet J, Lepault J, McDowell AW (1984) Cryo-electron microscopy of viruses. Nature (London) 308: 32–36

Arikawa J, Schmljohn AL, Dalrymple JM, Schmaljohn CS (1989) Characterization of Hantaan virus envelope glycoprotein antigenic determinants defined by monoclonal antibodies. J Gen Virol 70: 615–624

Baker TS, Drak J, Bina M (1988) Three-dimensional structure of SV40. Proc Natl Acad Sci USA 85: 422–426

Bishop DHL (1990) Bunyaviridae and their replication. Part I: Structure of bunyaviridae. In: Fields BN, Knipe DM (eds) Virology, vol 1, Raven, New York

Chiu W (1986) Electron microscopy of frozen hydrated biological specimens. Annu Rev Biophys Biophys Chem 15: 237–257

Crowther RA (1971) Procedures for three-dimensional reconstruction of spherical viruses by Fourier synthesis from electron micrographs. Phil Trans Roy Soc Lond B 261: 221–230

Dubochet J, Adrian M, Chang JJ, Homo JC, Lepault J, McDowall AW, Schultz P (1988) Cryo-electron microscopy of vitrified specimens. Q Rev Biophys 21: 129–228

Elliot RM (1990) Molecular biology of the Bunyaviridae. J Gen Virol 71: 501–522

Ellis DS, Shirodaria PV, Fleming E, Simpson DI (1988) Morphology and development of Rift Valley fever virus in Vero cell cultures. J Med Virol 24 (2): 161–174

Fazakerley JK, Gonzalez-Scarano F, Strickler J, Dietzschold B, Karush F, Nathanson N (1988) Organization of the middle RNA segment of snowshoe hare bunyavirus. Virology 167 (2): 422–432

Fuller SD (1987) The T = 4 envelope of Sindbis virus is organized by interactions with a complimentary T = 3 capsid. Cell 48: 923–934

Goldman N, Presser I Sreevalsan T (1977) California encephalitis virus: some biological and biochemical properties. Virology 76: 352

Gonzalez-Scarano F (1985) La Crosse virus G1 glycoprotein undergoes a conformational change at the pH of fusion. Virology 140: 209–216

Gonzalez-Scarano F, Nathanson N (1990) Bunyaviruses. In: Fields BN, Knipe DM (eds)Virology. Raven Press, New York

Gonzalez-Scarano F, Shope RE, Calisher CE, Nathanson N (1982) Characterization of monoclonal antibodies against the G1 and N proteins of La Crosse and Tahyna viruses, two California group bunyaviruses. Virology 120: 42–53

Gonzalez-Scarano F, Shope RE, Calisher CE, Nathanson N (1983) Monoclonal antibodies against the G1 and nucleocapsid proteins of La Crosse and Tahyna viruses. Prog Clin Biol Res 123: 145–156

Gonzalez-Scarano F, Pobjecky N, Nathanson N (1984) La Crosse bunyavirus can mediate pH-dependent fusion from without. Virology 132: 222–225

Gonzalez-Scarano F, Janssen RS, Najjar JA, Pobjecky N, Nathanson N (1985) An avirulent G1 glycoprotein variant of La Crosse-Bunyavirus with defective fusion function. J Virol 54 (3): 757–763

Gonzalez-Scarano F, Beaty B, Sundin D, Janssen R, Endres MJ, Nathanson N (1988) Genetic determinants of the virulence and infectivity of La Cross virus. Microbiol Pathog 4 (1): 1–7

Grady LJ, Kinch W (1985) Two monoclonal antibodies against La Crosse virus show host-dependent neutralizing activity. J Gen Virol 66: 2773–2776

Grady LJ, Sanders ML, Campbell WP (1983) Evidence for three separate antigenic sites on the G1 protein of La Crosse virus. Virology 126: 395–397

Harrison SC (1990) Principles of virus structure. In: Fields BN, Knipe DM, Chanock RM, Melnick JL, Roizman B, Shope RE (eds) Virology, vol 1. Raven, New York, pp 37–71

Hewlett MJ, Pettersson RF, Baltimore D (1977) Circular forms of Uukuniemi virion RNA: an electron microscopic study. J Virol 21: 1085–1093

Hogle JM, Chow M, Filman DJ (1985) Three dimensional structure of polivirus at 2.9 Å resolution. Science 229: 1358–1365

Jeng T-W, Chiu W (1987) High resolution cryo system designed for JEM 100CX electron microscope. Ultramicroscopy 23: 61–66

Jeng T-W, Talmon Y, Chiu W (1988) Containment system for the preparation of vitrified-hydrated virus specimen. J Electron Microsc Tech 8: 343–348
Jeng T-W, Crowther RA, Stubbs G, Chiu W (1989) Visualization of alpha-helices in tobacco mosaic virus by cryo-electron microscopy. J Mol Biol 205: 251–257
Keegan K, Collett MS (1986) Use of bacterial expression cloning to define the amino acid sequences of antigenic determinants on the G2 glycoprotein of Rift Valley fever virus. J Virol 58 (2): 263–270
Kingsford L (1984) Enhancement neutralization of La Crosse virus by the binding of specific pairs of monoclonal antibodies to the G1 glycoprotein. Virology 136: 265–273
Kingsford L, Hill DW (1983) The effect of proteolytic cleavage of La Crosse virus G1 glycoprotein on antibody neutralization. J Gen Virol 64: 2147–2156
Kingsford L, Ishizawa LD, Hill DW (1983) Biological activities of monoclonal antibodies reactive with antigenic sites mapped on the G1 glycoprotein of La Crosse virus. Virology 129: 443–455
Kuismanen E, Hedman K, Saraste J, Pettersson RF (1982) Uukuniemi virus maturation: accumulation of virus particles and viral antigens in the Golgi complex. Mol Cell Biol 2: 1444–1458
Lohman TM, Bujalowski W (1990) *E. coli* single strand binding protein: multiple single-stranded DNA binding modes and cooperativities. In: Revzin A (ed) The biology of non-specific DNA protein interactions CRC Press, Boca Raton (in press)
Ludwig GV, Christensen BM, Yuill TM, Schultz KT (1989) Enzyme processing of La Crosee virus glycoprotein G1: a bunyavirus-vector infection model. Virology 171 (1): 108–113
Lyons MJ, Heyduk J (1973) Aspects of the developmental morphology of California encephalitis virus in cultured vertebrate and arthropod cells and in mouse brain. Virology 54: 37–52
Marsh M, Helenius A (1980) Adsorptive endocytosis of Semliki Forest virus. J Mol Biol 142: 439–454
Murphy FA, Harrison AK, Whitfield SG (1973) Bunyaviridae: morphologic and morphogenetic similarities of Bunyamwera serologic super group viruses and several other arthropod-borne viruses. Intervirology 1: 297–316
Namba N, Stubbs G (1986) Structure of TMV at 3.6 Å resolution: implications for assembly. Science 231: 1401–1406
Najjar JA, Gentsch JR, Nathanson N, Gonzalez-Scarano F (1985) Epitopes of the G1 glycoprotein of La Crosse virus form overlapping clusters within a single antigenic site. Virology 144 (2): 426–432
Nermut MV, Hockley DJ, Gelderblom H (1987) Methods for the study of virus structure. In: Nermut MV, Stevens A (eds) Perspectives in medical virology, animal virus structure, vol 3. Amsterdam, pp 21–60
Obijeski JF, Bishop DHL, Palmer EL, Murphy FA (1976a) Segmented genome and nucleocapsid of La Crosse virus. J Virol 20: 664–675
Obijeski JG, Bishop DHL, Murphy FA, Palmer EL (1976b) Structural proteins of La Crosse virus. J Virol 19: 985–997
Pattersson RF, von Bonsdorf C-H (1975) Ribonucleoproteins of Uukuniemi virus are circular. J Virol 15: 386–392
Petterson RF, von Bonsdroff C (1987) Bunyaviridae. In: Nermut MV, Stevens A (eds) Perspectives in medical virology, animal virus structure, vol 3. Amsterdam, pp 147–157
Pobjecky N, Nathanson N, Gonzalez-Scarano F (1989) Use of resonance energy transfer assay to investigate the fusion function of La Cross virus. In: Kolakofsky D, Mahy BWJ (eds) Genetics and pathogenicity of negative strand viruses. Elsevier, Amsterdam, pp 24–32
Prasad BVV, Wang G-J Clerx JPM, Chiu W (1988) Three-dimensional structure of rotavirus. J Mol Biol 199: 269–275
Prasad BVV, Burns JW, Marietta E, Estes MK, Chiu W (1990) Localization of VP4 neutralization sites in rotavirus by three-dimensional cryo-electron microscopy. Nature 343: 476–478
Raju R, Kolakofsky D (1989) The ends of La Crosse virus genome and antigenome RNAs within nucleocapsids are base paired. J Viron 63 (1): 122–128
Richmond TJ, Finch JT, Rushton B, Rhodes D, Klung A (1984) Structure of the nucleosome core particle at 7 Å resolution, Nature 311: 532–537
Rossmann MG (1989) Neutralization of small RNA viruses by antibodies and antiviral agents. FASEB J 3: 2335–2343
Rossmann MG, Reuckert RR (1987) What does the molecular structure of viruses tell us about viral functions? Microbiol Sci 4: 206–214
Rossmann MG, Arnold E, Erickson JW, Frankenberger EA, Griffith JP, Hecht HJ, Johnson JR, Kramer G, Luo M, Mosser AG, Rueckert RR, Sherry B, Vriend G (1985) Structure of a human common virus and functional relationship to other picornaviruses. Nature (London) 317: 145–153
Samso AM, Bouloy M, Hannoun C (1975) Présence de ribonucléoproteins circulaire dans le virus Lumbo (bunyavirus). C R Acad Sci D280: 779–782

Samso A, Bouloy M, Hannoun C (1976) Mise en évidence de molécules d'acide ribonucleique circulaire dans le virus Lumbo (bunyavirus). C R Acad Sci D282: 1653–1655
Schmaljohn CS, Patterson JL (1990) Bunyaviridae and their replication. Part II. Replication of Bunyaviridae. In: Fields BN, Knipe DM (eds) Virology. Raven, New York, pp 1175–1194
Schrag JD, Prasad BVV, Rixon FJ, Chiu W (1989) Three-dimensional structure of HSV-1 nucleocapsid. Cell 56: 651–660
Smith JF, Pifat DY (1982) Morphogenesis of sandfly fever viruses (Bunyaviridae family). Virology 121: 61–81
Talmon Y, Prasad BV, Clerx JP, Wang GJ, Chiu W, Hewlett MJ (1987) Electron microscopy of vitrified-hydrated La Crosse virus. J Virol 61 (7): 2319–2321
Tycko B, Maxfield FR (1982) Rapid acidification of endocytic vesicles containing alpha-2-macroglobulin. Cell 28: 643–651
Vogel RH, Provencher SW, von Bonsdorff C-H, Adrian M, Dobochet J (1986) Envelope structure of Semliki Forest virus reconstructed from cryo-electron micrographs. Nature (London) 320: 533–535
von Bonsdorff C-H, Pettersson RF (1975) Surface structure of Uukuniemi virus. J Virol 16: 1296–1307

Bunyaviridae Genome Structure and Gene Expression

R. M. ELLIOTT[1], C. S. SCHMALJOHN[2], and M. S. COLLETT[3]

[1] Institute of Virology, University of Glasgow, Church Street, Glasgow G11 5JR, UK
[2] Virology Division, United States Army Medical Research Institute of Infectious Diseases, Fort Detrick, Frederick, MD 21701, USA
[3] MedImmune Inc., 19 Firstfield Road, Gaithersburg, MD 20878, USA

Current Topics in Microbiology and Immunology, Vol. 169

1 Introduction

More than 300 viruses are included in the family Bunyaviridae and hence show a number of unifying characteristics. However, with such a large number of members it is perhaps not unexpected that within the family there is significant diversity in genome structures and expression strategies. In this chapter we first present an overview of the unifying features as well as the diversity within the family. This will orientate the reader for the more detailed analyses of individual virus groups which follow.

At present the family Bunyaviridae contains five recognized genera—*Bunyavirus, Hantavirus, Nairovirus, Phlebovirus* and *Uukuvirus*—into which the majority of viruses are classified; a number of other viruses are included in the family but have yet to be assigned to a particular genus (BISHOP 1990). All currently recognized members of the family have been isolated from animal species (mammals, birds or arthropods), but recent data indicate that a thrip-transmitted plant virus, tomato spotted wilt (TSW) virus, fulfils the criteria for inclusion in the Bunyaviridae (DE HAAN et al. 1989, 1990) and thus is included in our discussion. The taxonomy and classification of the Bunyaviridae is a complex subject and outside the scope of this chapter; suffice it to say that the current classification must be regarded as fluid.

All members of the Bunyaviridae possess a genome comprising three segments of single-stranded RNA which are designated L (large), M (medium) and S (small); the RNA makes up 1%–2% of the virion (OBIJESKI et al. 1976a). The sizes of these RNA segments are presented in Table 1—either as estimated from the electrophoretic migration of the RNA in agarose or polyacrylamide gels or, for some segments, on nucleotide sequence determinations of cloned cDNA. The pattern of RNA segment size is distinctive for bunyaviruses, nairoviruses and TSW virus; there is less difference between the patterns of hanta-, phlebo- and uukuvirus RNAs.

A feature of the individual genome segments is the conservation of the 3′ terminal sequences (CLERX-VAN HAASTER and BISHOP 1980; OBIJESKI et al. 1980; CLERX-VAN HAASTER et al. 1982a, b; SCHMALJOHN and DALRYMPLE 1983; DE HAAN et al. 1989); these sequences are generally genus specific (Table 2), but those of phleboviruses and uukuviruses are identical. Note also that the bunyavirus and hantavirus consensus sequences show some similarity, the hantavirus differing in having an additional A residue at the 3′ end. The available sequence information shows that the 5′ terminal sequences are also conserved and are complementary to the 3′ termini (Table 2). A consequence of the terminal complementarity is that the ends of the RNAs may base-pair, which presumably accounts for the circular and/or panhandle forms of the RNAs which have been seen by electron microscopy (BOULOY et al. 1973/1974; SAMSO et al. 1976; HEWLETT et al. 1977; PARDIGON et al. 1982). It is postulated that the complementary sequences may provide signals for recognition by the virus-encoded polymerase or are involved in packaging of the viral genome. As the terminal sequences are the same for each of the three RNA segments in a given

Table 1. Sizes of the genome RNA segments of the Bunyaviridae[a]

RNA segment	Bunyaviridae genus					Tomato spotted wilt virus
	Bunyavirus	*Hantavirus*	*Nairovirus*	*Phlebovirus*	*Uukuvirus*	
L[b]	8100–9400	6500–8500	11 000–14 400	6500–8200	6500–8500	7900–8200
M[b]	5400–6900	4200–5700	4400–5700	3300–6600	3000–3900	5000–5400
S[b]	850–1500	1800–2300	1760–2050	1200–2400	1200–1900	2900–3400
L[c]	BUN: 6875					
M[c]	BUN: 4458 GER: 4534 LAC: 4526 SSH: 4527	HTN: 3616 NE: 3682 SR-11: 3651		PT: 4330 RVF: 3885	UUK:3231	
S[c]	BUN: 961 GER: 980 MAG: 945 LAC: 984 SSH: 982 Aino: 850	HTN: 1696 NE: 1785 PH: 1675 SR-11: 1769	DUG: 1712	PT: 1904 SFS: 1746	UUK:1720	TSW:2916

BUN, Bunyamwera; GER, Germiston; LAC, La Crosse; SSH, Snowshoe hare; MAG, Maguari; HTN, Hantaan 76–118; NE, nephropathia epidemica Hällnäs Bl; SR-11, Sapporo rat; PH, Prospect Hill; DUG, Dugbe; PT, Punta Toro; RVF, Rift Valley fever; SFS, sandfly fever–Sicilian; UUK, Uukuniemi; TSW, tomato spotted wilt.

[a] Sizes given in nucleotides.

[b] Estimates from electrophoresis of RNAs.

[c] Nucleotide sequence data from cloned cDNAs.

References:

L: BUN, ELLIOTT (1989b).

M: BUN, LEES et al. (1986); GER, PARDIGON et al. (1988); LAC, GRADY et al. (1987); SSH, ESHITA and BISHOP (1984); HTN, SCHMALJOHN et al. (1987b); YOO and KANG (1987); NE, GIEBEL et al. (1989); SR-11, ARIKAWA et al. (1990); PT, IHARA et al. (1985b); RVF, COLLETT et al. (1985); UUK, RONNHOLM and PETTERSSON (1987).

S: BUN, ELLIOTT (1989a); GER, GERBAUD et al. (1987b); MAG, ELLIOTT and McGREGOR (1989); LAC, AKASHI and BISHOP (1983); CABRADILLA et al. (1983); SSH, BISHOP et al. (1982); Aino, AKASHI et al. (1984); HTN, SCHMALJOHN et al. (1986a); NE, STOHWASSER et al. (1990); PH, PARRINGTON and KANG (1990); SR-11, ARIKAWA et al. (1990); DUG, WARD et al. (1990); PT, IHARA et al. (1984); SFS, MARRIOTT et al. (1989); UUK, SIMONS et al. (1990); TSW, DE HAAN et al. (1990).

Table 2. Consensus 3′ and 5′ terminal nucleotide sequences of Bunyaviridae genome RNA segments

Bunyavirus	3′	UCAUCACAUGA. . . . . . .UCGUGUGAUGA	5′
Hantavirus	3′	AUCAUCAUCUG . . . . .AUGAUGAU	5′
Nairovirus	3′	AGAGUUUCU . . . . AGAAACUCU	5′
Phlebovirus	3′	UGUGUUUC . . . . GAAACACA	5′
Uukuvirus	3′	UGUGUUUCUG . . . . CAGAAACACA	5′
Tomato spotted wilt virus	3′	UCUCGUUAG . . . . CUAACGAGA	5′

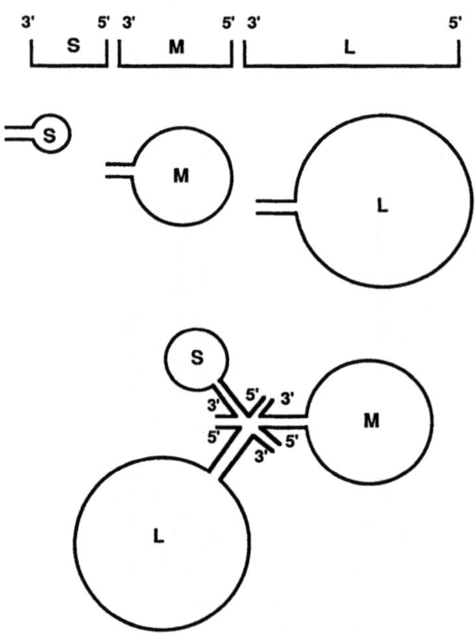

Fig. 1. Possible associations of the conserved complementary termini of Bunyaviridae RNA segments (not to scale). Based on Obijeski et al. 1980)

virus, the RNAs could be linked tandemly (Obijeski et al. 1980) although there is no experimental evidence for this. The possible associations of the RNA termini are depicted in Fig. 1.

For some viruses the mapping of viral proteins to genome segments has been achieved by genetic or biochemical methods (Bishop 1985); nucleotide sequence studies, along with N-terminal protein sequence data, have confirmed

Fig. 2. Coding strategies of Bunyaviridae genomes. *Thin lines*, genomic RNAs; *arrows*, mRNAs (■, 5′ end; ▶, 3′ end); *solid boxes*, gene products (not to scale)

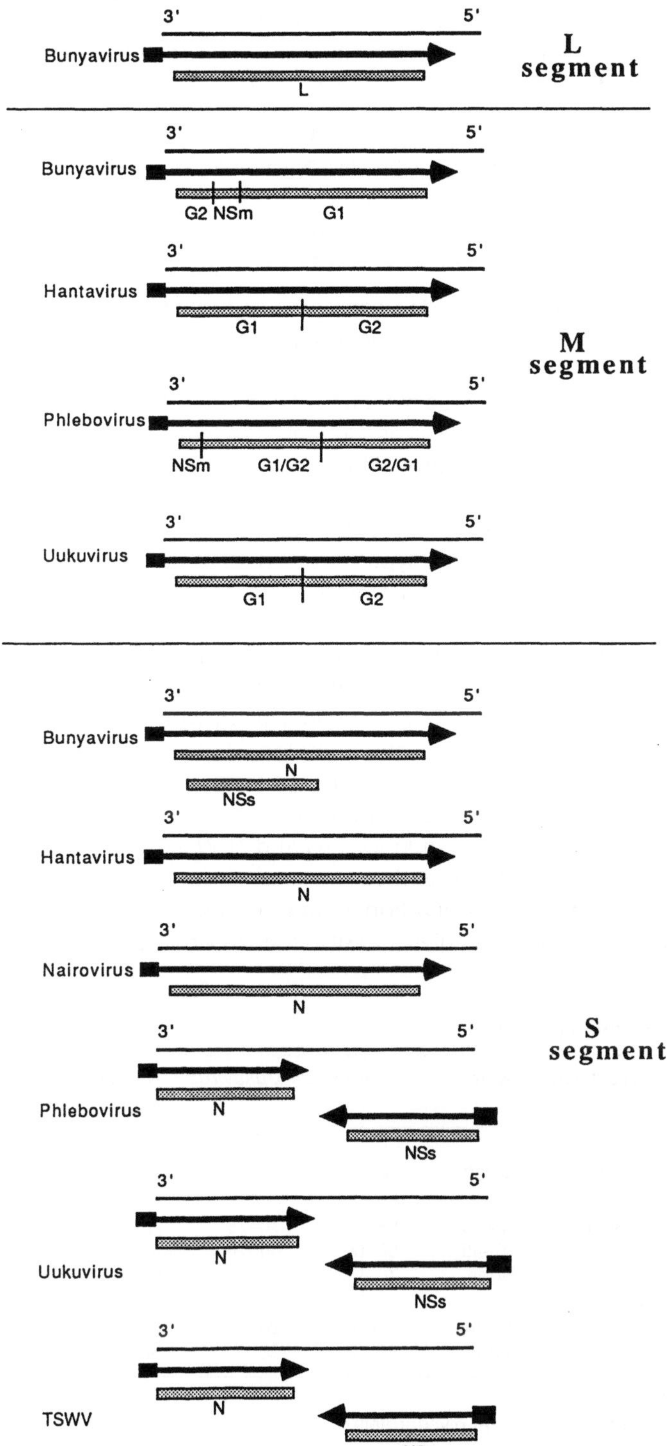
3'
5'
Bunyavirus
L
L
segment
Bunyavirus
G2 NSm
G1
Hantavirus
G1
G2
M
segment
Phlebovirus
NSm
G1/G2
G2/G1
Uukuvirus
G1
G2
Bunyavirus
N
NSs
Hantavirus
N
Nairovirus
N
S
segment
Phlebovirus
N
NSs
Uukuvirus
N
NSs
TSWV
N
NSs

these results and further have revealed the coding strategies employed by individual genome segments. These are shown in Fig. 2. The results of these studies can be generalized for the family as follows:

1. The L RNA segment encodes the L protein, the putative virion transcriptase.
2. The M RNA segment encodes the two virion glycoproteins G1 and G2 in the form of a (putative) precursor polypeptide which is probably processed by cotranslational cleavage. The glycoproteins are rich in Cys residues (> 5%).
3. The S RNA segment encodes the nucleocapsid protein, N.

The L and M RNA segments and the S RNA segments of bunya-, hanta- and nairoviruses are of negative polarity, i.e. the infecting genome segments must be transcribed into positive-sense mRNA for protein production. The S segments of phleboviruses, uukuviruses and TSW virus exhibit an ambisense strategy where the N protein open reading frame (ORF) is negative sense, but a second S segment ORF, encoding a nonstructural (NS) protein, NSs, is read in the genomic sense but is translated from a subgenomic mRNA (Fig. 2).

The M RNA segments of bunyaviruses and phleboviruses also encode an NS protein as part of the putative precursor polyprotein, and the bunyavirus S RNA segment encodes an NS protein in an overlapping reading frame to that of the N protein.

Table 3 summarizes the coding assignments and molecular weights of the Bunyaviridae proteins. The pattern of sizes of the structural proteins of phleboviruses and uukuviruses and of TSW virus is similar, whereas those of viruses in the other genera are distinct. No homology between equivalent proteins in different genera has been reported, with the exception of the N proteins and glycoproteins of uukuviruses and phleboviruses. In the detailed discussion which follows phleboviruses, uukuviruses and TSW virus are treated together in view of the overall similarity of these viruses.

One final, apparently general observation concerns the mode of transcription of the Bunyaviridiae. This resembles that of influenza viruses (KRUG et al.

Table 3. Proteins encoded by the Bunyaviridae[a]

RNA segment/ protein	*Bunyavirus*	*Hantavirus*	*Nairovirus*	*Phlebovirus*	*Uukuvirus*	Tomato spotted wilt virus
L segment						
L	259	>200	>200	>200	>200	>200
M segment						
G1	108–120	68–76	72–84	55–70	70–75	78
G2	29–41	52–58	30–45[b]	50–60	65–70	52–58
NSm	15–18	None	None	78 and 14	None	Unknown
S segment						
N	19–25	50–54	48–54	24–30	20–29	29
NSs	10–13	None	None	29–31	30	52

[a] Sizes given in kilodaltons

[b] Some nairoviruses may have three structural glycoproteins

1987) in that short (12–18 nucleotides) primers are cannibalized from cellular mRNAs to prime viral RNA synthesis. Hence additional nontemplated nucleotides are found at the 5′ ends of Bunyaviridae mRNA species transcribed from both negative and ambisense genomes (BISHOP et al. 1983; PATTERSON and KOLAKOFSKY 1984; ESHITA et al. 1985; COLLETT 1986; BOULOY et al. 1990). The 3′ termini of the mRNAs are truncated relative to the genomic RNAs and are probably not polyadenylated (ULMANEN et al. 1981; ABRAHAM and PATTNAIK 1983; PATTNAIK and ABRAHAM 1983; BOULOY et al. 1984; PATTERSON and KOLAKOFSKY 1984; ELLIOTT 1985; EMERY and BISHOP 1987; ESHITA et al. 1985; COLLETT 1986; BOULOY et al. 1990).

2 *Bunyavirus* Genus

The *Bunyavirus* genus is the largest in the family and includes over 160 viruses in 16 serogroups (BISHOP 1990). The prototype of the genus, and also of the family Bunyaviridae, is Bunyamwera (BUN) virus which was isolated in Uganda (SMITHBURN et al. 1946). Bunyaviruses are transmitted mostly by mosquitoes and infect a wide range of vertebrate species. Some members of the California serogroup (e.g. La Crosse, LAC, virus) cause encephalitis in man.

2.1 Genome Structure and Organization

BUN virus is the only member of the Bunyaviridae whose complete genome sequence has been reported (LEES et al. 1986; ELLIOTT 1989a, b). The entire genome comprises 12294 nucleotides (Table 1), of which 11718 (95.3%) encoded the viral proteins. As mentioned above, the terminal nucleotide sequences are conserved within a genus and the terminal sequences of the three BUN virus genome segments are displayed in Fig. 3. The terminal 11 bases at both the 3′ and 5′ ends are conserved, complementary among all sequenced genome segments, and characteristically show a mismatch at position 9 and −9. Following the bunyavirus-specific 11 bases, the next three or four nucleotides are conserved on a segment-specific basis, as judged from the available sequence data. Complementarity between the 3′ and 5′ ends is maintained thereafter for a variable number of residues in all the segments sequenced, but there is no significant homology between analogous segments of different viruses.

The genome RNA segments are encapsidated by the N protein (ca. 2000 copies per virion; OBIJESKI et al. 1976a). The nucleocapsids of bunyaviruses are circular (SAMSO et al. 1975; OBIJESKI et al. 1976b), and RAJU and KOLAKOFSKY (1989) have shown that the ends of the RNA are base-paired within the nucleocapsid. The signal for encapsidation by N protein probably resides within the complementary terminal sequences because, in general, genome negative-

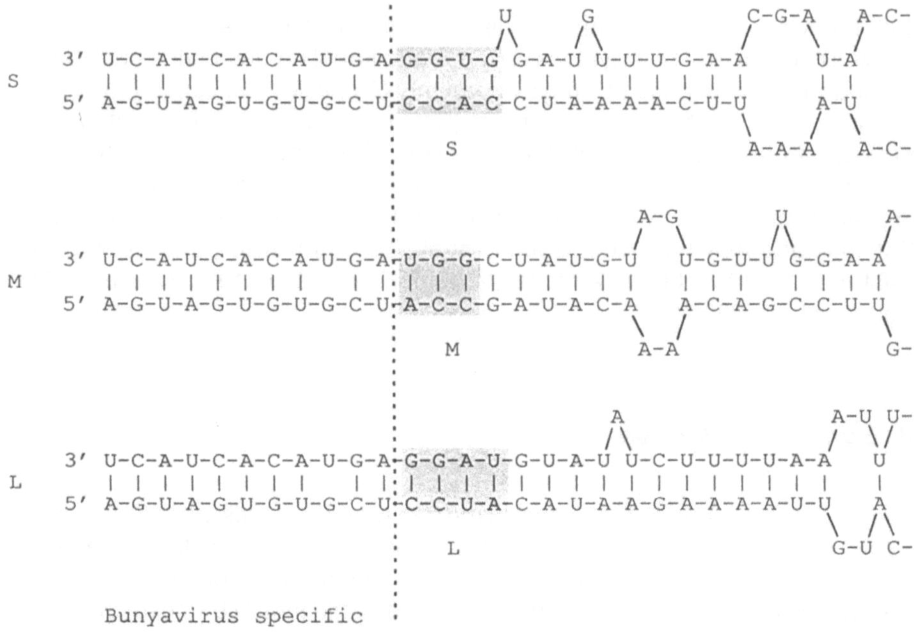

Fig. 3. Complementary sequences and possible base-paired structures between the 3′ and 5′ termini of Bunyamwera virus genomic RNA segments. The terminal 11 nucleotides are conserved in all genome segments; *shaded regions*, nucleotides which are conserved on a segment specific basis in all available sequenced bunyavirus RNA segments

sense RNA and full-length viral complementary positive-sense RNA (antigenome or replicative intermediate RNA) are encapsidated, whereas viral mRNA, which has a 5′-terminal extension and is truncated at the 3′ end, is not (RAJU and KOLAKOFSKY 1986). However RAJU and KOLAKOFSKY (1987a) reported that minor amounts of mRNA-like transcripts were encapsidated late in infection in mammalian cells, and HACKER et al. (1989) reported that in infected mosquito cells N protein controls its own synthesis by encapsidating its mRNA.

The BUN virus L RNA segment is 6875 nucleotides in length and has a base composition of 66.7% A + U residues; this is somewhat richer in A + U residues than the M (61.0%) or S (58.2%) segments, but the significance of this difference is not known. The L segment employs a straightforward negative-sense coding strategy (Fig. 2). At present only limited sequence information from the L RNA segments of other bunyaviruses is available; CLERX-VAN HAASTER et al. (1982a) and HACKER et al. (1990) have reported some sequences from snowshoe hare (SSH) and LAC viruses, and these show about 40% conservation with BUN virus L RNA.

Complete nucleotide sequences of four bunyavirus M segments, from viruses representing two serogroups, have been determined (Table 1, 4) and are similar in size (4458–4534 bases), base composition (61.0–63.2% A + U) and economy in the use of RNA to encode protein. Each segment encodes a single ORF in the

Table 4. Comparison of the M RNA segments of four bunyaviruses

Serogroup	Virus	RNA				Gene Products		
		Number of nucleotides				Number of amino acids	Number of potential glycosylation sites	
		Total	5′nc	3′nc	% A + U		G1	G2
Bunyamwera	BUN	4458	56	103	61.0	1433	2	2
	GER	4534	58	164	63.2	1437	2	2
California	LAC	4526	61	141	62.0	1441	2	2
	SSH	4527	61	142	61.0	1441	2	2

Table 5. Comparison of the S RNA segments of six bunyaviruses

Serogroup	Virus	RNA				Gene products	
		Number of nucleotides				Number of amino acids	
		Total	5′nc	3′nc	% A + U	N	NSs
Bunyamwera	BUN	961	85	174	58.2	233	101
	GER	980	87	194	57.6	233	109
	MAG	945	73	170	59.9	233	101
California	LAC	984	81	198	58.8	235	92
	SSH	982	79	198	56.3	235	92
Simbu	Aino	850	34	117	55.9	233	91

viral complementary RNA. Nucleotide sequence homology is greater between the M segments of viruses in the same serogroup (e.g. SSH virus versus LAC virus is 79%; GRADY et al. 1987) than between viruses in different serogroups (e.g. SSH virus versus BUN virus is 55%; LEES et al. 1986).

Six bunyavirus S segments have been sequenced, representing the Bunyamwera, California and Aino serogroups (Tables 1, 5). The S segments range from 850 to 984 nucleotides in length with a base composition of 55.9%–59.9% A + U residues. In each case the 3′ untranslated region is significantly longer than the 5′ untranslated region of the viral complementary RNA (Table 5).

2.2 Gene Expression Strategies

2.2.1 L Segment

The bunyavirus L RNA encodes the L protein (ELLIOTT 1989b; ENDRES et al. 1989), a large molecular weight protein which is synthesized in minor amounts in infected cells. There is a single AUG-initiated ORF in the BUN virus L segment viral complementary RNA flanked by 5′ and 3′ noncoding regions of 50 and 108 bases, respectively. This ORF contains 2239 codons; the next largest ORF is of 63 codons (ELLIOTT 1989b). In the genome-sense RNA there are three ORFs longer than 100 codons. The 5′ proximal AUG starts the longest of these, potentially of 129 amino acids, but there is no evidence that this encodes a *bona fide* gene product (ELLIOTT 1989b). Furthermore, some limited sequence analysis of the equivalent region of the LAC virus L segment indicates that the ORF is not maintained in this bunyavirus (HACKER 1990).

2.2.2 M Segment

The bunyavirus M RNA segment encodes, in the viral complementary RNA, the two virion glycoproteins G1 (M_r 108000–120000) and G2 (M_r 29000–41000), and a nonstructural protein termed NSm (M_r 15000–18000; GENTSCH and BISHOP 1979; FULLER and BISHOP 1982; ELLIOTT 1985). The four sequenced bunyavirus M segments each encode a single ORF of 1433–1441 amino acids, which contains a precursor to G1, G2 and NSm. The putative precursor has not been detected in infected cells (PENNINGTON et al. 1977; LEES et al. 1986), which suggests that cleavage may be a cotranslational event. It has not proved possible to translate bunyavirus M segment specific mRNA in vitro to analyse the cleavage events (ABRAHAM and PATTNAIK 1983; ELLIOTT 1985) although this methodology has been used to study phlebo- and uukuvirus glycoprotein biogenesis (see Sect. 5.2.2). There have been no reports thus far on the expression of bunyavirus M segment cDNAs in prokaryotic or eukaryotic systems.

FAZAKERLEY et al. (1988) used direct amino acid sequencing and specific peptide antibodies to determine the gene order to be 5′ G2–NSm–G1 3′ in the

SSH virus M segment complementary RNA. The other three bunyavirus precursor proteins can be readily aligned with that of SSH virus (ELLIOTT 1990) with the following observations: there is a putative signal peptide of 13–21 residues at the amino terminus of the precursor; G2 is generated by cleavage after a conserved Arg residue; although the amino terminus of G1 is poorly conserved, cleavage may occur after a common Ala residue; and the major trypsin-sensitive site in G1 (KINGSFORD and HILL 1983; GONZALEZ-SCARANO 1985) is conserved in all four proteins.

2.2.3 S Segment

The bunyavirus S segment specifies two proteins, the nucleocapsid protein N (M_r 19000–26000) and a nonstructural protein designated NSs (M_r 10000–12000), which are encoded in overlapping reading frames (GENTSCH and BISHOP 1978; CASH et al. 1979; BISHOP et al. 1982; FULLER and BISHOP 1982; AKASHI and BISHOP 1983; CABRADILLA et al. 1983; FULLER et al. 1983; AKASHI et al. 1984; BOULOY et al. 1984; ELLIOT 1985, 1989a; GERBAUD et al. 1987b; ELLIOTT and MCGREGOR 1989). Although one report claimed that multiple mRNAs were transcribed from the LAC virus S RNA segment, thereby generating different messages for N and NSs (PATTERSON et al. 1983), this is probably incorrect (RAJU and KOLAKOFSKY 1987b), and it is more likely that the two proteins are made from the same mRNA, the result of alternative initiation of translation (BISHOP et al. 1983; ELLIOTT and MCGREGOR 1989).

The S RNA segments of Germiston (GER) and Maguari (MAG) viruses contain a third AUG-initiated ORF, in the same frame as and downstream of, the NSs ORF (GERBAUD et al. 1987b; ELLIOTT and MCGREGOR 1989). The product of this third ORF can be translated in cell-free systems from RNA transcribed in vitro from cloned cDNA but has not been detected in vivo (ELLIOTT and MCGREGOR 1989). The significance of this ORF is thus questionable.

S segment cDNAs of SSH and MAG bunyaviruses have been cloned into recombinant baculoviruses (URAKAWA et al. 1988; ELLIOTT and MCGREGOR 1989), in both cases N protein, but not NSs, was efficiently expressed. The NSs protein can be detected in bunyavirus-infected mosquito cells (M.F. SCALLAN and R.M. ELLIOTT, unpublished results) so the failure to intiate translation at a downstream AUG is not a general feature of insect cells. Whether this observation has any implications for the control of NSs synthesis under normal conditions is unknown.

2.3 Proteins and Protein Function

2.3.1 L Segment Protein

The BUN virus L protein has a predicted molecular weight of 258656 (ELLIOTT 1989b), somewhat higher than the previous estimates of 145000–200000 which were based on the relative migration of L protein in SDS-polyacrylamide gels (PENNINGTON et al. 1977; MCPHEE and WESTAWAY 1981; SHORT et al. 1982). The protein is relatively rich in Ala and Gly residues and has an overall charge of + 14 at pH 7.

Some limited sequence data are available from the L segments of LAC and SSH viruses (CLERX-VAN HAASTER et al. 1982a; HACKER et al. 1990). Comparison of the predicted L protein sequences with the BUN virus L protein shows 63% identity of the amino terminal 46 residues and 32% identity of the carboxy terminal 108 residues; the amino termini of the LAC and SSH virus L proteins are 90% identical.

The L protein is presumed to be the virion-associated transcriptase or a component thereof. Transcriptase activity has been detected in detergent-disrupted preparations of Lumbo (BOULOY and HANNOUN 1976), LAC (PATTERSON et al. 1984), GER (GERBAUD et al. 1987a) and BUN (H. JIN and R.M. ELLIOTT, unpublished results) bunyaviruses. The enzymatic activity was weak compared to, for example, vesicular stomatitis virus, which has impeded detailed biochemical analysis of the enzyme. However PATTERSON et al. (1984) demonstrated stimulation of transcriptase activity by the dinucleotide ApG, by cap analogues such as $m^7GpppAm$, and by natural mRNAs; furthermore, they detected an endonuclease activity which cleaved capped mRNAs in vitro. These results are consistent with the notion that bunyavirus transcription, like that of influenza viruses, uses host-derived RNA fragments to prime mRNA synthesis.

Comparison of the BUN virus L protein with other negative-strand virus transcriptases failed to reveal any significant regions of homology, with the exception of an area of weak homology with the influenza virus PB1 protein (ELLIOTT 1989b). In fact, this region of the L protein (residues 950–1220) contains the four consensus motifs (dubbed the polymerase module) which POCH et al. (1989) have identified in 80 RNA-dependent polymerases and thus probably represents the polymerase catalytic domain.

2.3.2 M Segment Proteins

Bunyavirus glycoproteins have few oligosaccharide side-chains (VORNDAM and TRENT 1979; CASH et al. 1980; LEES et al. 1986; PARDIGON et al. 1988). The glycans are mostly of the endoglycosidase H (endo H) resistant, complex type (MADOFF and LENARD 1982), but some endo H-sensitive high mannose type glycans are also found. Comparison of the four glycoprotein sequences shows the presence of two strictly conserved N-linked glycosylation sites (Asn-X-Ser/Thr) in G2 and a single conserved site in G1. A second site in G1 is conserved between viruses in the same serogroup, and an additional glycosylation site is predicted in G2 of GER virus (ELLIOTT 1990). It is not known how many of these sites are actually glycosylated.

When the four proteins were compared, an overall homology of 60% was observed for G2, 50% for NSm and 40% for G1 (ELLIOTT 1990); G1 proteins are more conserved in the carboxy half than the amino half. Notably, however, the positions of the Cys residues are strictly maintained, which suggests that the glycoproteins may adopt a similar structure after folding. The hydropathy profiles of the four precursors are very similar (Fig. 4). G2 has a rather long hydrophobic sequence (residues 187–245) followed by a charged stop-transfer sequence, but the orientation of the protein in the viral membrane is not known. The hydrophobic domain in G2 was suggested by POBJECKY et al. (1989) to be responsible for the fusion of cells by bunyaviruses under acidic conditions (GONZALEZ-SCARANO et al. 1984). The role of NSm in the multiplication cycle of the virus is unknown, but examination of the amino acid sequences and hydropathy profiles (Fig. 4) suggests that it is a membrane-bound protein. The G1 protein appears to be a typical type I glycoprotein with its carboxy terminus anchored in the membrane and the amino terminus external. Available data indicate therefore that the antigenic differences that distinguish individual bunyaviruses are located in the external and more variable half of G1.

The gene products of the bunyavirus M segment have been implicated in many of the important biological properties of the viruses such as tissue tropism, virulence, neutralization, haemagglutination, etc. These are discussed in detail elsewhere in this volume.

2.3.3 S Segment Proteins

The six deduced N protein sequences are very similar in length (233 or 235 residues) and can be readily aligned with the minimum of inserted gaps (ELLIOTT 1989a). Overall, the N proteins show 40% similarity, but within a serogroup the

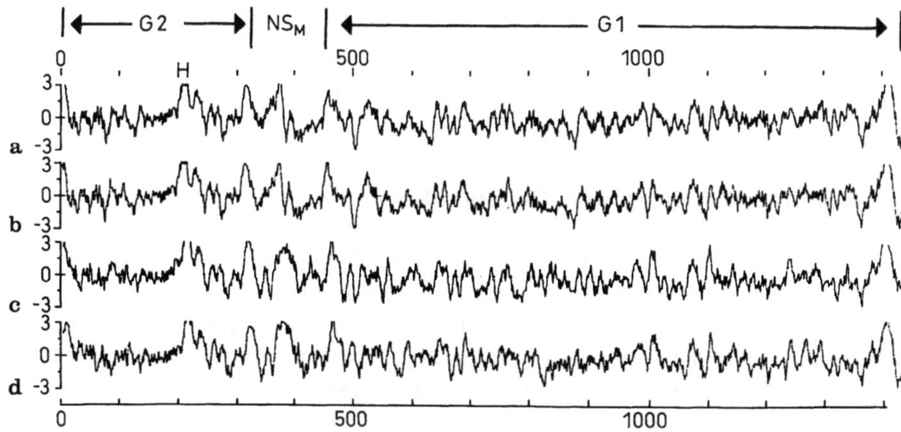

Fig. 4. Hydropathy profiles of SSH (*a*), LAC (*b*), BUN (*c*) and GER (*d*) virus glycoprotein precursors. The boundaries of G2, NSm and G1 are indicated. *H*, hydrophobic domain in G2 suggested to be involved in the fusion activity (see text). (From ELLIOTT 1990)

N protein similarity rises to over 80%. Certain regions of the N proteins are particularly well conserved, such as between residues 62 and 102, residues 123 and 169 and the carboxy terminal 15 residues. These conserved regions are presumably of functional significance to the protein (perhaps binding to viral RNA), and may also be antigenic determinants for the complement fixing antibodies which cross-react throughout the *Bunyavirus* genus (BISHOP 1990).

The NSs proteins, which are more variable in length (Table 5), can also be aligned with a minimum of gaps (ELLIOTT 1989a) but show greater diversity than the N proteins—25% similarity overall and 70% within serogroups. The difference in length of the NSs proteins is accounted for by variability at the carboxy termini of the proteins, and the most conserved region is between residues 71 and 91, which corresponds to the carboxy terminus of the shortest NSs protein. As the function of NSs has not been elucidated, the significance of the conserved residues is unknown.

3 *Hantavirus* Genus

The *Hantavirus* genus was added to the Bunyaviridae family in 1987 to accommodate a newly described group of rodent-borne viruses. The prototype virus of the genus, Hantaan (HTN) virus, and serologically related viruses, cause a variety of diseases which have been collectively termed haemorrhagic fever with renal syndrome (HFRS). Hantaviruses have been detected in indigenous rodents throughout the world, both in areas where HFRS occurs and in those where such diseases have not been reported. Nucleotide sequence information is currently availble for the M and/or S genome segments of four serologically distinct hantaviruses: HTN virus, Sapporo rat (SR-11) virus and nephropathia epidemica (NE) virus (Hällnäs strain), which are aetiological agents of severe, moderate and mild HFRS, respectively; and Prospect Hill (PH) virus, an American isolate not known to cause HFRS.

3.1 Genome Structure and Organization

Like those of other members of the family, the L, M and S genome segments of all hantaviruses examined to date display consensus sequences at their 3′ termini which are distinct from those described for viruses in the other four Bunyaviridae genera (SCHMALJOHN and DALRYMPLE 1983, 1984; SCHMALJOHN et al. 1985). Complementary 5′ terminal sequences have been identified by analyses of cDNA copies of viral genomes (SCHMALJOHN et al. 1986a, 1987b; GIEBEL et al. 1989; STOHWASSER et al. 1990; ARIKAWA et al. 1990; PARRINGTON and KANG 1990). As previously suggested for other viruses in the family, the complementarity of the terminal nucleotides may allow noncovalent circularization of the individual genome segments via stable, base-paired, panhandle structures (Fig. 5). It is possible that these structures are involved in viral replication, perhaps by serving as a recognition site for the virion-associated polymerase.

Nucleotide sequence analyses of cDNA representing the S genome segments of the four representative hantaviruses demonstrated that a strict negative-sense coding strategy is observed. A single, long ORF in the virus-complementary sense RNA appears to encode the sole gene product: the nucleocapsid (N) protein (Fig. 6). Although a small ORF, which could potentially code for a 6-kDa polypeptide, was identified in the same reading frame immediately following the termination codon of HTN N, a protein of this size has not been found in HTN virus-infected cells (SCHMALJOHN et al. 1986a). Moreover, this ORF is not conserved in the S genome segments of NE, PH or SR-11 viruses. Similarly, although an ORF was detected in an overlapping reading frame with that encoding N of NE and PH viruses, no protein of the predicted size (approximately 10 kDa) has been detected in infected cell cultures, and the reading frame is not conserved in the S segments of either

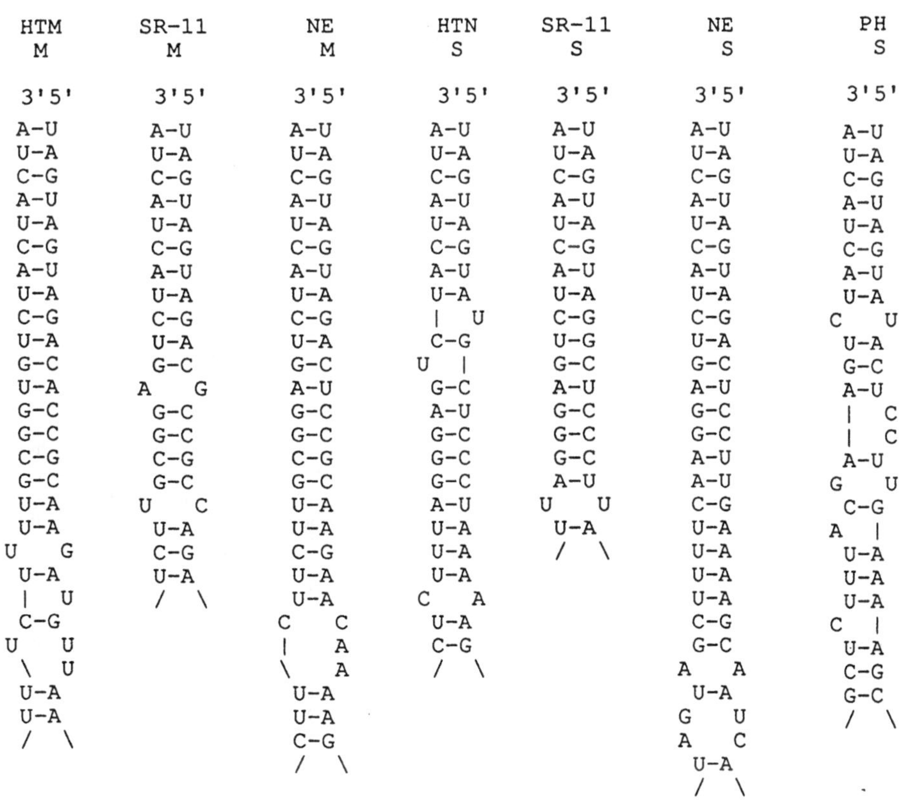

Fig. 5. Possible base-pairing of the terminal nucleotides of the M and the S genomic segments of hantaviruses

HTN or SR-11 viruses (SCHMALJOHN et al. 1986a; STOHWASSER et al. 1990; PARRINGTON and KANG 1990; ARIKAWA et al. 1990; Fig. 6). Thus unlike the S segments of viruses in the *Bunyavirus, Phlebovirus* and *Uukuvirus* genera, the S genome segments of hantaviruses do not appear to encode an NSs gene product. Consequently, approximately one-fourth of the hantavirus S segment is noncoding (Table 6). Because the function of the NSs protein has not yet been determined for any virus in the family, it is not known whether hantaviruses replicate in the absence of the function provided by NSs or perform that function with their much larger N protein. The same sort of S segment coding strategy and large noncoding region was reported for Dugbe virus, a member of the *Nairovirus* genus (WARD et al. 1990; see below).

Like the S segments of hantaviruses, the M genome segments do not appear to encode nonstructural proteins. Only one significant ORF, with a coding capacity of approximately 126 kDa, was detected in the virus-complementary sense RNA of the M segments of HTN, SR-11 and NE viruses (SCHMALJOHN et al. 1987b; YOO and KANG 1987; GIEBEL et al. 1989; ARIKAWA et al. 1990; Table 6, Fig. 6). Amino-terminal sequence analysis of the G1 and G2 envelope glycoproteins of HTN and SR-11 viruses demonstrated a gene order of 5′ G1–G2 3′ with respect to virus-complementary sense RNA (SCHMALJOHN et al. 1987b;

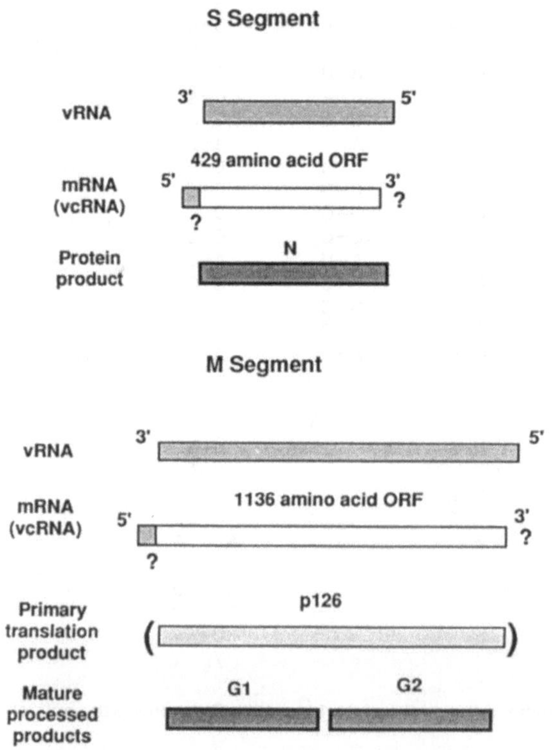

Fig. 6. Gene expression strategies of the M and S segments of Hantaan virus. Viral genomic RNA (vRNA) is transcribed to virus-complementary RNA (vcRNA) which serves as messenger RNA (mRNA). The termini of hantaviral mRNAs have not been defined (indicated by ?). *Open boxes* in the mRNA represent ORFs; *shaded areas*, possible 5′, nontemplated extension similar to those described for other viruses in the family. The S segment ORF is translated to yield the nucleocapsid protein. The M segment ORF is translated to yield the two envelope glycoproteins G1 and G2. Although the M segment has the capacity to yield a polyprotein with a calculated molecular weight of 126 kDa (p126), this protein has not been observed in vivo

Table 6. Coding capacity of hantavirus genomic segments[a]

Virus	RNA segment	Nucleotide sequences			Number of amino acids	Predicted M_r	Recognized gene products
		Total	Noncoding[b] 5′	3′			
HTN	M	3616	40	168	1136	126300	G1, G2
SR-11	M	3651	46	203	1134	125500	G1, G2
NE	M	3682	40	195	1148	126800	
HTN	S	1696	36	370	429	48100	N
SR-11	S	1769	42	437	429	48100	N
NE	S	1785	42	453	433	49400	N
PH	S	1675	42	331	434	49000	N

[a] Information is presented only for molecularly cloned genomic segments
[b] Noncoding sequences at the 5′ and 3′ termini are listed with respect to virus-complementary sense RNA

ARIKAWA et al. 1990). Synthetic peptides representing amino acids encoded by the exact end of the HTN M segment ORF were found to elicit antibodies in rabbits which recognized HTN G2 protein. Therefore, the carboxy terminus of G2 extends to the end of the coding region of the M segment. Similarly, synthetic peptides representing the last predicted hydrophilic region prior to the amino terminus of G2 (amino acids 588–614 of the ORF) were able to induce antibodies in rabbits that recognized HTN virus G1, indicating that the carboxy terminus of G1 extends at least to amino acid residue 588. Consequently, if an intergenic region is present, it could not encode a protein of more than 6 kDa (SCHMALJOHN et al. 1987b). Only a small amount of noncoding information is present at either the 3′ or 5′ terminus of the hantavirus M genome segments (Table 6).

Of the hantavirus genes sequenced to date, the S and the M segments of each virus share approximately the same degree of homology with cognate gene segments of another hantavirus. Although limited in scope, these data appear to differ from those obtained with influenza viruses, which display a higher rate of mutation in genes encoding the haemagglutinin and neuraminidase surface proteins than in genes encoding internal proteins such as the matrix protein and the nucleocapsid protein (SMITH and PALESE 1989). Whether this indicates that the surface glycoproteins of hantaviruses are not greatly influenced by immunological pressures or that surface and internal proteins are equally influenced by host factors awaits further studies on hantaviral evolution and gene sequence diversions.

Comparisons of the S segment nucleotide sequences of NE, PH or SR-11 virues to those of HTN virus revealed overall homologies of 60%, 57% and 70%, respctively. The M segments of NE and SR-11 viruses had nucleotide sequence homologies of 61% and 70%, respectively, with the M segment of HTN virus. Each of these four viruses was isolated originally from a rodent: HTN virus from an *Apodemus* field mouse in Korea, SR-11 virus from a laboratory rat in Japan, NE virus from a *Clethrionomys* vole in Scandinavia, and PH virus from a *Microtus* vole in the United States. Much higher sequence homologies (approximately 95%) were reported among the M segment of HTN virus and those of two viruses isolated from Korean haemorrhagic fever patients (SCHMALJOHN et al. 1988). Thus, despite their isolation from dissimilarly affected hosts (i.e. persistently infected rodents versus acutely infected humans) and over a 6-year time frame, the viruses apparently maintained a high degree of genetic stability in nature.

3.2 Gene Expression Strategies

3.2.1 L Segment

Preliminary sequence analysis of cDNA representing the L genome segment of HTN virus (SCHMALJOHN, unpublished data) indicates a coding strategy similar to that reported for the L segment of BUN virus (ELLIOTT 1989b). That is, a single, long ORF, detected in the virus-complementary sense RNA, has sufficient capacity to encode the viral L protein, which has an estimated molecular weight of 200 kDa (ELLIOTT et al. 1984). No other ORFs appear to be of significance, although more detailed analysis will be required to eliminate the possibility that minor ORFs play a role in coding. No obvious nucleotide sequence homology could be detected between the L segment of HTN virus and that of BUN virus.

3.2.2 M Segment

The hantaviral M segment encodes the G1 and G2 envelope glycoproteins. Hybridization with strand-specific probes to Northern blots of total RNA from HTN virus-infected cells, demonstrated that virus-complementary sense M segment RNA is similar in size to genomic M RNA (SCHMALJOHN et al. 1987a). Thus, subgenomic RNAs do not serve as messages for G1 and G2. Although it is assumed that the messenger RNA encoding G1 and G2 is like that of other viruses in the family, in that it has a 5′ extension consisting of host-derived, capped sequences and is truncated at the 3′ terminus, no direct evidence to support the assumption has yet been reported.

Two potential translation initiation codons precede sequences that code for the amino terminus of HTN virus G1. Only the first of these is conserved in the M segment ORFs of SR-11 and NE viruses (SCHMALJOHN et al. 1987b; GIEBEL et al. 1989; ARIKAWA et al. 1990). The exclusive use of this ATG for initiation of HTN virus G1 synthesis, as well as that of the other hantaviruses, is supported by the presence of more favourable flanking sequences surrounding the first ATG than the second and the inability to express HTN G1 in a baculovirus system when the first ATG is not present (probably because the seven amino acids following the ATG do not constitute an efficient signal sequence; SCHMALJOHN et al. 1987b, 1989b). In contrast, the 18 amino acids which extend from the first ATG to the amino terminus of HTN G1 and the corresponding

16 amino acids preceding those of SR-11 virus G1 constitute typical signal sequences (VON HEIJNE 1983). This stretch of amino acids probably dictates insertion of the nascent protein into membranes and is subsequently removed from the mature glycoprotein.

Generation of the hantaviral G2 protein is presumed to ensue upon cleavage of a polyprotein precursor. No polypeptide of a size corresponding to that encoded by the complete M segment ORF has been detected in hantavirus-infected cell cultures, suggesting either very rapid post-translational cleavage or, more likely, co-translational cleavage. Neither the cleavage site between HTN virus G1 and G2 nor factors effecting the cleavage have been identified, but it is believed that cleavage is mediated by host cell proteases.

An intriguing alternative mechanism of G2 synthesis is internal initiation of translation. An in-frame ATG codon followed by sequences encoding 16 amino acids which conform to criteria established for signal sequences (VON HEIJNE 1983) were identified preceding sequences encoding the G2 proteins of HTN and SR-11 viruses (SCHMALJOHN et al. 1987b; ARIKAWA et al. 1990). Expression studies with baculovirus and vaccinia virus vectors demonstrated that this initiation codon could function efficiently to produce HTN virus G2 in the absence of any additional initiation codons (SCHMALJOHN et al. 1989b, 1990). Similarly, baculovirus expression of cDNA, truncated such that the two ATG initiation codons preceding G1 were removed, but the rest of the gene (including the entire G1 and G2 coding sequences) remained, resulted in detectable synthesis only of G2 (SCHMALJOHN, unpublished data). Evidence that the ATG immediately upstream of G2 may be required for G2 synthesis was obtained by mutagenesis of cDNA representing the complete M segment such that the ATG was changed to ATC (isoleucine). Baculovirus expression of the mutated gene resulted in synthesis of detectable G1 but not G2 (SCHMALJOHN, unpublished). Whether this codon functions as the sole source of translation initiation of G2, and whether these results obtained by baculovirus expression correlate with M segment expression of authentic Hantaan virus, are interesting topics for further study.

3.2.3 S Segment

The expression strategy of the hantaviral S genome segment appears to be remarkably simple in comparison to those of bunyaviruses, phleboviruses and uukuviruses. The N protein is translated from a virus-complementary sense messenger RNA which is indistinguishable in size from genomic length S RNA (SCHMALJOHN et al. 1986a, 1987a). Like those of the M segment, the termini of the S segment messenger RNAs have not been characterized. As mentioned above, nonstructural protein coding information apparently is not included in the S segments of hantaviruses. The N protein of HTN virus was the only detectable product obtained by cell-free translation of authentic HTN S segment messenger RNA, or RNA transcribed from cDNA representing the viral S segment (SCHMALJOHN et al. 1986a, 1987a). Similarly, baculovirus or vaccinia virus recombinants containing cDNA representing the HTN S segment expressed only one detectable HTN-specific polypeptide: the N protein (SCHMALJOHN et al. 1989b, 1990).

3.3 Proteins and Protein Function

3.3.1 L Segment Protein(s)

The large ORF, which was found in the virus-complementary sense sequence of the L segment of HTN virus, has the capacity to encode a protein of approximately 240 kDa (SCHMALJOHN, unpublished data). A protein of similar size has been detected in Hantaan virion protein preparations (ELLIOTT et al. 1984). This large protein (L) is believed to be the RNA-dependent RNA polymerase, an activity which has been found in association with purified HTN virions (SCHMALJOHN and DALRYMPLE 1983). No information is available concerning the L segment messenger RNAs, or of possible processing of the L protein. Further characterization of functional aspects of this protein await in vitro expression of the L gene.

3.3.2 M Segment Protein(s)

The G1 and G2 proteins of hantaviruses are similar to those of other viruses in the family: they are glycosylated, are rich in cysteine residues and induce neutralizing antibody responses in animals (see below). Hydropathy plots of the predicted amino acids within the M segment ORFs of HTN, NE and SR-11 viruses are nearly superimposable (Fig. 7), despite overall amino acid sequence homologies of only 43% for G1 and 55% for G2 among the three viruses. A number of the observed amino acid differences can be attributed to conservative substitutions. Discounting these differences, the G1 and G2 proteins of the three viruses, respectively, display 58% and 71% amino acid sequence homologies (ARIKAWA et al. 1990). Deduced amino acid sequences encoded in the SR-11 virus M segment ORF are more closely related to those of HTN virus (78% identical) than to those of NE virus (53% identical). These amino acid sequence homologies are reflective of the closer serological relationship of HTN and SR-11 viruses as compared to HTN and NE viruses (SCHMALJOHN et al. 1985). Amino acid sequence homologies in excess of 98% have been reported for the G1 and G2 proteins of HTN virus and two isolates from Korean haemorrhagic fever patients, again reflecting the indistinguishable serological characteristics of these isolates (SCHMALJOHN et al. 1988).

Amino-terminal sequence analysis of the G1 proteins of HTN and SR-11 viruses revealed that the terminal residue of both proteins is leucine. Very little direct conservation of amino acids in the preceding sequences was observed although both had overall characteristics typical of signal sequences (ARIKAWA et al. 1990). The G2 amino-terminal residues of HTN and SR-11 viruses were

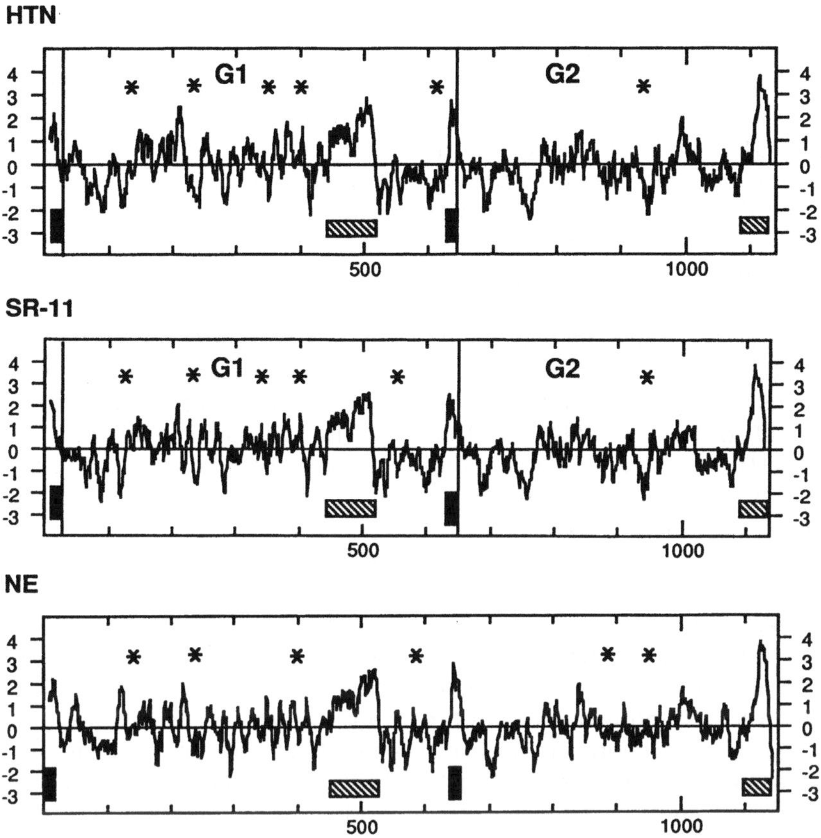

Fig. 7. KYTE-DOOLITTLE (1982) hydropathy plots of the M segment ORFs of hantaviruses performed by using the DNA Strider program (MARCK 1988). The hydropathic profiles of the deduced amino acids of the M segment ORFs of HTN, SR-11, and NE viruses are displayed. Where protein sequence information is available, *vertical lines* are added to represent the position of the amino terminus of the indicated proteins. In all cases, the mature glycoprotein coding sequences are preceded by a hydrophobic signal sequence (*black box*). Similarly, at or near the carboxy terminus of the first glycoprotein and at the end of the second protein, possible hydrophobic transmembrane anchor sequences are identified (*shaded boxes*). *Asterisks*, potential asparagine-linked glycosylation sites

found to be serine and alanine, respectively. Although amino-terminal sequence data have not been obtained for the G2 protein of NE virus, the five amino acids immediately preceding G2, which are identical for HTN and SR-11 viruses, are also conserved in the same position in the NE virus ORF, suggesting that the amino terminal residue of the G2 protein of NE virus is glutamic acid.

Both HTN and SR-11 viruses have 630 codons between the amino terminus of G1 and that of G2, which could potentially encode G1 proteins of 70 kDa. Based upon polyacrylamide gel electrophoresis, the G1 proteins of these viruses were estimated to be 68 kDa and 72 kDa, respectively (ELLIOTT et al. 1984; SCHMALJOHN and DALRYMPLE 1984; ARIKAWA et al. 1990). Although the carboxy-terminal amino acid sequence of HTN G1 was not determined, it was suggested that a short intergenic region between G1 and G2, encoding a polypeptide which may be removed during processing of G1 and G2, might explain the discrepancy between calculated and observed molecular weights of the HTN G1 protein (SCHMALJOHN et al. 1987b). If this is indeed so, then a shorter intergenic region between SR-11 virus G1 and G2 than between HTN virus G1 and G2 might account for the apparently larger G1 protein of SR-11 virus. Alternatively, differences in post-translational modifications of HTN and SR-11 G1 proteins, such as glycosylation, might also affect the apparent molecular weights of these proteins.

A comparison of potential asparagine-linked glycosylation sites in the G1 proteins of SR-11 and HTN viruses (Asn-X-Ser/Thr) revealed five sites in each. The positions of the first four sites were conserved, but the fifth site was located at amino acid positions 560–562 for SR-11 virus G1 and at amino acid positions 609–611 for HTN G1 (Fig. 7). Three of these four conserved glycosylation sites were also maintained in the NE virus G1 protein. The entire codon for the fourth glycosylation site, which is conserved between HTN and SR-11 viruses, is absent in that position of the NE virus M segment ORF. A potential glycosylation site located near the carboxy terminus of the G1 protein of NE virus was not conserved with either HTN or SR-11 viruses, and is probably not used due to the proline residue located between the asparagine and serine residues (BAUSE 1983). The single available glycosylation site in HTN G2 (928–930) was conserved in the G2 proteins of both SR-11 and NE viruses. This is the only potential G2 glycosylation site for HTN and SR-11 viruses; however, an additional site was identified in NE virus G2 (GIEBEL et al. 1989; Fig. 8). The HTN G2 glycosylation site is apparently used, as evidenced by the ability to radiolabel G2 with [^{3}H]glucosamine or [^{3}H]mannose and also by a reduction in the apparent molecular weight of G2, of about 3 kDa, after treatment with endoglycosidases (SCHMALJOHN et al. 1986b). The carbohydrate moieties attached to HTN virus G1 account for about 7 kDa of the apparent molecular weight, but it is not known how many of the available glycosylation sites are actually used. The oligosaccharides on both G1 and G2 were found to be mostly, but not entirely, of the high-mannose type (SCHMALJOHN et al. 1986b), perhaps reflecting the Golgi-associated site of maturation of these viruses (HUNG et al. 1985).

The intracellular transport of hantaviral proteins has not yet been clearly defined. It has been relatively difficult to observe viral morphogenesis, probably because of the low-titer persistent mode of replication which these viruses exhibit in cell culture. Antigen is generally not detected at the cell surface, and viruses are believed to mature in intracellular vesicles as described for other viruses in the family (HUNG et al. 1985; MURPHY et al. 1973). Although the factors dictating subcellular localization of viral proteins are not known, it is clear that the envelope proteins possess whatever signal(s) are necessary to be dispatched to the Golgi without additional requirements for other viral proteins or nucleic acid. This has been demonstrated by following the transport of

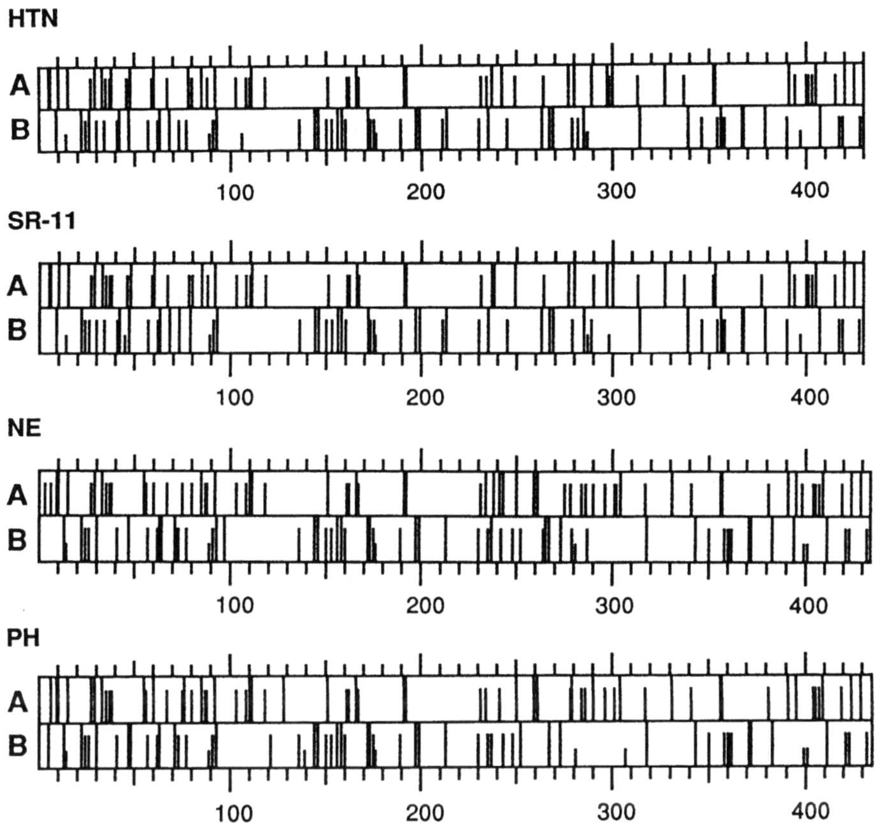

Fig. 8. Distribution of acidic and basic residues in the N protein of HTN, SR-11 and NE viruses. Using the DNA Strider sequence analysis program (MARCK 1988), the distribution of acidic (*A*) and basic (*B*) amino acid residues within the S segment ORFs are indicated. For *A*, *full vertical lines* represent glutamic acid residues and *intermediate lines* represent aspartic acid residues. For *B*, *full lines* represent the positions of arginine residues, *intermediate lines* are lysines and *short lines* histidines. *Numbers* represent amino acid residue position within the ORF

vaccinia virus-expressed HTN G1 and G2 in infected mammalian cell cultures (PENSIERO et al. 1988; R. PETTERSSON, personal communication) or of baculovirus-expressed HTN G1 and G2 in infected insect cell cultures (J. WHITE, personal communication).

As noted above, the envelope proteins of viruses in the Bunyaviridae family have been found to be cysteine rich, and the G1 and G2 proteins of hantaviruses are no exceptions. The positions of almost all of the Cys residues in the G1 and G2 proteins of HTN, SR-11 and NE-viruses are conserved, suggesting that the intramolecular bonds formed between these amino acids may be important to the structural integrity of these proteins (ARIKAWA et al. 1990).

The role of the envelope glycoproteins in protection from hantaviral infection is indisputable. Monoclonal antibodies to both the G1 and G2 proteins of HTN virus have been found to neutralize viral infectivity and passively to protect animals from infection (DANTAS et al. 1986; ARIKAWA et al. 1989; SCHMALJOHN et al. 1990). Such a humoral response is probably also important for cross-protection among hantaviruses, as evidenced by at least one neutralizing epitope which is shared by the G2 proteins of HTN and SR-11 viruses (ARIKAWA et al. 1989). Whether one of the two envelope proteins induces a better protective immune response than the other is not known; however, protection studies performed with baculovirus or vaccinia virus recombinants that express G1 and G2 coding regions separately, or those that express the entire M segment ORF (i.e. both G1 and G2), demonstrated that both proteins were required to achieve efficient protection from HTN virus challenge (SCHMALJOHN et al. 1990).

Monoclonal antibodies to HTN virus G1 and G2 have also been used to define antigenic relationships among the viruses in the genus. Haemagglutination inhibition assays and immunofluorescent antibody staining revealed conserved sites on the G2 proteins of HTN, SR-11, NE and PH viruses. Only one HTN G1-specific monoclonal antibody was identified which recognized antigens of both HTN and SR-11 viruses, and none was found which was cross-reactive with the G1 proteins of NE or PH viruses (ARIKAWA et al. 1989). These results, however, may be due simply to the smaller number of monoclonal antibodies available to G1 than to G2 rather than to a complete absence of hantaviral cross-reactive sites on G1. Similar results were obtained with monoclonal antibodies to the G1 and G2 proteins of a rat-borne hantavirus, B-1 virus (DANTAS et al. 1986).

3.3.3 S Segment Protein

The nucleocapsid proteins of hantaviruses and nairoviruses are considerably larger than those of viruses in the other four genera of Bunyaviridae. The functional significance of this (if any) is not known. Comparisons of the predicted amino acid sequence of the N proteins of SR-11, NE and PH viruses to that of HTN virus N revealed homologies of 83%, 61%, and 62%, respctively (ARIKAWA et al. 1990; PARRINGTON and KANG 1990; STOHWASSER et al. 1990). Interestingly, both the amino-terminal and carboxy-terminal residues are more highly conserved than the molecule as a whole. For example, STOHWASSER et al. (1990) reported a 73% conservation of amino acids 1–210 of NE and HTN N proteins and a 94% homology of amino acids 302–433. In contrast, the central portions of the viral N proteins (from amino acids 211–301) displayed only 13% homology. The same sort of comparisons can be made between the N proteins of SR-11 and PH viruses and that of HTN virus. These results suggest that the both ends of the nucleocapsid protein may have some role in viral replication, perhaps in interaction with the viral RNA. An overall conservation in the acidic and basic amino acids among all four viral N proteins further suggests structural similarities of these proteins (Fig. 8).

Factors influencing the assembly of nucleocapsids in hantavirus-infected cultures have not yet been defined. It is presumed that some sort of specific interaction with viral nucleic acid is required for generation of ribonucleocapsid structures. An intriguing observation with baculovirus-expressed HTN virus N suggests that such a signal may not be prerequisite to assembly. That is, expressed N was found to form structures which resembled authentic nucleocapsid structures when examined by immuno-electron microscopy (SCHMALJOHN, unpublished data). Whether nucleic acid is required for this assembly, or whether the structures consist only of self-assembled protein molecules remains to be determined.

As with other internal antigens of viruses, the hantaviral N protein does not induce a neutralizing antibody response in animals. Studies with baculovirus-expressed HTN virus proteins, however, demonstrated that N alone can protect animals from HTN virus infection (SCHMALJOHN et al. 1990). Although not proven, these results suggest that a cell-mediated immune response to HTN virus N may be protective. This hypothesis is consistent with previous findings by ASADA et al. (1987, 1988, 1989) which indicated that although neutralizing epitopes conserved between HTN and NE viruses or SR-11 and NE viruses have not been identified, spleen cells primed with NE virus were found to cross-react with HTN virus-infected target cells. Similarly, passively transferred T lymphocytes primed by NE virus were found to reduce the presence of HTN virus in nude mice. Because these studies were performed with authentic viral preparations, it was not possible to determine whether the cell-mediated response was specifically to N. To associate N conclusively with these results, it will be necessary to repeat these experiments using expressed or purified nucleocapsid proteins.

4 *Nairovirus* Genus

The *Nairovirus* genus contains 34 viruses which are divided into seven serogroups (CASALS and TIGNOR 1980; ZELLER et al. 1989). The prototype of the genus is Nairobi sheep disease (NSD) virus. Nairoviruses are transmitted mostly by ticks, and the group includes a severe human pathogen, Crimean-Congo hemorrhagic fever (CCHF) virus. At the molecular level the nairoviruses have so far received limited attention, and only one RNA segment—Dugbe (DUG) virus S RNA (WARD et al. 1990)—has been sequenced.

4.1 Genome Structure and Organization

Analysis of the sizes of the genome RNA segments (Table 1) indicates a distinctive pattern for nairoviruses, with the L RNA segment being substantially larger (perhaps twice the size) than the L RNAs of other Bunyaviridae. Some estimates of the size of the M segment indicate that this is also larger than that in other genera (CLERX and BISHOP 1981; CLERX et al. 1981; WATRET and ELLIOTT 1985; WARD et al. 1990). The coding assignments of the L and M segments are not known.

The S segment of DUG virus has been cloned as cDNA and its sequence determind (WARD et al. 1990). The segment is 1712 nucleotides long and has a base composition of 30.6% A, 20.1% C, 23.4% G and 25.9% U. Sixteen of the 17 terminal nucleotides at the 3′ and 5′ ends are complementary. No nucleotide sequence homology was found with representatives of the other Bunyaviridae genera. A large ORF was identified in the virus-complementary RNA encoding the N protein with an M_r of approximately 50000.

Cloned cDNAs derived from the M and S segments of DUG virus (NSD serogroup) have been used as hybridization probes to investigate sequence relationships within the *Nairovirus* genus (MARRIOTT et al. 1990). Some hybridization was noted with the S segments of CCHF virus and Hazara (HAZ) viruses (CCHF serogroup), but not with representatives of the other serogroups. These results are consistent with antigenic data which showed cross-reactivity by complement fixation or immunofluorescent antibody tests between viruses in these two serogroups (see CASALS and TIGNOR 1980).

4.2 Gene Expression Strategies

Nothing is known about the expression strategies of the nairovirus L and M segments. Assuming that the nairovirus L protein (which has a similar size to the analogous proteins of viruses in the other genera; BISHOP 1990) is encoded by the L RNA, then the coding capacity of this segment may be grossly underutilized.

Analyses of nairovirus glycoproteins present a complicated picture; estimates of M_r 72000–84000 for G1 and M_r 30000–45000 for G2 have been reported (CLERX and BISHOP 1980; CLERX et al. 1981; CASH 1985; WATRET and ELLIOTT 1985), but for HAZ virus three virion glycoproteins of M_r 84000, 45000 and 30000 were described (FOULKE et al. 1981). Examination of nairovirus-infected cells by pulse-chase experiments has identified a number of glycoprotein precursors which are related to the structural glycoproteins, and it appears that the mature G2 protein is produced during the final stages of virion maturation (CLERX and BISHOP 1980; CASH 1985; WATRET and ELLIOTT 1985. Thus, the glycoprotein expression strategy resembles that of other Bunyaviridae in gross terms by involving cleavage of precursor molecules, but no details of the cleavage events are known.

The DUG virus S RNA encodes a protein of predicted molecular weight 49400 in the virus-complementary RNA (WARD et al. 1990). This ORF is flanked by 3′ and 5′ untranslated regions of 49 and 337 nucleotides, respectively. A DUG virus N protein monoclonal antibody reacted with a fusion protein of β-galactosidase and the DUG virus ORF, thus confirming the identity of the predicted protein as N. Downstream of the N ORF a short (50 codon) ORF was observed which could encode a protein of M_r 5900. However, no evidence for the synthesis of this protein in DUG virus-infected cells was obtained. The layout of the DUG virus S RNA is, therefore, very similar to that of HTN virus (Sect. 3.1).

4.3 Protein Structure and Function

Little is known about the structure and function of nairovirus proteins, and we can only suggest that they resemble those of viruses in the other genera. As stated earlier (Table 1), the pattern of protein sizes of nairoviruses is distinctive. No homology was found between the DUG virus N protein and the N proteins specified by viruses in the other genera (WARD et al. 1990).

5 *Phlebovirus* and *Uukuvirus* Genera

Phleboviruses and uukuviruses are currently classified in separate genera. This has been based largely on their lack of serological cross-reactivity. However, more recent molecular genetic analyses, most notably sequencing data of molecularly cloned genomic RNAs, have suggested an evolutionary relationship between members of these two groups. These similarities are reviewed here. Also included in this section is TSW virus, a likely first plant-infecting member of the family (MILNE and FRANCKI 1984; DE HAAN et al. 1990). Although TSW virus has not yet been formally classified within the Bunyaviridae family, and certainly not assigned a generic status (though the names Phytophlebovirus and Phythrivirus have been suggested for a new genus), its inclusion here derives from recent sequence data indicating that it possesses certain unique and distinguishing features characteristic of phleboviruses and uukuviruses.

5.1 Genome Structure and Organization

The nucleotide sequence for the complete genome of a phlebovirus or of an uukuvirus has not yet been reported. The deficiency lies with the lack of sequence data for the L segment of these viruses. Studies have so far focused on the M and S segments (Table 1). The nucleotide sequences of the M segments of two phleboviruses (Punta Toro, PT, and Rift Valley fever, RVF) and of Uukuniemi (UUK) virus have been reported (Table 1). Mention of sequencing results for the M segment of TSW virus has been made (DE HAAN et al. 1988), but the data have not yet been published. Complete S segment sequences for two phleboviruses (PT and sandfly fever-Sicilian, SFS), for UUK virus, and for TSW virus, have been reported (Table 1).

As mentioned earlier, a consistent feature of the RNA for all viruses in the family Bunyaviridae is the conservation of the nucleotide sequence at the 3′ end of the three genomic segments of a virus. Futhermore, these 3′ terminal sequences are largely complementary to the 5′ end sequence of each segment. For phleboviruses, uukuviruses and TSW virus, the 3′ sequence conservation was noted from data derived from direct sequencing of the 3′ ends of viral RNA segments (Parker and Hewlett 1981; Collett et al. 1985; De Haan et al. 1989). These results were confirmed by sequence data from molecular clones, which further established the complementarity of the segment 5′ ends (Ihara et al. 1984, 1985b; Collett et al. 1985; Rönnholm and Pettersson 1987; Marriott et al. 1989; Simions et al. 1990). Table 7 summarizes these data where comparisons among the segment terminal sequences of a virus can be made.

Since sequences at segment termini are complementary, the possibility exists that they base-pair, resulting in either a tandem association of genomic segments or circular, panhandled structures for individual segments. This latter intramolecular RNA structure is most often considered, and seems supported by the electron microscopic observation of circular forms of UUK virus RNA (Hewlett et al. 1977). Possible energetically favourable base-paired structures of the termini of RNA segments of the viruses under discussion in this section are shown in Fig. 9. When predicting such possible base-paired structures, most analyses have taken into consideration only the segment sequences at the RNA termini. The body of the RNA may, however, influence the possible secondary structure at the termini. To investigate this, one study applied RNA folding algorithms to the entire M segment sequence of RVF (Collett et al. 1985). Although such an analysis is theoretical, a prominent feature of the

Table 7. End sequence conservation within the genomic segments of phleboviruses, Uukuniemi and tomato spotted wilt viruses

```
RVF   L    3' UGUGUUUCUG
              ||||||||||
RVF   M    3' UGUGUUUCUG
              |||||||| |
RVF   S    3' UGUGUUUCGG

PT    M    3' UGUGUUUCUGCCGUGCUACU.............UCUACACGGCCAGAAACACA   5'
              ||||| || |  |  ||  |             |  |  |  |||||||||||
PT    S    3' UGUGUAUC-GAGGACUUUU..............UUAAGUCCCCCAGAAACACA   5'

UUK   L    3' UGUGUUUCUG
              ||||||||||
UUK   M    3' UGUGUUUCUGCCGAUUGUACCA........UGGUACAUCGGCACAGAAACACA   5'
              ||||||||||| | ||| | |            |||| |  | ||||||||||
UUK   S    3' UGUGUUUCUGCAGGUUGAAUCG........AAUUACAACCUCCCAGAAACACA   5'

TSW   M    3'  CUCGUUAGUUACGUAGUUCU............ UCAACAUGGACUAACGAG    5'
               |||||||      ||                  | || || ||||||||
TSW   S    3' UCUCGUUAACACAGUCUUAAA.............UUUAACUGUGCUAACGAGA   5'
```

folded RNA model was the stem structure depicted in Fig. 9. In fact, the predicted structure involved 51base pairs.

Not only is the extensive base-pairing obvious in Fig. 9, but each segment seems to have a unique pattern of mismatched bases. These perhaps strategically positioned breaks in the duplex probably confer a unique secondary structure to the terminus of each segment. The combination of conserved terminal nucleotide sequence, yet an RNA secondary structure involving both termini of each segment that is unique, may provide signaling mechanisms for performance of processes common to all three genomic segments in a similar fashion (such as initiation of RNA transcription and replication), yet also allow individual L, M and S segment recognition for such activities as RNA segment packaging and virus assembly. Studies designed to investigate the functional significance of terminal sequence conservation and complementarity are needed in order to advance our understanding of the molecular mechanisms of replication of these viruses.

Another possible secondary structure has been discussed regarding the S segments of PT, UUK and TSW viruses. RNA folding analysis of sequences from the central region of these segments revealed inverted complementary sequences capable of forming a stem-loop structure (EMERY and BISHOP 1987; SIMONS et al. 1990; DE HAAN et al. 1990). For all three viruses, this sequence is A

```
 RVF       PT        UUK       TSW       SFS       PT        UUK       TSW
  M         M         M         M         S         S         S         S

 3'5'      3'5'      3'5'      3'5'      3'5'      3'5'      3'5'      3'5'

 U-A       U-A       U-A       C-G       U-A       U-A       U-A       U-A
 G-C       G-C       G-C       U-A       G-C       G-C       G-C       C-G
 U-A       U-A       U-A       C-G       U-A       U-A       U-A       U-A
 G-C       G-C       G-C       G-C       G-C       G-C       G-C       C-G
 U-A       U-A       U-A       U-A       U-A       U-A       U-A       G-C
 U-A       U-A       U-A       U-A       U-A      A   A      U-A       U-A
 U-A       U-A       U-A       A-U       U-A       U-A       U-A       U-A
 C-G       C-G       C-G       G-C       C-G       C-G       C-G       A-U
 U-A       U-A       U-A       U-A      C   A     G   A      U-A      A   C
 G-C       G-C       G-C      U   G     A   C     A   C      G-C       C-G
 |  C      |  C      |  A     A   /      G-C       G-C      C   C      A-U
 C-G       C-G       |  C      C-G       G-C       G-C      \   C      C-G
 C-G       C-G       C-G      G   U      G-C       G-C       A-U       A-U
 A-U       G-C       C-G       U-A      A   C     A   C      G-C       G-C
 C-G       U-A       G-C      A   |     U   U     C   U      G-C       U-A
 G-C       G-C       A-U       G-C      C   A     U   G      U-A      C  |
 U-A      C  |       U-A       U-A      A   G      U-A       U-A      U  |
 |  A      U-A      U  |       U-A       A-U       U-A       G-C       U-A
 |  C      A-U       G-C      C   C     U   C     U   U     A   A      A-U
 A-U      C   C      U-A      U   U      U-A       A-U       A-U       A-U
 A-U      U   U      A-U      |   |      / \       A-U      U   U      A-U
U   C     |   |      C-G                           A-U      |   |      A-U
 U-A                 C-G                           / \                 C-G
 / \                 A-U                                               A-U
                     / \                                               / \
```

Fig. 9. Possible base-pairing of the terminal sequences of the M and S segments of phleboviruses, Uukuniemi virus and tomato spotted wilt virus

+ U rich, and for PT virus involves about 200 nucleotides, for UUK virus about 60–70 nucleotides, and for TSW virus some 250 bases. However, the analogous region in the S RNA of SFS virus is very C rich and cannot be folded into an obvious base-paired structure (MARRIOTT et al. 1989). The possible functional significance of the predicted structure of PT virus has been hypothesized to be involved in transcription termination (see below). Comparative RNA folding analyses involving the complete sequence of S segments should be carried out to explore further the possible ramifications of these studies. If should be noted that whether any of these predicted secondary structures actually exist in vivo or have functional significance or consequence for virus replication, remains to be experimentally established.

When the terminal sequences of the RNA segments of phleboviruses, UUK and TSW viruses are compared to one another, clear and strong homologies are revealed between the former two viruses. Although much weaker, there may be some terminal sequence identity between the M segments of TSW and RVF viruses (Table 8); otherwise, the termini of TSW virus appear distinct.

Analyses of the currently sequenced phlebovirus, UUK and TSW virus M segments reveal that all possess a single AUG-initiated open reading in the vcRNA extending the length of the molecule. The 5′ noncoding regions are short (16–20 nucleotides), while the 3′ noncoding portions are longer (187–372 nucleotides; see Table 9). However, the actual mRNA serving the M segment ORF has ends distinct from virus-complementary RNA (see below). Within this ORF, the positions of the coding regions for two protein products of the M segment, glycoproteins G1 and G2, have been determined. For RVF virus, amino acid 154 of the ORF marks the amino-terminal residue of G2, and position 691 the start of G1 (COLLETT et al. 1985). For PT virus, the start for the first glycoprotein G1 is at

Table 8. Homologies among genomic segment end sequences of phleboviruses, Uukuniemi and tomato spotted wilt viruses

```
RVF M 3' UGUGUUUCUGCCA-CGUAAUU...............UCAACG-UGGC-CAGAAACACA  5'
         ||||||||||||   |  | |               || ||   ||| ||||||||||
PT  M 3' UGUGUUUCUGCCG-UGCUACU...............UCUACA-CGGC-CAGAAACACA  5'
         ||||||||||||| |  |||                  |||| |||| ||||||||||
UUK M 3' UGUGUUUCUGCCGAUUGUACC...............GGUACAUCGGCACAGAAACACA  5'
            |||   |     |    |                 |  |   | ||   |||
TSW M 3' CUCGUU--AGUUACGUAGUUCU................UCAACAUGGACU--AACGAG  5'
            |||   |  ||||| ||                  ||||| ||| |   |||
RVF M 3' UGUGUUUCUGCCACGUAAUUUA................UCAACGUGGCCAGAAACACA  5'


SFS S 3' UGUGUUUCCAGGGAUCAAUU..................ACUGAUCCCCCAGAAACACA  5'
         ||||| || |||||    |                        |||||||||||||||
PT  S 3' UGUGUAUCGAGGGACUUUUA..................UUAAGUCCCCCAGAAACACA  5'
         ||||| ||    |  |                          | | ||||||||||||
UUK S 3' UGUGUUUCUGCAGGUUGAAU..................UACAACCUCCCAGAAACACA  5'
         | |  ||      ||   |                   |   | ||     ||  | |
TSW S 3' UCUCGUUAACACAGUCUUAA..................UUUUAACUGUGUUAACGAGA  5'
         | |  ||       ||                      ||    |      ||  | |
SFS S 3' UGUGUUUCCAGGGAUCAAUU..................UUAAGUCCCCCAGAAACACA  5'
```

Table 9. Coding capacity of phlebovirus, UUK and TSW virus genomic segments[a]

Virus	RNA segment	ORF polarity	5′NC[b]	3′NC[b]	IR[c]	Number of amino acids	Predicted M_r	Recognized protein products
RVF	M	vcRNA	20	284		1197	133 000	G2, G1, 78-kDa, 14-kDa (NSm)
PT	M	vcRNA	16	372		1313	146 000	G1, G2, (NSm?)
UUK	M	vcRNA	17	187		1008	114 000	G1, G2
TSW	M	vcRNA	—	—		—	—	G1, G2
SFS	S	vcRNA	41		250	219	24 800	N
	S	vRNA	20			268	30 400	NSs
PT	S	vcRNA	28		360	250	26 900	N
	S	vRNA	34			243	29 100	NSs
UUK	S	vcRNA	34		74	254	28 500	N
	S	vRNA	25			273	32 000	NSs
TSW	S	vcRNA	153		508	259	28 800	N
	S	vRNA	87			465	52 400	(NSs?)

[a] Information for only molecularly cloned genomic segments is represented.
[b] NC, Noncoding region; number of nucleotides.
[c] IR, Intergenic region; number of nucleotides

amino acid 271 of the ORF and that for G2 at residue 810 (HARA et al. 1985b). In contrast to the two phleboviruses, the first glycoprotein G1 of UUK virus begins at amino acid 18 of its M segment ORF, followed by G2 at residue 514 (RÖNNHOLM and PETTERSSON 1987). In all cases, the estimated carboxy end of the first glycoprotein approximates the position of the N-terminus determined for the second glycoprotein. However, precise definition of the carboxy termini of these proteins has not yet been established by amino acid sequencing.

The glycoprotein gene order within the M segment mRNA for these viruses is therefore: for RVF virus, 5′ G2-G1 3′; for PT virus, 5′ G1-G2 3′; and for UUK virus, 5′ G1-G2 3′. That the first glycoprotein of the ORF of RVF virus is designated G2 and that of PT and UUK viruses G1 reflects only a nomenclature system based on protein gel mobility and should not be taken to imply differences in protein gene arrangement. In fact, as is discussed below, the first glycoproteins in all these viruses appear to be analogous and show sequence homologies. Similarly, the second glycoproteins bear some sequence relatedness with one another.

From the positioning of the viral envelope glycoproteins within the ORF, there exists considerable "preglycoprotein" coding potential in the two phlebovirus ORFs (about 17 kDa for RVF and 30 kDa for PT). For PT virus, this region has been suggested to encode a polypeptide termed NSm (IHARA et al. 1985b), although to date this putative gene product has not been described. The existence of proteins possessing the coding sequences of the preglycoprotein region in RVF virus has been reported (KAKACH et al. 1988) and is discussed in detail below. In contrast to the phleboviruses, the UUK virus M segment ORF does not possess a preglycoprotein coding region. This represents a significant difference between the two virus groups.

Inspection of the S segments of phleboviruses, UUK and TSW viruses reveals in all cases two significant AUG-initiated ORFs, one in the vRNA polarity and one in the vcRNA (Table 9). In their respective polarities, both ORFs are positioned near the 5′ end of the RNA, the phlebovirus and UUK virus 5′ noncoding regions being slightly shorter (20–41 nucleotides) than those of TSW (87 and 153 nucleotides). Both the viral and virus-complementary ORFs extend about one-half the length of the S segment and do not overlap one another. Each ORF encodes a single polypeptide. The ORF of the virus-complementary polarity codes for the viral nucleocapsid protein N; that in the viral-sense codes for a nonstructural protein designated NSs. This arrangement in which both polarities of the same RNA segment encode viral gene products has been termed ambisense (IHARA et al. 1984).

5.2 Gene Expression Strategies

5.2.1 L Segment

No reports have yet appeared on the complete cloning or sequencing of a phlebovirus or uukuvirus L segment. In the absence of such data, it has generally been assumed the L segment employs a simple negative sense coding and expression strategy. In support of this, a virus-complementary L RNA approximately the size as the genomic segment has been identified in RVF and UUK virus-infected cells (PARKER et al. 1984; PETTERSSON et al. 1985). These data, along with the observed size of the L protein (> 200 kDa) and the results from the cloning and sequencing of the L segment of BUN bunyavirus (ELLIOTT 1989b), makes the above assumption the probable reality.

5.2.2 M Segment

Since the phlebovirus, UUK and TSW virus M segments all possess a single ORF spanning the length of the virus-complementary RNA, it also appears that a simple negative-sense gene expression mechanism is operative for this segment. However, whereas the single large ORF of the L segment presumably encodes a single large polypeptide, the single ORF of the M segment of phleboviruses, UUK and TSW viruses (as well as of all Bunyaviridae) encodes at least two proteins, the viral envelope glycoproteins G1 and G2, and in one extensively investigated case (RVF virus) four polypeptides. Thus, superimposed on the negative-sense expression mode for the M segments are co- and/or post-translational processing and modification events.

The virus-complementary mRNA of the M segment of RVF and UUK viruses lacks a 3′ poly (A) tract and is similar in size to their respective genomic RNA (PARKER et al. 1984; PETTERSSON et al. 1985). However, closer examination of the RVF virus message revealed it to be truncated at its 3′ end by approximately 112 nucleotides with respect to the genomic sequence and extended beyond the 5′ genomic end by 12–14 nucleotides of heterogeneous sequence (COLLETT 1986). These characteristics of the ends of this phlebovirus segment mRNA (5′ end extension and 3′ end truncation relative to the genomic segment) appear to be common for all Bunyaviridae mRNAs examined to date. Several interesting consequences are suggested by these features. As a result of the 3′ end truncation, the mRNAs lack sequences complementary to the 5′ end of the RNA. Such molecules would therefore be unable to form the circular structures discussed above. The mRNA would thus be distinct from genomic (or anti-genomic) RNA species, and as a possible consequence would be excluded from participation in the processes of RNA replication and encapsidation. The mechanism by which these truncated molecules are generated is unknown but must involve some type of mRNA transcription termination signal recognized by the viral transcription enzyme complex. Inspection of nucleotide sequences in and around the estimated 3′ ends of Bunyaviridae mRNA has lead to speculation that purine-rich tracts might be involved (COLLETT 1986; RAJU and KOLAKOFSKY 1986). Further investigations will be necessary to define the actual (functional) transcription termination signals for mRNA synthesis.

With regard to initiation of mRNA transcription, the presence of added nucleotides at the 5′ end of the Bunyaviridae mRNA is reminiscent of orthomyxovirus mRNA, where mRNA biogenesis involves cap scavenging of sequences from host cell mRNA for use in the initiation of viral mRNA transcription (KRUG et al. 1987).

The large ORF of the M segment mRNAs of phleboviruses, UUK and TSW viruses is capable of encoding a polypeptide of 114–146 kDa (Table 9). M segment-specific polypeptides representing this full-length size have not been detected in phlebovirus or uukuvirus-infected cells. Since this ORF encodes at least two viral proteins, it is likely that the putative polyprotein is processed co-translationally. From cell-free translation studies, this appears to be the case. UUK virus M segment mRNA isolated from virus-infected cells when translated in a rabbit reticulocyte lysate gave rise to a primary translation product of about 110–114 kDa. When the reaction was carried out in the presence of microsomes, the 110–114 kDa protein was not observed. Rather, immunoprecipitable polypeptides migrating near the positions of the viral glycoproteins were found (ULMANEN et al. 1981). Similar in vitro translation results were obtained with RVF virus M segment mRNA-like transcripts produced through recombinant DNA means (SUZICH and COLLETT 1988).

The current view for protein translation from the UUK virus M segment mRNA is as follows (Fig. 10): translation initiation begins from the first AUG codon of the ORF; the initial 17 amino acids serve as a membrane insertion sequence, which is subsequently cleaved from glycoprotein G1; translation proceeds to the second glycoprotein which may also be preceded by a signal sequence, requiring that two proteolytic cleavage events occur, one at the carboxy terminus of G1 and another at the amino end of G2. The nature of the proteinases that affect these co-translational cleavages is unknown, although they are undoubtedly host cell derived and presumably microsome associated.

For phleboviruses, M segment expression is complicated by the fact that there is additional protein coding potential proceeding the mature glycoproteins. The coding capacity and gene expression strategy employed by this segment has been examined in detail in the case of RVF virus. Not only does the RVF virus M segment encode glycoproteins G2 and G1, but it also codes for two additional protein products: a glycosylated 78-kDa protein and a nonglycosylated 14-kDa protein (KAKACH et al. 1988; WASMOEN et al. 1988). The sequences represented in the 78-kDa protein begin from the first AUG codon of the ORF and extend to the carboxy end of the first glycoprotein. Those of the 14-kDa protein encompass the amino acids between the second in-phase AUG codon (located 37 amino acids donwstream from the first AUG codon) and the start of the mature glycoprotein G2. Attempts to demonstrate in virus-infected cells a precursor-product relationship between the 78-kDa protein and the two proteins whose sequences it encompasses (the 14-kDa protein and glycoprotein G2) were unsuccessful (COLLETT et al. 1989). Using site-directed mutagenesis and a faithful surrogate expression system (recombinant vaccinia virus), it was shown that the 78-kDa protein initiated at the first AUG codon, and that the 14-kDa protein initiated independently at the second in-phase translational site (KAKACH et al. 1988; COLLETT et al. 1989; SUZICH et al. 1990). Investigation of the

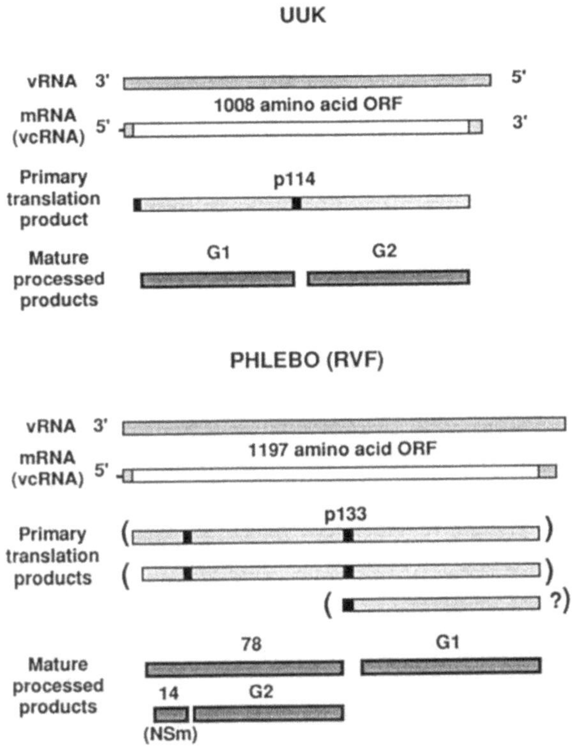

Fig. 10. M segment gene expression. Viral genomic RNA (vRNA) is transcribed to mRNA. The M segment mRNA is truncated at its 3′ end by about 100–150 bases and extended at its 5′ end by 11–14 nucleotides. *Open box* in the mRNA represents the ORF, *shaded areas*, noncoding. Based on cell-free translation studies, the mRNA directs the synthesis of primary translation products which in the presence of microsomes are processed co-translationally to yield mature protein products. *Black boxes*, putative signal sequences preceding the mature glycoprotein coding regions. The primary products (*in parentheses*) are not observed in virus-infected cells. In the case of RVF, there are two, and possibly three, such primary translation products; one initiating from the first AUG of the ORF, another from the second AUG, and a putative third from an as yet undefined initiation site. See text for details

AUG codon requirements for biogenesis of the two glycoproteins yielded additional surprising results. The first glycoprotein (G2) appeared to require a translation start site within the preglycoprotein region but not the first AUG of the ORF and likely the same AUG employed for the 14-kDa protein synthesis (the second in-phase AUG). A portion of the production of the second glycoprotein (G1) also involved one or more of the five AUG codons within the preglycoprotein region. However, about half the production of the G1 glycoprotein was independent of all AUG codons in the preglycoprotein region (SUZICH et al. 1990).

From these data, the following scheme for RVF virus M segment expression can be envisioned (Fig. 10): primary translation results in at least two polyprotein products initiated from AUG codons within the preglycoprotein region of the large ORF; one putative precursor, starting from the first methionine codon and extending to the end of the ORF (1197 amino acids), would serve as the precursor to the 78-kDa protein and possibly also glycoprotein G1; a second primary translation product, beginning from the second AUG and continuing to the ORF termination codon (1159 amino acids), would ultimately yield the 14-kDa protein and both glycoproteins. Finally, a significant portion (50%) of glycoprotein G1 must arise by a pathway distinct from those presented above. The mechanism by which this portion of the second glycoprotein is synthesized independently of the other three M segment gene products is unclear at present but certainly may involve internal translation initiation.

Thus, the expression strategy of this phlebovirus M segement is surprisingly complex. Not only does it involve multiple translational initiation events, but the biosynthetic pathways for the two major viral glycoproteins appear to be distinct from one another. These results are provocative. Their consideration in light of data from studies of UUK virus glycoprotein transit further increases the intrigue of M segment gene expression. KUISMANEN (1984) and PERSSON et al. (1989) showed that the kinetics of synthesis and transport of the two glycoproteins of UUK virus were different, and that they were apparently transported independently before their dimerization and incorporation into virions. Finally, data from expression studies of the M segment of HTN virus suggest independent production of the two glycoproteins of this virus is also possible (see Sect. 3.2.2). Further investigations will be necessary to clarify and establish the mechanisms of glycoprotein biogenesis by these viruses.

The complexity of such an expression strategy suggests that it might have some regulatory role. There is a suggestion that this may be the case for the use of the first and second in-phase AUG codons of the RVF virus M segment. The common glycosylation site that exists between the 78-kDa and 14-kDa proteins is used only by the former (KAKACH et al. 1989). Translational initiation from the first AUG codon appears to predetermine utilization of this N-linked glycosylation site and furthermore precludes the proposed proteolytic cleavage at the preglycoprotein-glycoprotein G2 junction (KAKACH et al. 1989). Conversely, translation from the second AUG codon results in the failure to use the preglycoprotein glycosylation site but does allow proteolytic cleavage at the amino terminus of mature glycoprotein G2 to yield the 14-kDa protein. The putative proteolytic cleavage at this junction was found to be independent of glycosylation (KAKACH et al. 1989), implicating the amino acid sequence between the first and second AUG codons as influencing subsequent protein glycosylation and proteolytic processing. Thus, use of the two-site translational initiation expression strategy serves as a mechanism for controlling co- and post-translational protein modifications. Unfortunately, the roles of the 78-kDa and 14-kDa proteins in virus replication are not known.

5.2.3 S Segment

Since the S segments of phleboviruses, UUK and TSW viruses possess functional ORFs in each the vRNA and vcRNA polarities, their expression strategy has been designated ambisense (IHARA et al. 1984). Analyses of RNA species for S segment sequences in virus-infected cells revealed that, in addition to full-length S vRNA and S vcRNA, two subgenomic RNAs were present (PARKER et al. 1984; IHARA et al. 1985a; PETTERSSON et al. 1985; MARRIOTT et al. 1989; DE HAAN et al. 1990). Cell-free translation of these subgenomic RNA species showed them to encode the viral N and NSs proteins (PARKER et al. 1984; IHARA et al. 1985a; PETTERSSON et al. 1985). Further studies of the PT virus S segment mRNAs demonstrated that the N protein was encoded by the subgenomic viral-complementary mRNA and the NSs protein by the subgenomic viral-sense mRNA (IHARA et al. 1985a). Analyses of the 5′ ends of the two PT mRNAs revealed them to be heterogeneous in nucleotide sequence and extended beyond the end of the genomic sequence by some 11–18 bases (IHARA et al. 1985a).

The overall size of the S segment subgenomic mRNAs indicated that they terminated at their 3′ ends near the middle of the genomic segment, within a so-called intergenic region (IR) between the two ORFs. Detailed analyses of the 3′ termini of the PT S mRNAs showed them both to map to the same 40 nucleotide region within the IR (EMERY and BISHOP 1987). An interesting feature of the PT virus S segment IR is that it is A + U rich and can potentially form an energetically stable hairpin structure involving some 200 nucleotides (EMERY and BISHOP 1987). It is suggestive that such an RNA secondary structure might be involved in

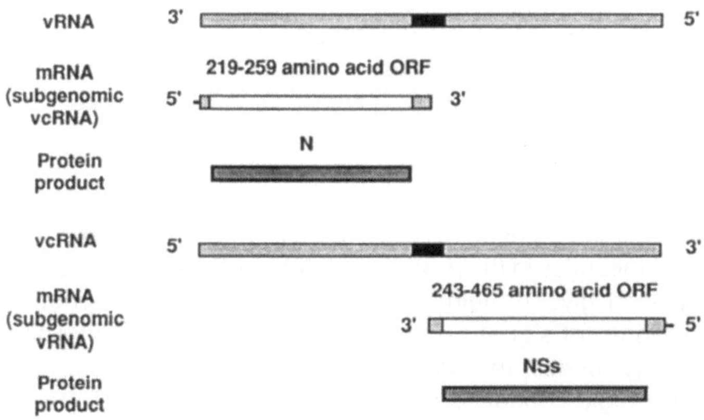

Fig. 11. S segment gene expression. Viral genomic RNA (vRNA) is transcribed to yield a subgenomic virus-complementary mRNA roughly half the size of genomic RNA. After genomic RNA replication, full-length vcRNA is used to produce a second subgenomic mRNA of viral-sense. Both mRNAs possess added sequences at their 5′ ends. *Open box* in the mRNAs represents the ORF; *shaded areas*, noncoding. *Black box* depicted in the full-length RNAs represents the intergenic region between the two ORFs of the S segment sequence. Translation of the individual mRNAs yields their respective protein products: N from the viral-complementary mRNA, and NSs from the viral-sense mRNA

transcription termination. Interest in this possibility is further provided by the fact that the S segment IRs of UUK and TSW viruses are also A + U rich and are predicted to form a stem-loop structure (SIMONS et al. 1990; DE HAAN et al. 1990). However, the S RNA of SFS virus has a predominantly C-rich IR which cannot be folded into a base-paired structure (MARRIOTT et al. 1989). Understanding the transcription termination signals employed by these viruses will require further investigation.

The specific details of S segment mRNA biogenesis are discussed elsewhere in this volume (see chapter by Kolakofsky and Hacker). Schematically, gene expression of the S segment RNA of phleboviruses, UUK and TSW viruses is summarized in Fig. 11. The viral genomic RNA serves as template for production of the subgenomic virus-complementary sense mRNA, which is then translated to produce the N protein. For production of the subgenomic viral-sense mRNA to be used for NSs protein synthesis, a full-length vcRNA template is used.

5.3 Proteins and Protein Function

5.3.1 L Segment Protein(s)

Little is known about the gene product(s) encoded by the phlebo- and uukuvirus L segment RNA. Assuming features analogous to those revealed by the BUN virus L segment sequence (ELLIOTT 1989b), this RNA codes for only the L protein. This polypeptide ($M_r > 200\,000$) is readily detected in virus-infected cells and in purified virions. The L protein is most certainly, and at least, a component of the viral RNA-dependent RNA polymerase involved in RNA transcription and replication. It is likely to be a multifunctional polypeptide possessing distinct domains responsible for specific activities. Good biochemical and enzymological investigations on this important protein are sorely needed.

5.3.2 M Segment Proteins

The hydropathy plots of the M segment ORF of PT, RVF and UUK viruses are displayed in Fig. 12. The similarities in the profiles are striking. Several details are noteworthy. Preceding each glycoprotein is a stretch of 17–20 hydrophobic amino acids that may represent signal or membrane insertion sequences for translocation of the polypeptides through the endoplasmic reticulum. An additional hydrophobic region near the carboxy-terminal end of each glycoprotein may represent transmembrane anchor sequences. The region between the putative transmembrane domain of the first glycoprotein and the signal sequence of the second may represent a "cytoplasmic tail" of the first glycoprotein (RONNHOLM and PETTERSSON 1987). However, since the precise carboxy terminus of the first glycoprotein has not been determined for any of these viruses, it is still possible that a small intergenic peptide is removed from between the mature coding sequence of the two envelope glycoproteins.

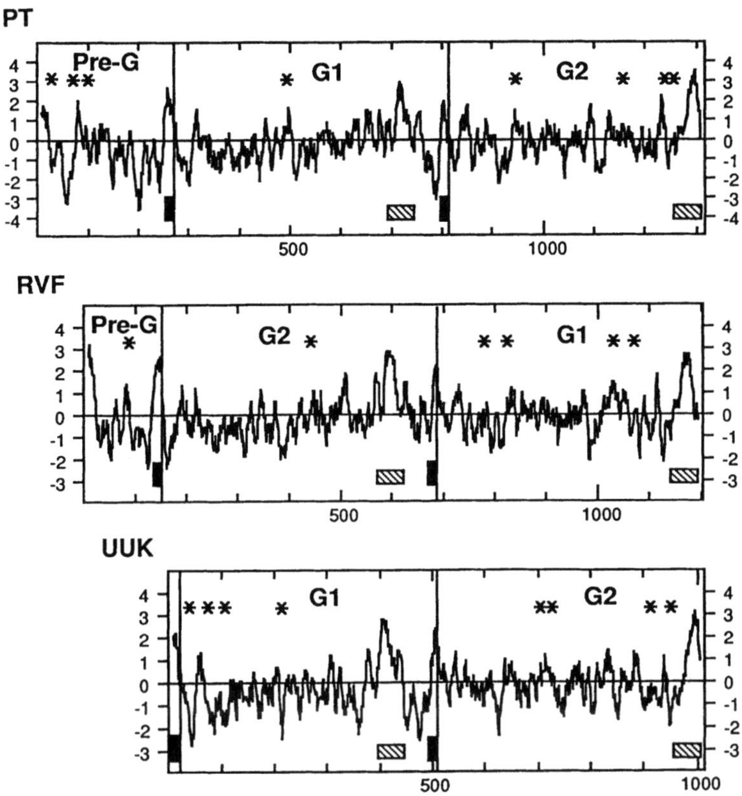

Fig. 12. Hydropathy plots of the M segment ORFs of phleboviruses and Uukuniemi virus. Using the program of MARCK (1988), the hydropathic profile of the amino acid sequence of the M segment ORF of PT, RVF, and UUK viruses are displayed. *Vertical lines*, position of the amino terminus of the indicated proteins. *Pre-G*, protein coding region preceding that of the first envelope glycoprotein. In all cases, the mature glycoprotein coding sequences are preceded by a hydrophobic signal sequence (*black box*). Similarly, toward the carboxy end of the first glycoprotein and at the end of the second, there appear hydrophobic putative transmembrane anchor sequences (shaded boxes). *Asterisks*, potential asparagine-linked glycosylation sites

Inspection of Fig. 12 also graphically displays the apparent N-terminal addition to the phlebovirus ORFs (or deletion from the UUK virus ORF) of the preglycoprotein (NSm) region.

Sequence comparisons among the M segment proteins of these viruses have been made. In these comparisons, percentages of aligned sequence identity of less than 18%–20% are not likely to represent significant relatedness. No apparent sequence homology (8% identity) exists between the preglycoprotein regions of the PT and RVF virus M segment ORFs. However, there is a clear and obvious relationship between the analogous glycoproteins of these two viruses (35% aligned amino acid identity between the first glycoproteins of PT and RVF viruses; 49% between the second glycoproteins). Although no apparent identity is revealed between the first glycoprotein of UUK and the phleboviruses

(15%–18% identity), there appears to be a weak overall sequence relatedness between the second glycoproteins of these viruses (identity level of 25% between UUK and PT viruses; 22% between UUK and RVF viruses). Confirmation of this is provided by analyses of conservative substitution homology and of the existence of local regions of higher sequence identity (RÖNNHOLM and PETTERSSON 1987).

As with all members of the Bunyaviridae, the glycoproteins of phlebo- and uukuviruses are rich in Cys residues (5%–7% of the total amino acid content). The positions of most (70%) of the Cys residues within both glycoproteins are conserved among these three viruses (IHARA et al. 1985b; RÖNNHOLM and PETTERSSON 1987), suggesting not only some important role in protein structure of function but further indicating a relatedness between phleboviruses and uukuviruses.

Within the first glycoprotein of these viruses, there is a single site for asparagine-linked glycan attachment in RVF virus G2 and four sites in PT virus G1 and UUK virus G1. In all three viruses, the second glycoprotein possesses four such sites. Within the PT and RVF virus preglycoprotein region, there exist three sites and one site, respectively, for potential N-linked glycosylation. Based on gel mobility differences between glycosylated and unglycosylated forms of the envelope glycoproteins, it is generally thought that most sites are occupied with carbohydrate. More direct analyses in the RVF virus system have shown this indeed to be the case for the two glycoproteins (ALONSO-CAPLAN and DALRYMPLE, personal communication 1988). Furthermore, the RVF virus 78-kDa protein was shown to be glycosylated at both the preglycoprotein region site and the site that it shares with glycoprotein G2 (KAKACH et al. 1989). Interestingly, the same preglycoprotein site was not utilized in the 14-kDa protein. Detailed structure of the carbohydrate associated with these glycoproteins has been reported for UUK virus (PESONEN et al. 1982). Glycoprotein G1 contains only complex, endoglycosidase H-resistant glycans, while G2 has mostly high mannose, endoglycosidase H-sensitive sugars.

There are no published reports of O-linked glycosylation of phlebo- or uukuvirus glycoproteins. Investigation of the RVF virus glycoproteins for this modification indicated there was none (KAKACH and COLLETT, unpublished data). Another potential secondary protein modification of these glycoproteins for which there are no reports is fatty acid acylation. Given the large number of Cys residues in these proteins and their processing pathway, such a modification could be expected.

A characteristic of the glycoproteins of viruses in the Bunyaviridae family is their ability to accumulate and be retained in the Golgi complex, the site of virus maturation. Generally, only trace amounts, if any, of the glycoproteins are detectable at the cell surface (MATSUOKA et al. 1988). However, in one report, virus maturation was observed at the surface of RVF virus-infected primary rat hepatocytes (ANDERSON and SMITH 1987). The Golgi-specific property of these glycoproteins has been shown for UUK and RVF viruses to be independent of virus replication and of viral gene products from the other genomic segments (GAHMBERG et al. 1986; WASMOEN et al. 1988). These observations suggest that

Golgi localization and/or retention signals are contained within the viral glycoproteins themselves. Studies to investigate these putative signals further for the RVF virus glycoproteins have employed recombinant vaccinia viruses for surrogate gene expression. A recombinant virus was constructed containing only the first glycoprotein (G2) gene and the 22 amino acid hydrophobic sequence immediately preceding it (DALRYMPLE et al. 1989). Authentic-appearing G2 glycoprotein was made and found to be largely Golgi complex-localized (WASMOEN and COLLETT, unpublished data). However, when a similar construct was made in which the 76 amino acids immediately preceding the mature second (G1) glycoprotein, and therefore at the carboxy end of glycoprotein G2, were deleted (virus vSC7–1860; DALRYMPLE et al. 1989), the localization of the resultant carboxy end truncated G2 protein changed from the compact Golgi site to a diffuse reticular pattern (WASMOEN and COLLETT, unpublished data). Interestingly, this deletion removed sequences up to, but not including, the putative carboxy end transmembrane anchor domain. Thus, the sequences within the putative cytoplasmic tail of RVF virus glycoprotein G2 may be important either for transit from the endoplasmic reticulum to the Golgi complex or for Golgi localization and/or retention, although other interpretations are possible.

Recombinant viruses expressing only the second RVF virus glycoprotein (G1) have also been constructed. However, interpretation of the results of their analysis has been difficult. For example, in one recombinant (unpublished), glycoprotein G1 expressed alone was found to be mostly Golgi associated, while in another construct in which glycoprotein G1 was expressed in the total absence of G2 (virus 9; KAKACH et al. 1988), G1 was found exclusively in the plasma membrane exposed to the cell surface (WASMOEN and COLLETT, unpublished data). Additional studies are necessary before these sets of data can be reconciled. However, they do indicate that under at least some circumstances, when expressed alone, the RVF glycoprotein G1 can localize outside the Golgi complex.

The above results suggest that distinct or multiple mechanisms may be involved in the Golgi localization of these proteins. Considerations of the means for Golgi retention must take into account features of protein sequence, structure, transit kinetics and capability, and protein-protein associations. Clearly, further investigations are needed before we will understand the nature of the signals involved in the Golgi complex localization of Bunyaviridae glycoproteins.

The major envelope glycoproteins are likely involved in virus-cell attachment. However, the specific interactions involved and the nature of cellular receptors for virus infection remain to be elucidated. Virus penetration and endosomal membrane fusion probably represent additional activities mediated by these glycoproteins. It is clear that both glycoproteins of PT and RVF viruses are able to induce and interact with virus-neutralizing antibodies. Polyclonal, monospecific antisera to either G2 or G1 are able to neutralize virus infectivity in vitro (DALRYMPLE et al. 1981; SMITH, MEEGAN and DALRYMPLE, unpublished data).

Furthermore, monoclonal antibodies (MAbs) directed against the two glycoproteins have been isolated that are also virus-neutralizing (PIFAT et al. 1988; SMITH et al., unpublished data). However, investigations in both the RVF and PT virus systems have shown that only antibodies to the first glycoprotein are able to confer protection against lethal virus challenge in a mouse model (PIFAT et al. 1988; SMITH and DALRYMPLE, unpublished data). Neutralizing MAbs or antisera to the second glycoprotein were not protective when passively administered to animals. Thus, in this system, there is a distinction between in vitro virus neutralization and animal protection.

In active immunization experiments using RVF virus glycoprotein immunogens produced in either recombinant *Escherichia coli*, vaccinia virus, or baculovirus expression systems, glycoprotein G2 was found to be necessary and sufficient for induction of a protective immune response (COLLETT et al. 1987; DALRYMPLE et al. 1989; SCHMALJOHN et al. 1989a). The amino acid sequences of several epitopes defined by MAbs that are both virus neutralizing and protective have been identified in the RVF glycoprotein G2 (KEEGAN and COLLETT 1986).

The M segment coding potential preceding the first mature glycoprotein in RVF and PT viruses is 17 kDa and 30 kDa, respectively. For PT virus, the existence of polypeptides possessing these sequences (a putative NSm protein) has not been demonstrated. As mentioned above, sequences from this region of the RVF virus M segment ORF are found in two viral proteins: the glycosylated 78-kDa and nonglycosylated 14-kDa proteins (KAKACH et al. 1988). Comparison of the amino acid sequences from the preglycoprotein region in PT and RVF viruses reveal no apparent homology. Similarly, no homology was found with sequences of the M segment ORFs of other viruses of Bunyaviridae, including the putative NSm protein of bunyaviruses.

Studies of the 78-kDa and 14-kDa proteins of RVF virus showed the former to be exclusively localized to the Golgi complex of cells, while the latter was found in the Golgi as well as reticular structures (WASMOEN et al. 1988). That the 78-kDa protein is glycosylated, and the 14-kDa is not (KAKACH et al. 1989) suggests that the processing and transit of these two products may be distinct. Based on Triton X-114 extraction experiments, both proteins partition exclusively in the detergent phase, indicating their inherent capability for integral membrane association (KAKACH and COLLETT, unpublished data).

The functional roles of these two proteins is not known. They do not appear to be necessary for the normal synthesis, processing and transport of the major envelope glycoproteins (KAKACH et al. 1988; WASMOEN et al. 1988). The 78-kDa protein, but not the 14-kDa protein, can be found in purified virion preparations. Sera from convalescent animals infected in RVF virus do not possess antibodies to the 14-kDa protein. Interestingly, when just the 14-kDa protein was expressed in transfected mammalian cells, it proved to be extremely toxic, causing formation of giant syncytia and entire cell culture death (WASMOEN and COLLETT, unpublished data). These results are reminiscent of the inherent cytopathogenicity of the matrix protein of vesicular stomatitis virus (BLONDEL et al. 1990), raising the possibility that the RVF 14-kDa protein might be involved in virus assembly and/or maturation. However, the fact that no sequence similarities exist with those of other viruses in the family, and that this coding region is plainly absent in the related virus UUK virus (Fig. 12) suggests that the 14-kDa (NSm) protein may play a role more specific to phleboviruses. Possibilities for such a specific function include aspects of virus cell tropism and virus-vector transmission.

5.3.3 S Segment Proteins

Homology analyses of the sequences of the N proteins of the phleboviruses and UUK virus show that they are all clearly related (MARRIOTT et al. 1989; SIMONS et al. 1990). Levels of sequence identity between virus N proteins are: 47% for PT and SFS; 35% for PT and UUK; and 32% for SFS and UUK viruses. If conserved amino acid substitutions are allowed, these percentages nearly double. The N protein sequence of TSW virus shows no apparent sequence relatedness to those of the above viruses (identity levels of 12%–18%; DE HAAN et al. 1990). When the NSs protein sequences of all these viruses are similarly compared to one another, no significant similarities are revealed (12%–18% identities), except for a 26% identity level between the NSs of PT and SFS viruses.

The N protein functions in formation of the individual L, M and S nucleocapsids and therefore must interact directly with viral RNA. As viral mRNA is not found in nucleocapsid structures, there must be specific recognition signals or factors that determine N protein-RNA interaction. Details of this interaction and of the recognition sequences involved have not been elucidated. However, inspection of the charge distribution along the amino acid sequence of these N proteins suggests possibilities. Although less pronounced for the TSW virus N protein, the phlebovirus and UUK virus N proteins possess an obvious acidic amino-terminal domain of about 30 residues which is followed by a 70 residue region with distinctly basic character (Fig. 13). It is also likely the N protein is involved in the regulation of mRNA transcription, vRNA replication, and virion assembly and maturation, although investigations directly addressing these postulated activities are lacking.

As with the viral-glycoproteins, the N protein accumulates specifically in the Golgi complex, probably in the form of nucleocapsids (KUISMANEN et al. 1982, 1984). How this accumulation occurs, whether it is an intrinsic feature of N or of nucleocapsids, or whether the N protein interacts specifically with a Golgi-retained structure or protein (e.g. the viral glycoproteins), remain areas for future study.

Although a potent immune response is generated against the N protein in virus-infected animals, antibodies (polyclonal or monoclonal) to this protein are neither virus neutralizing in vitro nor protective in vivo (PIFAT et al. 1988; SMITH and DALRYMPLE, unpublished data).

The NSs protein of RVF virus is a phosphoprotein (STRUTHERS et al. 1984) and can be found in the nuclei of infected cells (SWANEPOEL and BLACKBURN 1977;

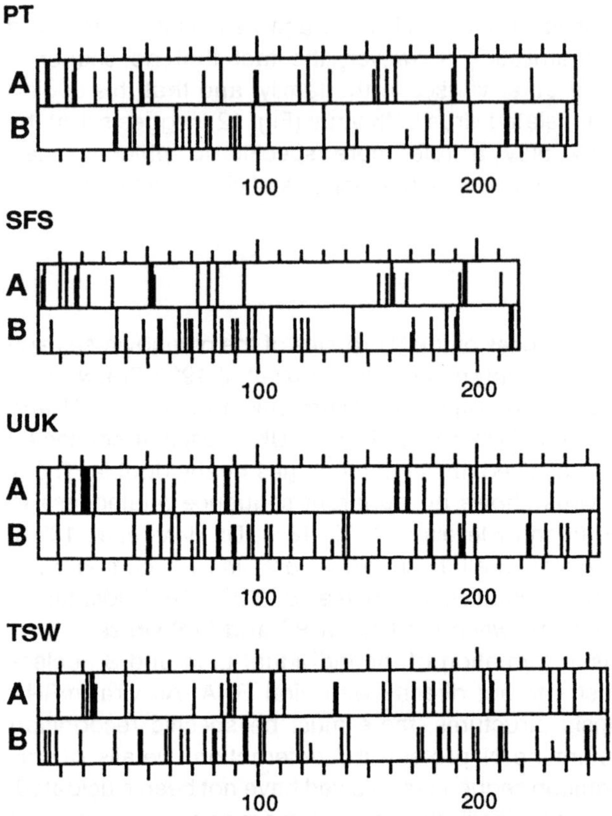

Fig. 13. Distribution of acidic and basic residues in the N protein of phleboviruses, Uukuniemi virus, and TSW virus. Using the program of MARCK (1988), the distribution of acidic (*A*) and basic (*B*) amino acid residues within the ORF encoding the N protein of PT, SFS, UUK, and TSW viruses are displayed. For *A*, *full vertical lines* represent aspartic acid residues and *intermediate lines* glutamic acids. For *B*, *full lines* represent the positions of arginine residues, *intermediate lines* are lysines and *short lines* histidines. *Numbers* represent amino acid residue position within the ORF

STRUTHERS and SWANEPOEL 1982). However, the functional role of the NSs protein in virus replication is not known. As a consequence of the ambisense expression strategy of the S RNA segment, the NSs protein would not be expected to be made until after RNA replication (or secondary transcription). This, coupled with the fact the NSs protein is not found in virions, suggests that this polypeptide may not function in the early processes of virus replication (e.g. primary transcription) but rather acts in late events. Given this scenario, possible roles include regulation of transcription and replication activities (e.g. attenuation of primary transcription) or virion morphogenesis. However, the detection of full length vcRNA in UUK virions raises the possibility that NSs mRNA and protein could be produced early in infection (SIMONS et al. 1990). And, in fact, the NSs protein of

RVF virus is found very early after infection, where it appears to accumulate as intranuclear inclusions. Later in the infection, NSs can also be found in the cytoplasm (DRIER and SMITH, personal communication).

Recently, a RVF virus mutant was isolated which proved to be defective in NSs protein expression (SALUZZO and SMITH, personal communication). Analysis of the genomic RNA of this virus revealed the S segment to be shorter than that of other RVF virus isolates. Further investigation showed that the S segment possessed a 600 nucleotide deletion within the coding region of NSs (LOFTS and SMITH, personal communication). This virus replicated to levels indistinguishable from those of wild-type virus in Vero cells, producing all viral gene products normally, with the exception of NSs. However, replication in other cell types (e.g. MRC-5 cells) was abortive. The virus did replicate to wild-type levels in mosquitoes and was transmissible. When mice and hamsters were inoculated, this virus demonstrated an attenuated phenotype and elicited a solidly protective immune response (SMITH, personal communication). The story of the NSs protein, its manner and kinetics of biogenesis and its role in virus life cycle remains a puzzling yet intriguing one.

6 Concluding Remarks

Through the application of molecular techniques, particularly gene cloning and nucleotide sequencing, much has been learned over the past few years about the genome structures and gene expression strategies of viruses in the family Bunyaviridae. These studies have revealed considerable diversity within this single taxonomic grouping. However, as we have indicated throughout this chapter, out knowledge in these areas is far from complete. More extensive molecular analyses of additional viral genomes are required before we can truely understand the phylogenetic relationships among the members of the Bunyaviridae.

In addition, we must move beyond the descriptive and cataloging phase of bunyavirology. So very little is known about gene function. Efforts aimed at understanding the roles of viral RNA sequences and protein products in virus replication need to be intensified. As several members of the family are important pathogens, the roles of virus gene products in pathogensis and in immunity must be further elucidated. Development of a system for the genetic manipulation of members of this virus group will greatly accelerate advancement in these areas.

References

Abraham G, Pattnaik AK (1983) Early RNA synthesis in Bunyamwera virus infected cells. J Gen Virol 64: 1227–1290

Akashi H, Bishop DHL (1983) Comparison of the sequences and coding of La Crosse and snowshoe hare bunyavirus S RNA species. J Virol 45: 1155–1158

Akashi H, Gay M, Ihara T, Bishop DHL (1984) Localized conserved regions of the S RNA gene products of bunyaviruses are revealed by sequence analyses of the Simbu serogroup Aino virus. Virus Res 1: 51–63

Anderson GW, Smith JF (1987) Immunoelectron microscopy of Rift Valley fever viral morphogenesis in primary rat hepatocytes. Virology 161: 91–100

Arikawa J, Schmaljohn AL, Dalrymple JM, Schmaljohn CS (1989) Characterization of HTN virus envelope glycoprotein antigenic determinants defined by monoclonal antibodies. J Gen Virol 70: 615–624

Arikawa J, LaPenotiere HF, Iacono-Connors L, Wang W, Schmaljohn CS (1990) Coding properties of the S and the M genome segments of Sapporo rat virus: comparison to other causative agents of hemorrhagic fever with renal syndrome. Virology 176: 114–125

Asada H, Tamura M, Kondo K, Okuno Y, Takahashi Y, Dohi Y, Nagai T, Kurata T, Yamanishi K (1987) Role of T lymphocyte subsets in protection and recovery from Hantaan virus infection in mice. J Gen Virol 68: 1961–1969

Asada H, Tamura M, Kondo K, Dohi Y, Yamanishi K (1988) Cell-mediated immunity to virus causing haemorrhagic fever with renal syndrome: generation of cytotoxic T lymphocytes. J Gen Virol 69: 2179–2188

Asada H, Balachandra K, Tamura M, Kondo K, Yamanishi K (1989) Cross-reactive immunity among different serotypes of virus causing haemorrhagic fever with renal syndrome. J Gen Virol 70: 819–825

Bause E (1983) Structural requirements of N-glycosylation of proteins. Biochem J 209: 331–336

Bishop DHL (1985) Replication of arenaviruses and bunyaviruses. In: Fields BN (ed) Virology. Raven, New York, pp 1083–1110

Bishop DHL (1990) Bunyaviridae and their replication. I. Bunyaviridae. In: Fields BN, Knipe DM, Chanock RM, Hirsch MS, Melnick JL, Monath TP, Roizman B (eds) Fields Virology. Raven, New York, pp 1155–1173

Bishop DHL, Gould KG, Akashi H, Clerx-van Haaster CM (1982) The complete sequence and coding content of the snowshoe hare bunyavirus small (S) viral RNA species. Nucleic Acids Res 10: 3703–3713

Bishop DHL, Gay ME, Matsuoko Y (1983) Nonviral heterogeneous sequences are present at the 5′ ends of one species of snowshoe hare bunyavirus S complementary RNA. Nucleic Acids Res 11: 6409–6418

Blondel D, Harmison GG, Schubert M (1990) Role of matrix protein in cytopathogenesis of vesicular stomatitis virus. J Virol 64: 1716–1725

Bouloy M, Hannoun C (1976) Studies on Lumbo virus replication. I. RNA dependent RNA polymerase associated with virions. Virology 69: 258–268

Bouloy M, Krams-Ozden S, Horodniceanu F, Hannoun C (1973,/1974) Three segment RNA genome of Lumbo virus (bunyavirus). Intervirology 2: 173–180

Bouloy M, Vialat P, Girard M, Pardigon N (1984) A transcript from the S segment of the Germiston bunyavirus is uncapped and codes for the nucleoprotein and a nonstructural protein. J Virol 49: 717–723

Bouloy M, Pardigon N, Vialet P, Gerbaud S, Girard M (1990) Characterization of the 5′ and 3′ ends of viral messenger RNAs isolated from BHK 21 cells infected with Germiston virus (*Bunyavirus*). Virology 175: 50–58

Cabradilla CD, Holloway BP, Obijeski JF (1983) Molecular cloning and sequencing of the La Crosse virus S RNA. Virology 128: 463–468

Casals J, Tignor GH (1980) The *Nairovirus* genus: serological relationships. Intervirology 14: 144–147

Cash P (1985) Polypeptide synthesis of Dugbe virus, a member of the *Nairovirus* genus of the Bunyaviridae. J Gen Virol 66: 141–148

Cash P, Vezza AC, Gentsch JR, Bishop DHL (1979) Genome complexities of the three mRNA species of snowshoe hare bunyavirus and in vitro translation of S mRNA to viral N polypeptide. J Virol 31: 685–694

Cash P, Hendershot L, Bishop DHL (1980) The effects of glycosylation inhibitors on the maturation and intracellular polypeptide synthesis induced by snowshoe hare bunyavirus. Virology 103: 235–240

Clerx JPM, Bishop DHL (1981) Qalyub virus, a member of the newly proposed *Nairovirus* genus (Bunyaviridae). Virology 108: 361–372

Clerx JPM, Casals J, Bishop DHL (1981) Structural characteristics of nairoviruses (genus *Nairovirus*, Bunyaviridae) J Gen Virol 55: 165–178
Clerx-van Haaster C, Bishop DHL (1980) Analysis of the 3′ terminal sequences of snowshoe hare and La Crosse bunyaviruses. Virology 105: 564–574
Clerx-van Haaster CM, Akashi H, Auperin DD, Bishop DHL (1982a) Nucleotide sequence analyses and predicted coding of bunyavirus genome RNA species. J Virol 41: 119–128
Clerx-van Haaster CM, Clerx JPM, Ushijima H, Akashi H, Fuller F and Bishop DHL (1982b) The 3′ terminal sequences of Bunyaviruses and Nairoviruses (Bunyaviridae): evidence of end sequence generic differences within the virus family. J Gen Virol 61: 289–292
Collett MS (1986) Messenger RNA of the M segment RNA of Rift Valley fever virus. Virology 151: 151–156
Collett MS, Purchio AF, Keegan K, Frazier S, Hays WM, Anderson DK, Parker MD, Schmaljohn C, Schmidt J, Dalrymple JM (1985) Complete nucleotide sequence of the M RNA segment of Rift Valley fever virus. Virology 144: 228–245
Collett MS, Keegan K, Hu S-L, Sridhar P, Purchio AF, Ennis WH, Dalrymple JM (1987) Protective subunit immunogens to Rift Valley fever virus from bacteria and recombinant vaccinia virus. In: Mahy B, Kolakofsky D (eds) The biology of negative strand viruses. Elsevier Science, Amsterdam, pp 321–329
Collett MS, Kakach LT, Suzich JA, Wasmoen TL (1989) Gene products and expression strategy of the M segment of the phlebovirus Rift Valley fever virus. In: Mahy B, Kolakofsky D (eds) Genetics and pathogenicity of negative strand viruses. Elsevier Biomedical Press, Amsterdam, pp 49–57
Dalrymple JM, Peters CJ, Smith JF, Gentry MK (1981) Antigenic components of Punta Toro virus. In: Bishop DHL, Compans RW (eds) The replication of negative strand viruses. Elsevier, New York, pp 167–172
Dalrymple JM, Kakach LT, Hasty S, Collett MS (1989) Mapping protective detrerminants of Rift Valley fever virus using recombinant vaccinia viruses. In: Brown F, Chanock RM, Ginsberg HS, Lerner RA (eds) Vaccines 89. Cold Spring Harbor Laboratory, New York, pp 371–375
Dantas JR, Okuno Y, Asada H, Tamura M, Takahashi M, Tanishita O, Takahashi Y, Kurata T, Yamanishi K (1986) Characterization of glycoproteins of viruses causing hemorrhagic fever with renal syndrome (HFRS) with monoclonal antibodies. Virology 151: 379–384
de Haan P, Wagemakers L, Goldbach R, Peters D (1988) Tomato spotted wilt virus: a bunyavirus in plants. Virus Res. [Suppl] 2: 83
de Haan P, Wagemakers L, Peters D, Goldbach R (1989) Molecular and terminal sequence determination of the S and M RNAs of tomato spotted wilt virus. J Gen Virol 70: 3469–3473
de Haan P, Wagemakers L, Peters D, Goldbach R (1990) The S RNA segment of tomato spotted wilt virus has an ambisense character. J Gen Virol 71: 1001–1007
Elliott LH, Kiley MP, McCormick JB (1984) Hantaan virus: identification of virion proteins. J Gen Virol 65: 1285–1293
Elliott RM (1985) Identification of non-structural proteins encoded by viruses of the Bunyamwera serogroup (family Bunyaviridae). Virology 143: 119–126
Elliott RM (1989a) Nucleotide sequence analysis of the S RNA segment of Bunyamwera virus, the prototype of the family Bunyaviridae. J Gen Virol 70: 1281–1285
Elliott RM (1989b) Nucleotide sequence analysis of the large (L) genomic RNA segment of Bunyamwera virus, the prototype of the family Bunyaviridae. Virology 173: 426–436
Elliott RM (1990) Molecular biology of the Bunyaviridae. J Gen Virol 71: 501–522
Elliott RM, McGregor A (1989) Nucleotide sequence and expression of the small (S) RNA segment of Maguari bunyavirus. Virology 171: 516–524
Emery VC, Bishop DHL (1987) Characterization of Punta Toro S mRNA species and identification of an inverted complementary sequence in the intergenic region of Punta Toro phlebovirus ambisense S RNA that is involved in mRNA transcription termination. Virology 156: 1–11
Endres MJ, Jacoby DR, Janssen RS, Gonzalez-Scarano F, Nathanson N (1989) the large viral RNA segment of California serogroup bunyaviruses encodes the large viral protein. J Gen Virol 70: 223–228
Eshita Y, Bishop DHL (1984) The complete sequence of the M RNA of snowshoe hare bunyavirus reveals the presence of internal hydrophobic domains in the viral glycoprotein. Virology 137: 227–240
Eshita Y, Ericson B, Romanowski V, Bishop DHL (1985) Analyses of the mRNA transcription processes of snowshoe hare bunyavirus S and M RNA species. J Virol 55: 681–689

Fazakerley JK, Gonzalez-Scarano F, Stricler J, Dietzschold B, Karush F, Nathanson N (1988) Organization of the middle RNA segment of snowshoe hare bunyavirus. Virology 167: 422–432
Foulke RS, Rosato PR, French GR (1981) Structural polypeptides of Hazara virus. J Gen Virol 53: 169–172
Fuller F, Bishop DHL (1982) Identification of virus-coded non-structural polypeptides in bunyavirus-infected cells. J Virol 41: 643–648
Fuller F, Bhown AS, Bishop DHL (1983) Bunyavirus nucleoprotein, N, and a non-structural protein, NSs, are coded by overlapping reading frames in the S RNA. J Gen Virol 64: 1705–1714
Gahmberg N, Kuismanen E, Keranen S, Pettersson RF (1986) Uukuniemi virus glycoproteins accumulate in and cause morphological changes in the Golgi complex in the absence of virus maturation. J Virol 57: 899–906
Gentsch JR, Bishop DHL (1978) Small viral RNA segment of bunyaviruses codes for viral nucleocapsid protein. J Virol 28: 417–419
Gentsch JR, Bishop DHL (1979) M viral RNA segment of bunyaviruses codes for two glycoproteins, G1 and G2. J Virol 30: 767–770
Gerbaud S, Pardigon N, Vialet P, Bouloy M (1987a) The S segment of the Germiston bunyavirus genome: coding strategy and transcription. In: Mahy BWJ, Kolakofsky D (eds) The biology of negative strand viruses. Elsevier Biomedical Press, Amsterdam, pp 191–198
Gerbaud S, Vialat P, Pardigon N, Wychowski C, Girard M, Bouloy M (1987b) The S segment of Germiston virus RNA genome can code for three proteins. Virus Res 8: 1–13
Giebel LB, Stohwasser R, Zöller L, Bautz EKF, Darai G (1989) Determination of the coding capacity of the M genome segment of nephropathia epidemica strain Hällnäs B1 by molecular cloning and nucleotide sequence analysis. Virology 172: 498–505
Gonzalez-Scarano F (1985) La Crosse virus G1 glycoprotein undergoes a conformational change at the pH of fusion. Virology 140: 209–216
Gonzalez-Scarano F, Pobjecky N, Nathanson N (1984) La Crosse bunyavirus can mediate pH-dependent fusion from without. Virology 132: 222–225
Grady LJ, Sanders ML, Campbell WP (1987) The sequence of the M RNA of an isolate of La Crosse virus. J Gen Virol 68: 3057–3071
Hacker D, Raju R, Kolakofsky D (1989) La Crosse virus nucleocapsid protein controls its own synthesis in mosquito cells by encapsidating its mRNA. J Virol 63: 5166–5174
Hacker D, Rochat S, Kolakofsky D (1990) Anti-mRNAs in La Crosse virus infected cells. J Virol 64: 5051–5057
Hewlett MJ, Pettersson RF, Baltimore D (1977) Circular forms of Uukuniemi virion RNA: an electron microscopic study. J Virol 21: 1085–1093
Higgins DG, Sharp PM (1988) CLUSTAL: a package for performing multiple sequence alignment on a microcomputer. Gene 73: 237–244
Hung T, Chou Z, Zhao T, Xia S, Hang C (1985) Morphology and morphogenesis of viruses of hemorrhagic fever with renal syndrome (HFRS): some peculiar aspects of the morphogenesis of various strains of HFRS virus. Intervirology 23: 97–108
Ihara T, Akashi H, Bishop DHL (1984) Novel coding strategy (ambisense genomic RNA) revealed by sequence analyses of Punta Toro phlebovirus S RNA. Virology 136: 293–306
Ihara T, Matsuura Y, Bishop DHL (1985a) Analyses of the mRNA transcription-processes of Punta Toro phlebovirus (Bunyaviridae). Virology 147: 317–325
Ihara T, Smith J, Dalrymple JM, Bishop DHL (1985b) Complete sequences of the glycoprotein and M RNA of Punta Toro phlebovirus compared to those of Rift Valley fever virus. Virology 144: 246–259
Kakach LT, Wasmoen TL, Collett MS (1988) Rift Valley fever virus M segment: use of recombinant vaccinia viruses to study phlebovirus gene expression. J Virol 62: 826–833
Kakach LT, Suzich JA, Collett MS (1989) Rift Valley fever virus M segment: phlebovirus expression strategy and protein glycosylation. Virology 170: 505–510
Keegan K, Collett MS (1986) Use of bacterial expression cloning to define the amino acid sequences of antigenic determinants on the G2 glycoprotein of Rift Valley fever virus. J Virol 58: 263–270
Kingsford L, Hill DW (1983) The effect of proteolytic cleavage of La Crosse virus G1 glycoprotein on antibody neutralization. J Gen Virol 64: 2147–2156
Krug RM, St Angelo C, Broni B, Shapiro G (1987) Transcription and replication of influenza virion RNA in the nucleus of infected cells. Cold Spring Harbor Symp Quant Biol 52: 353–358
Kuismanen E (1984) Posttranslational processing of Uukuniemi virus glycoproteins G1 and G2. J Virol 51: 806–812
Kuismanen E, Hedman K, Saraste J, Pettersson RF (1982) Uukuniemi virus maturation, accumulation of virus particles and viral antigens in the Golgi complex. Mol Cell 2: 1444–1458

Kuismanen E, Bang B, Hurme M, Pettersson RF (1984) Uukuniemi virus maturation: immunofluorescence microscopy with monoclonal glycoprotein-specific antibodies. J Virol 51: 137–146
Kyte J, Doolittle RF (1982) A simple method for displaying the hydropathic character of a protein. J Mol Biol 157: 105–132
Lees JF, Pringle CR, Elliott RM (1986) Nucleotide sequence of the Bunyamwera virus M RNA segment: conservation of structural features in the bunyavirus glycoprotein gene product. Virology 148: 1–14
Madoff DH, Lenard J (1982) A membrane glycoprotein that accumulates intracellularly: cellular processing of the large glycoprotein of La Crosse virus. Cell 28: 821–829
Marck C (1988) 'DNA Strider' a 'C' program for the fast analysis of DNA and protein sequences on the Apple Macintosh family of computers. Nucleic Acids Res 16: 1829–1836
Marriott AC, Ward VK, Nuttall PA (1989) The S RNA segment of sandfly fever Sicilian virus: evidence for an ambisense genome. Virology 169: 341–345
Marriott AC, Ward VK, Higgs S, Nuttal P (1990) RNA probes detect nucleotide sequence homology between members of two different nairovirus serogroups. Virus Res 16: 77–81
Matsuoka Y, Ihara T, Bishop DHL, Compans RW (1988) Intracellular accumulation of Punta Toro virus glycoproteins expressed from cloned cDNA. Virology 167: 251–260
McPhee DA, Westaway EG (1981) Proteins and glycoproteins specified by Bunyamwera virus and by Belmont virus, a possible bunyavirus, in mammalian cells. J Gen Virol 54: 149–159
Milne RG, Francki RIB (1984) Should tomato spotted wilt virus be considered as a possible member of the family Bunyaviridae? Intervirology 22: 72–76
Murphy FA, Harrison AK, Whitfield SG (1973) Morphologic and morphogenetic similarities of Bunyamwera serological supergroup viruses and several other arthropod-borne viruses. Intervirology 1: 297–316
Obijeski JF, Bishop DHL, Murphy FA, Palmer EL (1976a) Structural proteins of La Crosse virus. J Virol 19: 985–987
Obijeski JF, Bishop DHL, Palmer EL, Murphy FA (1976b) Segmented genome and nucleocapsid of La Crosse virus. J Virol 20: 664–675
Obijeski JF, McCauley J, Skehel JJ (1980) Nucleotide sequences at the termini of La Crosse virus RNAs. Nucleic Acids Res 8: 2431–2438
Pardigon N, Vialet P, Gerbaud S, Girard M, Bouloy M (1988) Nucleotide sequence of the M segment of Germiston virus: comparison of the M gene products of several bunyaviruses. Virus Res 11: 73–85
Pardigon N, Vialet P, Girard M, Bouloy M (1982) Panhandles and hairpin structures at the termini of Germiston virus RNAs (Bunyavirus). Virology 122: 191–197
Parker MD, Hewlett MJ (1981) The 3'-terminal sequences of Uukuniemi and Inkoo virus RNA genome segments. In: Bishop DHL, Compans RW (eds) Replication of negative strand viruses. Elsevier, New York, pp 125–145
Parker MD, Smith JF, Dalrymple JM (1984) Rift Valley fever virus intracellular RNA: a functional analysis. In: Compans RW, Bishop DHL (eds) Segmented negative strand viruses. Academic Press, Orlando, pp 21–28
Parrington MA, Kang CY (1990) Nucleotide sequence analysis of the S genomic segment of Prospect Hill virus: comparison with the prototype hantavirus. Virology 175: 167–175
Patterson JL, Kolakofsky D (1984) Characterization of La Crosse virus small genome transcripts. J Virol 49: 680–685
Patterson JL, Holloway B, Kolakofsky D (1984) La Crosse virions contain a primer-stimulated RNA polymerase and a methylated cap-dependent endonuclease. J Virol 52: 215–222
Patterson JL, Cabradilla C, Holloway BP, Obijeski JF, Kolakofsky D (1983) Multiple leader RNAs and messenger RNAs are transcribed from the La Crosse virus small genome segment. Cell 33: 791–799
Pattnaik AK, Abraham G (1983) Identification of four complementary RNA species in Akabane virus-infected cells. J Virol 47: 452–462
Pennington TH, Pringle CR, McCrae MA (1977) Bunyamwera virus-induced polypeptide synthesis. J Virol 24: 397–400
Pensiero MN, Jennings GB, Schmaljohn CS, Hay J (1988) Expression of the Hantaan virus M genome segment by using a vaccinia virus recombinant. J Virol 62: 696–702
Persson R, Wikstrom L, Pettersson RF (1989) Dimerization and intracellular transport of two Golgi-specific bunyavirus membrane glycoproteins. In: Compans RW, Helenius A, Oldstone MBA (eds) Cell biology of virus entry, replication, and pathogenesis. Alan R Liss, New York, pp 229–241

Pesonen M, Kuismanen E, Petterson RF (1982) Monosaccharide sequence of protein-bound glycans of Uukuniemi virus. J Virol 41: 390–400

Pettersson RF, von Bonsdorff C-H (1975) Ribonucleoproteins of Uukuniemi virus are circular. J Virol 15: 386–392

Pettersson RF, Kuismanen E, Rönnholm R, Ulmanen I (1985) mRNAs of Uukuniemi virus, a bunyavirus. In: Becker Y (ed) Viral messenger RNA: transcription, processing, splicing and molecular structure. Nijhoff, Boston, pp 283–300

Pifat DY, Osterling MC, Smith JF (1988) Antigenic analysis of Punta Toro virus and identification of protective determinants with monoclonal antibodies. Virology 167: 442–450

Pobjecky N, Nathanson N, Gonzalez-Scarano F (1989) Use of resonance energy transfer assay to investigate the fusion function of La Crosse virus. In: Kolakofsky D, Mahy BWJ (eds) Genetics and pathogenicity of negatvie strand viruses. Elsevier, Amsterdam, pp 24–32

Poch O, Sauvaget I, Delarue M, Tordo N (1989) Identification of four conserved motifs among RNA-dependent polymerase encoding elements. EMBO J 8: 3867–3874

Raju R, Kolakofsky D (1986) Inhibitors of protein synthesis inhibit both La Crosse virus S-mRNA and S genome syntheses in vivo. Virus Res 5: 1– 9

Raju R, Kolakofsky D (1987a) Unusual transcripts in La Crosse virus-infected cells and the site for nucleocapsid assembly. J Virol 61: 667–672

Raju R, Kolakofsky D (1987b) Translational requirements of La Crosse virus S-mRNA synthesis: in vivo studies. J Virol 61: 96–103

Raju R, Kolakofsky D (1989) The ends of La Crosse virus genome and antigenome RNAs within the nucleocapsid are base paired. J Virol 63: 122–128

Rönnholm R, Pettersson RF (1987) Complete nucleotide sequence of the M RNA segment of Uukuniemi virus encoding the membrane glycoproteins G1 and G2. Virology 160: 191–202

Samso A, Bouloy M, Hannoun C (1975) Presence de ribonucleoproteins circulaires dans le virus Lumbo (Bunyavirus). C R Acad Sci (III) D280: 779–782

Samso A, Bouloy M, Hannoun C (1976) Mise en evidence de molecules d'acide ribonucleique circulaire dans le virus Lumbo (Bunyavirus). C R Acad Sci (III) D282: 1653–1655

Schmaljohn CS, Dalrymple JM (1983) Analysis of Hantaan virus RNA: evidence for a new genus of Bunyaviridae. Virology 131: 482–491

Schmaljohn CS, Dalrymple JM (1984) Biochemical characterization of Hantaan virus. In: Compans RW, Bishop DHL (eds) Segmented negative strand viruses. Academic Press, Orlando, pp 117–124

Schmaljohn CS, Hasty SE, Dalrymple JM, LeDuc JW, Lee HW, von Bonsdorff C-H, Brummer-Korvenkontio M, Vaheri A, Tsai TF, Regnery HL, Goldgaber D, Lee PW (1985) Antigenic and genetic properties of viruses linked to hemorrhagic fever with renal syndrome. Science 227: 1041–1044

Schmaljohn CS, Jennings G, Hay J, Dalrymple JM (1986a) Coding strategy of the S genome segment of Hantaan virus. Virology 155: 633–643

Schmaljohn CS, Hasty SE, Rasmussen L, Dalrymple JM (1986b) Hantaan virus replication: effects of monensin, tunicamycin and endoglycosidases on the structural glycoproteins. J Gen Virol 67: 707–717

Schmaljohn CS, Jennings GB, Dalrymple JM (1987a) Identification of Hantaan virus messenger RNA species. In: Mahy BWJ, Kolakofsky D (eds) The biology of negative strand viruses. Elsevier, Amsterdam pp 116–121

Schmaljohn CS, Schmaljohn AL, Dalrymple JM (1987b) Hantaan virus M RNA: coding strategy, nucleotide sequence, and gene order. Virology 157: 31–39

Schmaljohn CS, Arikawa J, Hasty SE, Rasmussen L, Lee HW, Lee PW, Dalrymple JM (1988) Conservation of antigenic properties and sequences encoding the envelope proteins of prototype Hantaan virus and two virus isolates from Korean haemorrhagic fever patients. J Gen Virol 69: 1949–1955

Schmaljohn CS, Parker MD, Ennis WH, Dalrymple JM, Collett MS, Suzich JA, Schmaljohn AL (1989a) Baculovirus expression of the M genome segment of Rift Valley fever virus and examination of antigenic and immunogenic properties of the expressed proteins. Virology 170: 184–192

Schmaljohn CS, Arikawa J, Dalrymple JM, Schmaljohn AL (1989b) Expression of the envelope glycoproteins of Hantaan virus with vaccinia and baculovirus recombinants. In: Kolakofsky D, Mahy BWJ (eds) Genetics and pathogenicity of negative strand viruses. Elsevier, Amsterdam, pp 58–66

Schmaljohn CS, Chu Y-K, Schmaljohn AL, Dalrymple JM (1990) Antigenic subunits of Hantaan virus expressed by baculovirus and vaccinia virus recombinants. J Virol 64: 3162–3170

Short NJ, Meek AD, Dalgarno L (1982) Seven infection-specific polypeptides in BHK cells infected with Bunyamwera virus. J Virol 43: 840–843

Simons JF, Hellman U, Pettersson RF (1990) Uukuniemi virus S RNA segment: ambisense coding strategy, packaging of complementary strands into virions, and homology to members of the genus *Phlebovirus*. J Virol 64: 247–255

Smith FI, Palese P (1989) Variation in influenza virus genes. In: Krug RM (ed) The influenza viruses. Plenum, New York, pp 319–359

Smithburn KC, Haddow AJ, Mahaffy AF (1946) A neurotropic virus isolated from *Aedes* mosquitoes caught in the Semliki Forest. Am J Trop Med 26: 189–208

Stohwasser R, Giebel LB, Zöller L, Bautz EKF, Darai G (1990) Molecular characterization of the RNA S segment of nephropathia epidemica virus strain Hällnäs B1. Virology 174: 79–86

Struthers JK, Swanepoel R (1982) Identification of a major non-structural protein in the nuclei of Rift Valley fever virus-infected cells. J Gen Virol 60: 381–384

Struthers JK, Swanepoel R, Shepherd SP (1984) Protein synthesis in Rift Valley fever virus-infected cells. Virology 134: 118–124

Suzich JA, Collett MS (1988) Rift Valley fever virus M segment: cell-free transcription and translation of virus-complementary RNA. Virology 164: 478–486

Suzich JA, Kakach LT, Collett MS (1990) Expression strategy of a phlebovirus: biogenesis of protein from the Rift Valley fever virus M segment. J Virol 64: 1549–1555

Swanepoel R, Blackburn NK (1977) Demonstration of nuclear immunofluorescence in Rift Valley fever virus-infected cells. J Gen Virol 34: 557–561

Ulmanen I, Seppäla P, Pettersson RF (1981) In vitro translation of Uukuniemi virus-specific RNAs: identification of a nonstructural protein and a precursor to the membrane glycoproteins. J Virol 37: 72–79

Urakawa T, Small DA, Bishop DHL (1988) Expression of snowshoe hare bunyavirus S RNA coding proteins by recombinant baculoviruses. Virus Res 11: 303–317

von Heijne G (1983) Patterns of amino acids near signal-sequence cleavage sites. Eur J Biochem 133: 17–21

Vorndam AV, Trent DW (1979) Oligosaccharides of the California encephalitis viruses. Virology 95: 1–7

Ward VK, Marriot AC, El-Ghorr A, Nuttall PA (1990) Coding strategy of the S RNA segment of Dugbe virus (*Nairovirus*; Bunyaviridae). Virology 175: 518–524

Wasmoen TL, Kakach LT, Collett MS (1988) Rift Valley fever virus M segment: cellular localization of M segment-encoded proteins. Virology 166: 275–280

Watret GE, Elliott RM (1985) The proteins and RNAs specified by Clo Mor virus, a Scottish nairovirus. J Gen Virol 66: 2513–2516

Yoo D, Kang CY (1987) Nucleotide sequence of the M segment of the genomic RNA of Hantaan virus 76–118. Nucleic Acids Res 15: 6299–6300

Zeller HG, Karabatsos N, Calisher CH, Digoutte J-P, Cropp CB, Murphy FA, Shope RE (1989) Electron microscopic and antigenic studies of uncharacterized viruses. II. Evidence suggesting the placement of viruses in the family Bunyaviridae. Arch Virol 108: 211–227

Bunyavirus RNA Synthesis: Genome Transcription and Replication

D. KOLAKOFSKY and D. HACKER

1 Nucleocapsids, the Templates for RNA Synthesis

The three bunyavirus genome segments, termed small (S), medium (M), and large (L), are found both in virions and infected cells as viral nucleocapsids (NCs). These are large, possibly helical, structures in which the RNA is complexed with predominantly the viral N protein, with sedimentation constants ($S_{20,w}$) for La Crosse virus (LAC) of 115 S (L), 90 S (M), and 65 S (S) in sucrose gradients. Based on the chemical composition of the virus and the proportion of each protein by SDS-PAGE, OBIJESKI et al. (1976) estimated that there would be one N protein for every six nucleotides.

Early studies showed that the RNA in NCs was solubilized by high concentrations of RNase A (100 µg/ml, 37°; OBIJESKI et al. 1976), in contrast to the RNA in rhabdovirus or paramyxovirus NCs. However, PETTERSSON et al. (1971) reported that 1 µg/ml RNase (37 °C) did not solubilize the RNA. In a more recent study, in which the ratio of RNase to RNA was controlled, and digestion was measured by methods which would detect a single nick, it was found that no NC

Department of Microbiology, University of Genewa School of Medicine, CMU, 9 Ave de Champel, 1211 Geneva, Switzerland

Current Topics in Microbiology and Immunology, Vol. 169

RNA was digested at 22 °C, even when 25 times more RNase A or micrococcal nuclease was used than was necessary to digest completely the viral mRNA (HACKER et al. 1989). The RNA in bunyavirus NCs is thus protected from attack by reasonable concentrations of either RNase A or micrococcal nuclease but appears not to be as tightly complexed as the RNA in paramyxovirus and rhabdovirus NCs.

NCs are also sufficiently stable to high salt that they band as a complex in CsCl density gradients (OBIJESKI et al. 1976; PETTERSSEN et al. 1971; SAMSO et al. 1975). This latter property, which is common to all (−)RNA virus NCs except those of influenza virus, allows for easy separation of the encapsidated and unencapsidated viral RNAs. When cytoplasmic extracts of infected cells are centrifuged on CsCl gradients, virtually all of the genome and antigenome RNAs are found as NCs at a density of 1.31 g/ml, whereas virtually all of the mRNAs are found in the pellet fraction (PATTERSON and KOLAKOFSKY 1984; RAJU and KOLAKOFSKY 1987a). (An exception to this rule occurs in mosquito cells, discussed below). The banded NCs contain only the viral N protein, and we assume that other viral proteins, for example, L and possibly NS_S, have been removed during centrifugation. The buoyant density of the bunyavirus NCs in CsCl is very similar if not identical to that of rhabdovirus and paramyxovirus NCs, as well as tobacco mosaic virus (1.31 g/ml). These latter structures have all been found to contain 4% RNA by weight. For the *Bunyavirus* genus, whose N protein is approximately 25 kDa, this would correspond to one N monomer per three nucleotides.

When viewed in the electron microscope, all three NC segments were found to be circular (OBIJESKI et al. 1976; PETTERSSON and VON BONSDORFF 1975; SAMSO et al. 1975), even though the genomic RNA contains free 5′ (pppA; GENTSCH et al. 1977; OBIJESKI et al. 1976; PETTERSSON et al. 1977) and 3′ (hydroxyl) ends (OBIJESKI et al. 1976). When the naked RNA was viewed in the electron microscope under partially denaturing conditions (SAMSO et al. 1976; HEWLETT et al. 1977), it too was found to be circular. This suggested that the ends of each segment would contain complementary sequences, and this was soon confirmed (OBIJESKI et al. 1980). More recently, psoralen cross-linking studies on NCs both in virions and in infected cells showed that the ends of the RNA within NCs are indeed capable of base-pairing to form dsRNA, as they are in the naked RNA (RAJU and KOLAKOFSKY 1989). Moreover, the ends of the RNA within NCs are not the only region of the genome capable of base-pairing interactions. When oligonucleotides complementary to either genomes or antigenomes are added to cell extracts in the presence of RNase H, a variable fraction of the NC RNA, up to 50% depending on the oligonucleotide, is specifically cleaved, although at a much slower rate than the unencapsidated RNA (BELLOCQ and KOLAKOFSKY 1987; KOLAKOFSKY et al. 1987). Similar experiments with paramyxovirus NCs showed that its RNA was insensitive to such treatment (unpublished data). The RNA within bunyavirus NCs is thus more accessible to the milieu than that of paramyxo- or rhaboviruses, even though they all apparently contain the same ratios of protein and RNA. Although negatively stained bunyavirus NCs do not give images of high resolution, some authors (OBIJESKI et al. 1976) have detected structures that appear to be more like beads on a string than the disclike structures of TMV or paramyxovirus NCs. It is unclear whether the RNA is arranged on the outside of the protein, like nucleosomes, or within a groove created by the assembled N, as in TMV. In either case, one assumes that the N protein must be locally displaced from the RNA for the RNA to base-pair.

2 Viral RNA Synthesis

When five snowshoe hare virus (SSH) S segments, generated by primer extension of samples containing both mRNAs and antigenomes, were sequenced (BISHOP et al. 1983), a somewhat confusing picture emerged. Two were found to have the 5′ ends of the RNA at positions +2 or +3, relative to the 5′ end of the antigenome being set at position +1, and these were assumed to represent incomplete RTase transcripts of the S RNAs. Moreover, three clones contained 5′ ends at position −13. These extra sequences, which were heterogeneous, could not have been templated from the gemone, whose 3′ end is also defined as position +1. These authors concluded that one of the S RNA species, probably the mRNA, contained nontemplated nucleotides at its 5′ end. When LAC S antigenomes from CsCl banded material was directly examined by primer extension (without cloning), a single 5′ end at position +1 was found, as expected. However, extension on the CsCl pellet RNA gave a gaussian distribution of bands at positions −10 to −18. When ddNTP were included in these reactions, the sequence could be read to position +2 but became heterogenous beyond this point (PATTERSON and KOLAKOFSKY 1984).

These results suggested that the extra sequences on the mRNAs might be due to primers derived from host cell mRNAs used to initiate mRNA synthesis, in analogy with the cap-snatching mechanism of influenza virus mRNA synthesis (KRUG 1981, 1983). Further work has supported this suggestion (PATTERSON et al. 1984). The polymerase activity of purified LAC virions was found to be stimulated by oligonucleotides of the $(A)_nG$ series as well as cap groups (^mGpppA), and primer extension studies of the polymerase products showed that the oligonucleotides were indeed acting as primers. The most potent stimulator on a molar basis was a natural plant virus mRNA, and the 5′ ends of these products were at positions −12 and −13, similar to those of natural mRNAs. Like influenza virus, LAC was found to contain a methylated cap-dependent endonuclease, which cuts the plant virus mRNA 13 and 14 nt from its 5′ end. More recently, we have shown that the extra sequences on intracellular LAC mRNAs are in fact capped, because they can be specifically selected by anti-cap antibodies (Fig. 2) (HACKER et al. 1990). The mechanism of mRNA initiation of bunyaviruses and influenza viruses thus appears to be remarkably similar, except that bunyavirus mRNAs are made in the cytoplasm, and therefore use a

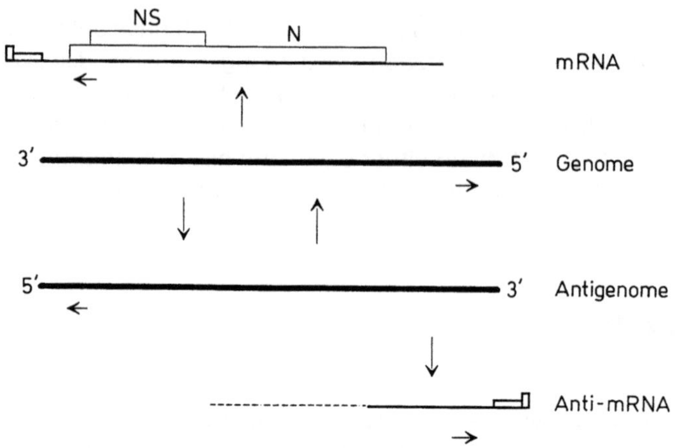

Fig. 1. A schematic representation of the LAC S segment RNAs. Genomes and antigenomes (*thick lines*) are found as NCs which band in CsCl gradients, mRNAs (with coding regions shown above), and anti-mRNAs (*thin lines*) are foud in the CsCl pellet fraction. The N and NS_s open reading frames are indicated on the mRNA. *Open boxes* at the 5′ end of these chains represent cap groups and host-derived sequences. *Horizontal arrows*, primers used for 5′ end analysis; *dotted line* at the 3′ end of the anti-mRNA indicates that this end is probably heterogenous and incomplete; *vertical arrows* show the direction of RNA synthesis. The S genome is 983 nt long. The S mRNA terminates at or around nt 886 but has an average of 14 additional nt at its 5′ end. The 3′ end of the anti-mRNA has never been detected

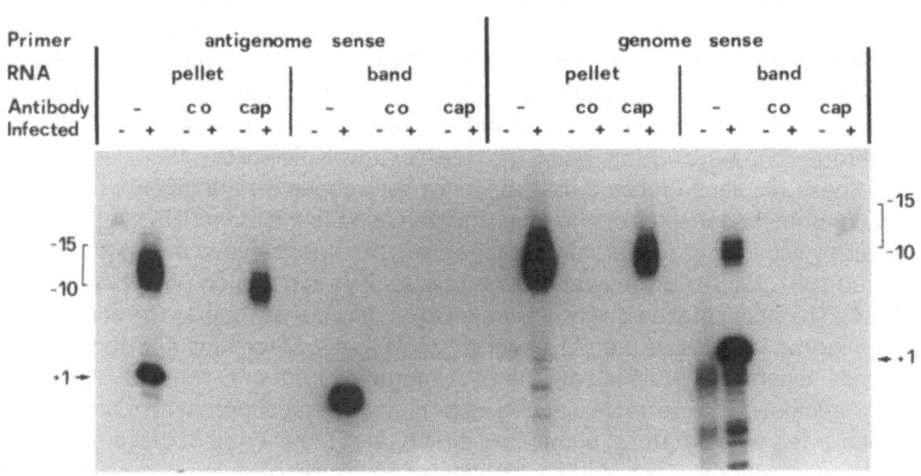

Fig. 2. The nontemplated sequences at the 5′ ends of both the LAC S mRNAs and anti-mRNAs are capped. S genome- and antigenome-sense primers were extended both on CsCl band and pellet RNA, from both mock (−) and LAC-infected (+) cells. The cDNA:RNA hybrids were either analyzed directly or after selection with protein-A sepharose beads coated with either anti-cap antibodies (*cap*) or control antiserum to a Sendai virus protein (*co*). Position + 1 marks the 5′ end of the genome and antigenome. Position − 10 to − 15 marks the 5′ ends of the mRNAs or anti-mRNAs. The antigenome-sense primer detects genomes and anti-mRNAs. The genome sense primer detects antigenomes and mRNAs

stable pool of mRNA transcripts as the source of the primers (ROSSIER et al. 1986). The mechanism for mRNA initiation would then be:

1. The viral polymerase, presumably attached to the 3′ end of the genome template, binds the cap group of a host mRNA, either directly or via a host cell cap-binding protein.
2. The polymerase cleaves this chain 10–18 nt downstream of the cap, depending on the particular sequence of the host mRNA, leaving a 3′ hydroxyl group on the capped fragment.
3. The 3′ base of the fragment is then aligned on the 3′ base of the template, not necessarily by base-pairing (BOULOY et al. 1990), and is used as a primer for viral mRNA synthesis. The first base incorporated by the polymerase would be position +2.

In contrast to mRNA synthesis, antigenomes would initiate with a triphosphate at position +1. Moreover, as the bunyavirus polymerase apparently does not have an activity which caps triphosphates, the 5′ end of the antigenome (or genome) remains unmodified.

The 3′ ends of the mRNAs never extend to the end of their templates. For the non-ambisense segments (Fig. 1), the ends of the mRNAs are missing only 50–100 nt, such that only for the S segment can mRNAs and antigenomes be separated on gels. An important consequence of these missing sequences is that the ends of the mRNA, unlike the antigenome, are no longer complementary and so cannot form circular structures, which might interfere with their translation. The template sequences corresponding to the 3′ ends of six mRNAs have been mapped by nuclease protection, and there appear to be two different but related consensus sequences. The LAC S (PATTERSON and KOLAKOFSKY 1984) and L (HACKER et al 1990) and Germiston (GER) M mRNAs (BOULOY et al. 1990) all end within the sequence 3′ GUUUUU 5′, which is identical to that used by influenza virus, whereas GER S mRNA (BOULOY et al. 1990) ends within an almost identical sequence 3′ GUUU<u>G</u>U 5′. The SSH (IHARA et al. 1985) and Rift Valley fever virus (COLLETT 1986) M mRNAs, however, end within the sequence 3′ ACCCC 5′. These consensus sequences are probably signals for mRNA termination, but one cannot exclude the possibility that the polymerase has proceeded past this site, and that the 3′ end was formed by an endonuclease. In this latter case, which we consider less likely, these sequences would be signals for cleavage. Interestingly, the mRNAs are not polyadenylated, but most have the potential to form stem-loop structures just upstream of their 3′ ends. This structure may substitute for the poly(A) tail in conferring mRNA stability.

The S segments of phlebo- (IHARA et al. 1984) and uukuviruses (SIMONS et al. 1990), as well as the recently elucidated plant bunyavirus, tomato spotted wilt virus (DEHAAN et al. 1990), are ambisense, i.e., they contain protein coding sequences at the 5′ ends of genome as well as antigenome sense RNAs. Two mRNAs of opposite polarity are transcribed from these segments (see ELLIOTT et al., this volume). That transcribed from the 3′ end of the genome is the N mRNA similar to non-ambisense segments, and that from the 3′ end of the antigenome codes for a nonstructural protein called NS_s. The ORFs for these proteins do not overlap but are separated by an intergenic region, which, except for sandfly fever Sicilian phlebovirus (MARRIOTT et al. 1989), has the potential to form a strong stem-loop structure. [An NS_s protein is also translated from an overlapping ORF contained within the N ORF of the S mRNA of members of the non-ambisense *Bunyavirus* genus (Fig. 1; CABRADILLA et al. 1983).] The position of the 3′ ends of the mRNAs, which like the non-ambisense mRNAs are also not polyadenylated, have been reported only for Punta Toro phlebovirus (EMERY and BISHOP 1987). These authors were unable to carry out S1 nuclease mapping because of the secondary structure in the intergenic region, and they instead used a series of oligonucleotides in RNA blots of the mRNAs. Curiously, the 3′ ends of the mRNAs mapped to very near the loop region rather than to the base of the stem. There are no obvious consensus sequences here for termination. We assume that the stem-loop structure also forms within NCs and somehow plays a role in terminating the mRNAs, but there is no experimental evidence for this as yet. The 5′ ends of the mRNAs are also similar to those of non-ambisense segments (IHARA et al. 1985; COLLETT 1986).

3 Anti-mRNAs in LAC-Infected Cells

As non-ambisense segments contain all of their protein coding sequences on the antigenome sense RNAs, one might have expected that genome-sense mRNAs would not be transcribed from these segments. In accordance with this, RNA blots of CsCl pellet material examined with antigenome-sense probes detect only a minor fraction of the full-length genomes, which contaminate the pellet fraction (RAJU and KOLAKOFSKY 1987a). However, when pellet RNA is examined for genome-sense mRNAs or "anti-mRNAs" by primer extension, not only is the minor fraction of genomes here detected (position +1), but also a series of bands which extend to positions −10 to −18 (Fig. 2). These latter bands are not only of the same length as those of the antigenome-sense mRNAs, but they are also capped. The initiation of anti-mRNA synthesis is thus probably the same as that for mRNAs. Anti-mRNAs have been found for both the LAC S and L segments by primer extension but not by Northern blotting, which suggests that their 3′ ends are heterogeneous, possibly because of premature termination due to the absence of concurrent translation (HACKER et al. 1990).

The anti-mRNAs are 15–30 times less abundant than the mRNAs. However, since their antigenome templates are also five times less abundant than gemones intracellularly (ROSSIER et al. 1986), this would suggest that the anti-mRNAs are initiated only three to six times less frequently on antigemone templates than are mRNAs on genome templates. Moreover, the NS_s or genome-sense mRNAs of phlebo- and uukuviruses are also considerably less abundant than the antigenome-sense N mRNAs (IHARA et al. 1984, 1985;

MARRIOTT et al. 1989; SIMONS et al. 1990). This is as expected for mRNAs which code for a minor nonstructural protein as compared to the most abundant viral protein. There are in fact as many anti-mRNAs relative to mRNAs in LAC-infected cells as there are, for example, NS_s mRNAs relative to N mRNAs in Uukumeimi-infected cells (SIMONS et al. 1990). The difference in mRNA synthesis between ambisense and non-ambisense segments is therefore not at the level of mRNA initiation from antigenome templates but rather whether a translatable transcript of defined length is produced. What role, if any, is played by the anti-mRNAs during infection is unclear.

4 Relative Amounts of S, M, and L Segment mRNAs

Experiments which measure the relative amounts of the different RNAs must deal with the problem of RNA degradation, which is particularly acute for LAC and possibly also the other bunyaviruses. LAC infection of mammalian cells induces a general mRNA instability approximately 5 h postinfection (hpi) (RAJU and KOLAKOFSKY 1988), and there is the added problem of degradation during isolation. Highly specific breaks in even the S mRNA transcripts are often found when these are separated by electrophoresis. These breaks were first thought to be due to the polymerase stopping and restarting, to yield a nested set of leader RNAs and mRNAs (PATTERSON et al. 1983). However, subsequent studies showed that the breaks were more likely the result of highly specific nicking of the mRNAs (RAJU and KOLAKOFSKY 1987a). Moreover, since the S and L RNAs are seven fold different in length, the L RNAs are much larger targets for degradation than the other RNAs. Therefore, methods such as electrophoresis under denaturing conditions cannot be used for their estimation, as these depend on the integrity of each chain.

We have carried out such experiments on LAC-infected BHK and C6/36 (mosquito) cells using slot blots and hybridization with riboprobes of each segment (ROSSIER et al. 1988). The CsCl pellet was examined for mRNAs, and the banded RNAs were then examined for genome-sense RNAs. The S, M, and L genomes of LAC are found in equimolar amounts in BHK cells and in roughly equimolar amounts in mosquito cells (M was three fold more abundant than S, with L in between), throughout the period when these RNAs accumulate. Nevertheless, the relative amounts of the S, M, and L mRNAs are clearly nonequimolar over the same period of time (Fig. 3). For BHK cells, the S mRNAs (five per S genome) were approximately ten fold more abundant than M mRNAs (0.5 per M genome), and M mRNA was ten fold more abundant than the L mRNAs (0.05 per L genome). Mosquito cells contain the same ratio of S mRNA per S genome but even lower ratios of M and L mRNAs than in BHK cells. These different ratios of mRNAs/template are unlikely to be due to differential degradation since they are found at 3 and 4 hpi in BHK cells, before mRNA

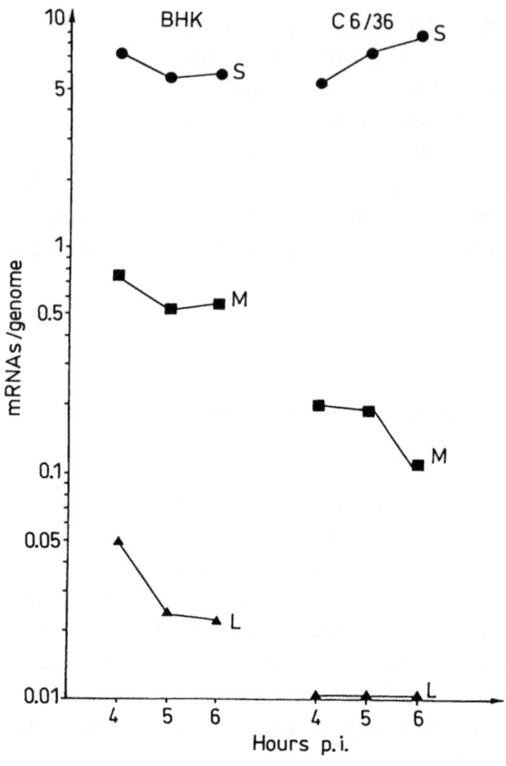

Fig. 3. Ratios of LAC mRNAs per genome template in BHK and mosquito (C6/36) cells. The relative amounts of genomes in CsCl-band RNA and mRNAs in CsCl-pellet RNA were estimated with segment-specific as well as strand-specific riboprobes. The ratios are plotted as a function of time post-infection. Only early times are used for this calculation, as at later times mRNA instability interferes with the measurement. The ratios are thought to reflect the frequency with which each genome segment is transcribed

instability becomes acute, and are similar in mosquito cells where mRNA instability is not induced.

The three genome segments therefore appear to transcribe their mRNAs at very different rates, depending on the requirements for their individual gene products. For the ambisense S segments, where there is also considerably less NS_s mRNA than N mRNA, part of this difference would be due to the lower amount of antigenomes relative to genomes. The remainder is presumably due to stronger rates of initiation of mRNAs from genomes than from antigenomes (i.e., promoter strength).

5 The Curious Coupling of mRNA Synthesis to Translation

In all negative-strand RNA virus infected cells, the addition of drugs which inhibit protein synthesis also blocks genome replication at all times during the infection, and this is also the case for bunyaviruses (PATTERSON and KOLAKOFSKY 1984). This requirement for on-going protein synthesis is thought not to reflect a direct coupling of genome replication and translation. More likely it reflects a coupling of replication and assembly. There is thus a requirement for a continuous supply of unassembled N protein, which is in limiting amounts, and must be continously replenished as genome replication proceeds and depletes this pool.

For the other negative-strand RNA viruses, the addition of these drugs either has no effect on mRNA synthesis or rather enhances it. For bunyaviruses in most mammalian cells, however, the addition of drugs such as cycloheximide not only blocks genome replication but mRNA synthesis as well. This translational requirement was found in BHK cells for Bunyamwera (ABRAHAM and PATTNAIK 1983), Akabane (PATTNAIK and ABRAHAM 1983), LAC (RAJU and KOLAKOFSKY 1986), and GER viruses (GERBAUD et al. 1987) but not for SSH virus (VEZZA et al. 1979; ESHITA et al. 1985). However, SSH virus in our BHK cells has the same translational requirement as LAC (unpublished data). Furthermore, we found only a very slight translational requirement for LAC mRNA synthesis in C6/36 cells (RAJU et al. 1989). The translational requirement is therfore cell type dependent, and not all lines of BHK cells appear to be the same in this respect.

The translation requirement for mRNA synthesis made little sense, since bunyavirions carry out RNA synthesis in vitro (BOULOY and HANNOUN 1976; GERBAUD et al. 1987; PATTERSON et al. 1984; RANKI and PETTERSSON 1975; SCHMALJOHN and DALRYMPLE 1983). However, when the length of the in vitro LAC transcripts were examined, virtually all of the S RNAs were found to have terminated prematurely, at nt 175, and drugs such as cycloheximide had no effect on this reaction. Full-length S mRNA (terminated at nt 886) could be made in vitro but only in the presence of a rabbit reticulocyte lysate which was active in translation. Addition of cycloheximide to this coupled system did not affect the rate of mRNA initiation but caused all of the transcripts to terminate prematurely, at nt 175. There is considerable homology between the sequences present near nt 175 and nt 886 (BELLOCQ et al. 1987). The presence of the nt 175 terminated S transcript could also be shown in vivo when the infected cells were treated with these drugs (RAJU and KOLAKOFSKY 1987a).

The translational requirement found in vitro, and in most cells in vivo, was not for mRNA initiation but apparently to prevent the polymerase from terminating prematurely. Further in vitro studies (BELLOCQ and KOLAKOFSKY 1987) showed that the polymerase could be induced progressively to read through nt 175 and to continue to nt 886, even in the absence of on-going protein synthesis, by progressively replacing the guanosines of the nascent chain with inosine. These studies also showed that nt 175 was only the first, and probably the strongest, of several sites where the polymerase would terminate prematurely in the absence of translation. Since replacement of uridines with Br-uridine had the opposite effect, this suggested that premature termination is due to base-pairing of the nascent mRNA chain and the template, and that concurrent translation of the nascent chain is required to break these interactions. During antigenome synthesis, we assume that concurrent assembly of this nascent chain with N protein would not only prevent premature termination before nt 886 but at nt 886 as well (KOLAKOFSKY et al. 1987).

The question then remains why the translational requirement is not present in all cells. The lack of a translational requirement for mRNA synthesis in mosquito cells could also be partially reproduced in vitro (RAJU et al. 1989). Genomes from mosquito cells were much less active than those of BHK cells, but their polymerases read through nt 175 efficiently without concurrent translation. To examine whether this difference might be due to a host protein or a host modification of a viral component, mixing experiments were carried out. The addition of uninfected mosquito cell extracts of BHK genomes had no effect of any kind. However, addition of uninfected BHK extracts to mosquito cell genomes both stimulated mRNA initiation and recreated the translational requirement for efficient read-through of nt 175. Moreover, simply washing the BHK genomes by pelleting them through sucrose cushions was found to increase their ability to read through nt 175 without concurrent translation, i.e., to remove the translational requirement. The translational requirement then appears to be due to a host component. In line with the model for this requirement discussed above, the host component could function by favoring the annealing between the nascent mRNA and its template during synthesis.

6 Persistent and Asymptomatic Infection of Mosquito Cells

Although arboviruses grow equally in mammalian and insect cells, the consequence of the infection is quite different for the two cell types, and in the organisms. For example, the natural host for LAC is the mosquito, *Aedes trisenatus* (see BEATY and CALISHER, this volume), in which the infection is persistent and apparently asymptommatic, and the same is true for the infection of mosquito cells in culture. Infection of mammals, on the other hand, can cause encephalitis (see GONZALEZ-SCARANO et al., this volume), and infection of mammalian cells in culture is highly cytopathic. At least one reason for this difference is the general mRNA instability induced in mammalian but not in mosquito cells (RAJU and KOLAKOFSKY 1988). However, there is clearly much more to this phenomenon.

When Bunyamwera virus or LAC infections of these two cell types are examined in parallel (NEWTON et al. 1981; ELLIOTT and WILKIE 1986; ROSSIER et al. 1988), one clear difference is that viral macromolecules accumulate more slowly in mosquito than in mammalian cells. This difference remains even when the two infections are carried out at the same temperature and culture conditions (ROSSIER et al. 1988) and is probably due in part to the lower levels of polymerase gene expression found in mosquito cells. Moreover, the mosquito cell infection remains asymptomatic even when the intracellular concentration of viral products reaches very similar levels to those found in the mammalian cell infection. The intracellular levels of the viral products per se cannot therefore account for this difference.

Except for the L mRNA, which remains close to or below the level of detection in mosquito cells, all the viral RNAs accumulate linearly until approximately 24 hpi but then level off and begin to decrease slowly over the next several days (ROSSIER et al. 1988). At 24 hpi, one finds that viral protein synthesis and genome replication become severely reduced, even though mRNA levels and rates of synthesis continue almost unchanged. This suggests that there is a form of translational control operating at this time in these infections, and recent studies have shown that this happens in a novel and unexpected manner.

We had previously found that early in BHK cell infections, before mRNA instability became acute, a small but measurable fraction (1%–2%) of the S mRNA was assembled into NCs (RAJU and KOLAKOFSKY 1987b). When the mosquito cell infection was examined for S mRNA encapsidation, at 12 hpi, less than 10% of this mRNA was encapsidated, as judged by either CsCl density gradient fractionation or immunoselection with an N-specific antibody. However, at 24–48 hpi, more than 75% of the mRNA was assembled with N protein, and this ratio remained high subsequently. Host cell mRNAs, on the other hand, were not associated with N protein. As expected, the assembled S mRNA could not be translated in vitro (HACKER et al. 1989).

Before 12 hpi, there appears to be little unassembled N protein present, as judged by the ability of drugs such as cycloheximide to efficiently prevent new genome replication. However, at 24 hpi, the pool of unassembled N must have grown in size, as cycloheximide is now much less efficient in blocking genome replication. The selective encapsidation of the S mRNA with time postinfection therefore appears to be due to the growing size of the pool of unassembled N intracellularly. Moreover, unlike genomes and antigenomes which appear as NCs at all times as soon as they are made, even those S mRNAs which are made late appear first as free mRNAs, and only much later are they assembled into NCs.

The following model was proposed to account for these unusual events (HACKER et al. 1989). Although genome replication depends on a supply of unassembled N, the down-regulation of genome replication in the mosquito cell infection cannot be due to the lack of unassembled N. Its pool, on the contrary, grows with time, and there is clearly sufficient N to encapsidate the majority of the S mRNAs. The down-regulation of genome replication must then be due to some other cause, possibly modification of the relatively low levels of the viral polymerase, which inhibits its function as a replicase but not as a transcriptase. What follows the reduction in genome replication appears more straightforward. The levels of unassembled N are controlled by the rate at which new N is made and the rate at which it is then removed from the pool by assembling with RNA. This assembly must be highly specific, as N normally assembles only with the viral genomes and antigenomes. When new genome synthesis is down-regulated, but mRNA synthesis continues, the primary target for N assembly is removed, while N continues to be made at the same rate. The pool of unassembled N then grows and at higher levels begins to associate with the viral mRNAs, which are secondary targets. Presumably all the viral mRNA are secondary targets, as

L mRNA is also assembled under these conditions (D. HACKER, unpublished data).

In this way, N controls its own synthesis in the absence of sufficient genome replication, and also down-regulates the expression of the other viral mRNAs. All of this takes place without N interfering with the expression of the cellular mRNAs, presumably because they have a much lower affinity for unassembled N. Without such a self-regulatory system, N would accumulate to the point that it becomes either insoluble or associated nonspecifically with host mRNAs, either of which would eventually lead to cytopathic effects. The ability of N specifically to assemble the viral RNAs under these circumstances then explains how the mosquito cell infection begins in an acute phase in which the viral products accumulate linearly but then switches to a persistent phase in which the viral products stop accumulating. The slow decrease in the level of the viral products which then ensues could be explained by the turnover of these products becoming greater than their rate of synthesis.

7 Specificity of N Protein–RNA Interactions

As mentioned above, in mosquito cells N protein can assemble over 75% of the S mRNA, and probably also the other viral mRNAs, without assembling host mRNAs. The assembly process is thought to be controlled at the level of initiation, by N protein interacting with a specific sequence on its target RNA. Subsequent addition of N would then make contact with both the RNA and the preexisting N in the complex, and the N–N contacts would thus permit any downstream sequence to be encapsidated. The sequence specific for the initiation of NC assembly must be contained on all genome and antigenome chains and is therefore likely to be a sequence that is conserved between them, i.e., the first and last 11 nt of each of these chains.

In LAC-infected BHK cells, a small amount of an S genome transcript is found which starts at position +1 like the antigenome, but whose 3′ end is at position 886 like the mRNA (RAJU and KOLAKOFSKY 1987b). This transcript, called 1 to 886, is found only as NCs. An even smaller amount of an S transcript, called −15 to 983, which starts like the mRNA but ends like the antigenome, is also found but predominantly as unassembled chains. These results argue that the site for the initiation of NC assembly is the conserved 11 nt at the 5′ ends of genomes and antigenomes. The viral mRNAs would of course also contain this sequence, not at the precise 5′ ends of these chains but downstream of the capped primer used to initiate the mRNAs. It is presumably the precise location of this sequence, and possibly also the presence of the cap group and the proteins which bind to it, which decreases the affinity of the N protein to the initiation sequence on the mRNAs. Thus, at low concentrations of N, only genomes and antigenomes would be assembled, and at higher concentrations viral but not cellular mRNAs would also be assembled. One assumes that at even higher concentrations, any RNA would be assembled. However, such concentrations are never reached intracellularly because of the manner in which unassembled N controls its own synthesis.

8 Genome Transcription and Replication

Two types of RNA synthesis take place on genome NCs: (a) polymerases start at the 3′ end of the template with a capped primer and stop short of the 5′ end of the template, i.e., respond to the mRNA termination signal, to make mRNAs, or (b) they start at position +1 with a triphosphate and continue to the precise end of the template, to make antigenomes (Fig. 1).

What controls this switch between transcription and replication, which presumably occurs at the level of chain initiation, is unclear. It is unlikely to be the concentration of unassembled N which drives the switch towards replication, as proposed for the nonsegmented negative-strand RNA viruses (VIDAL and KOLAKOFSKY 1989) since precisely the opposite result occurs during the mosquito cell infection. However, these results do not rule out the possibility that unassembled N still participates in the initiation of genome replication, only that it is not the determining factor. The mosquito cell infection is also noteworthy because it demonstrates that the transcriptase and replicase functions of the polymerase can be controlled independently, and that something else, which is not unassembled N, is required for replication but not transcription. The most likely candidates for this factor are modification of the L or NS_s proteins, about which we know very little.

Since genomes and antigenomes are found only as NCs, and the sites for the initiation of assembly appear to be located at the 5′ ends of these chains, it is logical that their synthesis and assembly take place concurrently. There are then two possible explanations why the transcriptase responds to the mRNA termination signals whereas the replicase does not. One is that they are simply preprogrammed in this way, by the same differences that account for the way in which they initiate their chains. The other is that the difference depends on the presence of assembling N protein on the nascent antigenome chain, which prevents interactions between this chain and the template required for termination. This would be analogous to the way in which concurrent translation of the nascent mRNAs prevents premature termination.

The above description would also hold for RNA synthesis from antigenome templates, from either ambisense or non-ambisense segments, but there are quantitative diffences of some importance. As there are five times as many genomes as antigenomes intracellularly, and each is the template for the other. This suggests that genomes are initiated on antigenome templates 5 × 5, or 25, times more frequently, on a per-template basis, than are antigenomes on genome templates. On the other hand, considerably less genome-sense mRNAs are initiated from antigenome templates than mRNAs from genome templates, regardless of whether the segment is ambisense. Genomes would therefore have strong promoters for mRNA synthesis and weak promoters for antigenome synthesis, whereas the reverse would be true for antigenome templates.

9 The Critical End Sequences of the Bunyavirus Segments

The sequences at the ends of the bunyavirus segments appear to be critical in the control of RNA synthesis during infection. For two viruses, Bunyamwera virus and LAC, all six ends have been determined (ELLIOTT 1990; HACKER et al. 1990). The LAC ends are shown in Fig. 4 as the panhandle structures thought to be present in genome NCs (see also ELLIOTT et al., this volume). G:U pairs are included, as they occur in rRNA and tRNA, and in somes cases are recognition sites for protein interactions (MCCLAIN et al. 1988; WYATT et al. 1989). Antigenome panhandles would simply be the complements of these, except that the G:U pairs are replaced with A:C pairs, which would break the helical stack. The circularity of the NCs seen in the electron microscope and the fact that the ends of the RNA within NCs can be cross-linked with psoralen is good evidence that these structures exist within NCs and suggests that they might play a role in RNA synthesis.

```
                   A    A  A
3' UCAUCACAUGAGGUGA CUUA UG AA
           :                            S genome
5' AGUAGUGUGCUCCACU GAAU AC UU
                           A

                        C
3' UCAUCACAUGAAGGUUCAUAU  UAUUGCA
           :                  :         M genome
5' AGUAGUGUGCUUCCAAGUAUA  AUAAUGU
                        AA

                          U
3' UCAUCACAUGAGGAUAGAUGUUU GAAUGUGUCA
           :               :   :        L genome
5' AGUAGUGUGCUCCUAUCUACAAA UUUAUACAGU
   .        .         .
   1       10        20
```

Fig. 4. Panhandle structures of the ends of the LAC segment RNAs. The complementary sequences at the ends of the S, M, and L segments (CABRADILLA et al. 1983; GRADY et al. 1987; OBIJESKI et al. 1980; HACKER et al. 1990) that can form ds-RNA are shown. Unmatched or mismatched bases are shown above or below the duplex. U:G pairs are indicated with *double dots*. The terminal 11 nt of each segment, indicated in *boldface*, show the sequences conserved among all members of the *Bunyavirus* genus

By comparing the end sequences of the three LAC segments and those of Bunyamwera virus, several points are worth noting. (a) Except for the first 11 nt, which are strictly conserved, none is at all conserved, either in sequence or in the position and nature of the loops and bulges. It is therefore not so surprising that if genomes contain mRNA promoters, so would antigenomes, even though these viruses do not employ ambisense coding stategies. (b) The longer the segment, the longer and stronger is the panhandle. The increased complementarity at the ends of the longer segments would help these ends to pair, as the length of the intervening sequence decreases the probability of the ends finding each other. The segments might then have equal probabilities of circularizing, in spite of their differences in length. (c) The longer the segment, the less mRNA (and apparently also anti-mRNA; HACKER et al. 1990) is made per template. Assuming that breaking the paired ends of the template are required for mRNA synthesis, there is then a correlation between the ease in opening the ends and the frequency of mRNA synthesis. This correlation does not hold for genome or antigenome synthesis, which is roughly equimolar for the three LAC segments intracellularly (ROSSIER et al. 1988). However, mRNA and genome synthesis clearly initiate by different mechanisms.

References

Abraham G, Pattnaik AK (1983) Early RNA synthesis in Bunyamwera virus-infected cells. J Gen Virol 64: 1277–1290

Bellocq C, Kolakofsky D (1987) Translational requirement for La Crosse virus S-mRNA synthesis: a possible mechanism. J Virol 61: 3960–3967

Bellocq C, Rzju R, Patterson JL, Kolakofsky D (1987) Translational requirement of La Crosse virus S-mRNA synthesis: in vitro studies. J Virol 61: 87–95

Bishop DHL, Gay ME, Matsuoko Y (1983) Nonviral heterogeneous sequences are present at the 5′ ends of one species of snowshoe hare bunyavirus S complementary RNA. Nucleic Acids Res 11: 6409–6418

Bouloy M, Hannoun C (1976) Studies on Lumbo virus replication. I. RNA dependent RNA polymerase associated with virions. Virology 69: 258–268

Bouloy M, Pardigon N, Vialat P, Gerbaud S, Girard M (1990) Characterization of the 5′ and 3′ ends of viral messenger RNAs isolated from BHK21 cells infected with Germiston virus (*Bunyavirus*). Virology 175: 50–58

Cabradilla CD, Holloway BP, Obijeski JF (1983) Molecular cloning and sequencing of the La Crosse virus S RNA. Virology 128: 463–468

Collett MS (1986) Messenger RNA of the M segment RNA of Rift Valley fever virus. Virology 151: 151–156

Elliott RM (1990) Molecular biology of the Bunyaviridae. J Virol 71: 501–522

Elliott RM, Wilkie ML (1986) Persistent infection of *Aedes albopictus* C6/36 cells by Bunyamwera virus. Virology 150: 21–32

Emery VC, Bishop DHL (1987) Characterization of Punta Toro S mRNA species and identification of an inverted complementary sequence in the intergenic region of Punta Toro Phlebo-virus ambisense S RNA that is involved in mRNA transcription termination. Virology 156: 1–11

Eshita Y, Ericson B, Romanowski V, Bishop DHL (1985) Analyses of the M RNA transcription processes of snowshoe hare bunyavirus S and M RNA species. J Virol 55: 681–689

Gentsch JR, Bishop DHL, Obijeski JF (1977) The virus particle nucleic acids and proteins of four bunyaviruses. J Gen Virol 34: 257–268

Gerbaud S, Pardigon N, Vialat P, Bouloy M (1987) The S segment of the Germiston bunyavirus genome: coding strategy and transcription. In: Mahy BWJ, Kolakofsky D (eds) The biology of negative strand viruses. Elsevier, Amsterdam, pp 191–198
Grady LJ, Sanders ML, Campbell WP (1987) The sequence of the M RNA of an isolate of La Crosse virus. J Gen Virol 68: 3057–3071
Haan de P, Wagemakers L, Peters D, Goldbach R (1990) The S RNA segment of tomato spotted wilt virus has an ambisense character. J Gen Virol 71: 1001–1007
Hacker D, Raju R, Kolakofsky D (1989) La Crosse virus nucleocapsid protein controls its own synthesis in mosquito cells by encapsidating its mRNA. J Virol 63: 5166–5174
Hacker D, Rochat S, Kolakofsky D (1990) Anti-mRNAs in La Crosse *Bunyavirus* infected cells. J Virol :
Hewlett MJ, Petterson RF, Baltimore D (1977) Circular forms of Uukuniemi virion RNA: an electron microscopic study. J Virol 21: 1085–1093
Ihara T, Akashi H, Bishop DHL (1984) Novel coding strategy (ambisense genomic RNA) revealed by sequence analyses of Punta Toro phlebovirus S RNA. Virology 136: 293–306
Ihara T, Matsuura Y, Bishop DHL (1985) Analyses of the mRNA transcription processes of Punto Toro phlebovirus (Bunyaviridae). Virology 147: 317–325
Kolakofsky D, Bellocq C, Raju R (1987) The translational requirement of La Crosse virus S-mRNA synthesis. Cold Spring Harbor Symp Quant Biol 52: 373–379
Krug RM (1981) Priming of influenza viral RNA transcription of capped heterologous RNAs. In: Shatkin AJ (ed) Initiation signals in viral gene expression. Springer, Berlin Heidelberg New York, pp 125–149 (Current topics in microbiology and immunology, vol 93)
Krug RM (1983) Transcription and replication of influenza viruses. In: Palese P, Kingsbury DW (eds) Genetics of influenza viruses. Springer, Vienna New York, pp 70–98
Marriott AC, Ward VK, Nuttall PA (1989) The S RNA segment of sandfly fever Sicilian virus: evidence for an ambisense genome. Virology 169: 341–345
McClain WH, Chen Y-M, Fross K, Schneider J (1988) Association of transfer RNA acceptor identity with a helical irregularity. Science 242: 1681–1684
Newton SE, Short NJ, Dalgardno L (1981) Bunyamwera virus replication in cultured *Aedes albopictus* (mosquito) cells: establishment of a persistent viral infection. J Virol 38: 1015–1024
Obijeski JF, Bishop DHL, Palmer EL, Murphy FA (1976) Segmented genome and nucleocapsid of La Crosse virus. J Virol 20: 664–675
Obijeski JF, McCanley J, Skeliel JJ (1980) Nucleotide sequence at the termini of La Crosse Virus RNA. Nucleic Acids Res 8: 2431–2438
Patterson JL, Kolakofsky D (1984) Characterization of La Crosse virus small-genome transcripts. J Virol 49: 680–685
Patterson JL, Cabradilla C, Holloway BP, Obijeski JF, Kolakofsky D (1983) Multiple leader RNAs and messenger RNAs are transcribed from the La Crosse virus small genome segment. Cell 33: 791–799
Patterson JL, Holloway B, Kolakofsky D (1984) La Crosse virions contain a primer-stimulated RNA polymerase and a methylated cap-dependent endonuclease. J Virol 52: 215–222
Pattnaik AK, Abraham G (1983) Identification of four complementary RNA species in Akabane virus-infected cells. J Virol 47: 452–462
Pettersson RF, von Bonsdorff CH (1975) Ribonucleoproteins of Uukuniemi virus are circular. J Virol 15: 386–392
Pettersson R, Kääriäinen L, von Bonsdorff CH, Okerblom N (1971) Structural components of Uukuniemi virus, a non-cubical tick-borne arborirus. Virology 46: 721–729
Pettersson RF, Hewlett MJ, Baltimore D, Coffin JM (1977) The genome of Uukuniemi virus consists of three unique RNA segments. Cell 11: 51–63
Raju R, Kolakofsky D (1986) Inhibitors of protein synthesis inhibit both La Crosse virus S-mRNA and S genome syntheses in vivo. Virus Res 5: 1–9
Raju R, Kolakofsky D (1987a) Translational requirement of La Crosse virus S-mRNA synthesis: in vivo studies. J Virol 61: 96–103
Raju R, Kolakofsky D (1987b) Unusual transcripts in La Crosse virus-infected cells and the site for nucleocapsid assembly. J Virol 61: 667–672
Raju R, Kolakofsky D (1988) La Crosse virus infection of mammalian cells induces mRNA instability. J Virol 62: 27–32
Raju R, Kolakofsky D (1989) The ends of La Crosse virus genome and antigenome RNAs within nucleocapsids are base paired. J Virol 63: 122–128

Raju R, Raju L, Kolakofsky D (1989) The translational requirement for complete La Crosse virus mRNA synthesis is cell-type dependent. J Virol 63: 5159–5165
Ranki M, Pettersson R (1975) Uukuniemi virus contains an RNA polymerase. J Virol 16: 1420–1425
Rossier CJ, Patterson JL, Kolakofsky D (1986) La Crosse virus small genome mRNA is made in the cytoplasm. J Virol 58: 647–650
Rossier C, Raju R, Kolakofsky D (1988) La Crosse virus gene expression in mammalian and mosquito cells. Virology 165: 539–548
Samso A, Bouloy M, Hannoun C (1975) Présence de ribonucléoproteines circulaires dans le virus Lumbo (*Bunyavirus*). CR Acad Sci (III) D280: 779–782
Samso A, Bouloy M, Hannoun C (1976) Mise en évidence de molécules d'acide ribonucléique circulaire dans le virus Lumbo (*Bunyavirus*). CR Acad Sci (III) D282: 1653–1655
Schmaljohn CS, Dalrymple JM (1983) Analysis of Hantaan virus RNA: evidence for a new genus of Bunyaviridae. Virology 131: 482–491
Simons JF, Hellman U, Pettersson RF (1990) Uukuniemi Virus S RNA Segment: Ambisense Coding Strategy, Packaging of Complementary Strands into Virions, and Homology to Members of the Genes Phlebovirus. J Virol 64/1: 247–255
Vezza AC, Repik PM, Cach P, Bishop DHL (1979) In vivo transcription and protein synthesis capabilities of bunyaviruses: wild-type snowshoe hare virus and its temperature-sensitive group I, group II and group I/II mutants. J Virol 31: 426–436
Vidal S, Kolakofsky D (1989) Modified model for the switch from Sendai virus transcription to replication. J Virol 63: 1951–1958
Wyatt JR, Puglisi JD, Tinoco I JR (1989) RNA folding: pseudoknots, loops and bulges. Bioessays 11: 100–106

Bunyavirus Protein Transport and Assembly*

Y. MATSUOKA[1], S. Y. CHEN[2], and R. W. COMPANS[2]

1 Introduction

One of the distinctive features of the Bunyaviridae is their intracellular maturation process, which occurs by budding at smooth-surfaced membranes in the Golgi region (MURPHY et al. 1973). Virions bud into the Golgi cisternae and are then transported and released at the cell surface probably by a vesicular transport process utilizing the exocytic pathway. Before assembly of the virus particles, all structural components including proteins as well as viral genome segments must reach the site where virus particles are formed. Recent studies of the envelope glycoproteins have emphasized their role in determining the site of virus maturation. Bunyavirus glycoproteins accumulate at the membranes of the Golgi apparatus prior to virus assembly, indicating that the glycoproteins may serve to direct other structural components to the site of maturation. Recently it has become evident that bunyavirus glycoproteins expressed from cloned cDNAs are retained in the Golgi complex and thus behave similarly to resident proteins of the Golgi complex.

* This research was supported by grant no. AI 12680 from the National Institute of Allergy and Infectious Diseases
[1] Secretech, Inc., 1025 18th Street South, Suite 201, Birmingham, AL 35205, USA
[2] Department of Microbiology, University of Alabama at Birmingham, Birmingham, AL 35294, USA

Current Topics in Microbiology and Immunology, Vol. 169

The present review focuses on the biogenesis of the glycoproteins, including their processing, transport, and possible mechanism of intracellular localization. A series of studies with Uukuniemi virus (UUK) of *Uukuvirus* genus, as well as recent molecular biological approaches using cDNA clones of two phleboviruses, Punta Toro virus (PTV) and Rift Valley fever virus (RVFV), have greatly contributed to understanding the biosynthesis and transport of bunyavirus glycoproteins. The limited information available on the transport of other virus-specific proteins is also summarized.

2 Biogenesis of Proteins Encoded by the M Genome Segment

2.1 Protein Synthesis

Two glycoproteins designated G1 and G2 and, in some genera a nonstructural protein designated NS_M, are encoded by the M genome segment. Sequence analysis (COLLETT et al. 1985; ESHITA and BISHOP 1984; GRADY et al. 1987; FAZAKERLEY et al. 1988; IHARA et al. 1985; LEE et al. 1986; RONNHOLM and PETTERSSON 1987; SCHMALJOHN et al. 1987) as well as the study of mRNA species in virus-infected cells (ESHITA et al. 1985; PARKER et al. 1984) have indicated that these proteins are translated from a single mRNA species as a precursor polyprotein and subsequently cleaved into individual polypeptides. It was also suggested from studies of UUK glycoprotein processing, by synchronized initiation of protein synthesis and short pulse-labeling, that the proteolytic cleavage occurs cotranslationally (KUISMANEN 1984). The designations G1 and G2 have been used for the larger and smaller cleavage products, respectively. However, because of differences in the location of cleavage sites, the gene order of G1 and G2 on the M segment differs among various members of the Bunyaviridae. In the discussion which follows, we therefore sometimes employ the terms "first glycoprotein" and "second glycoprotein" to specify the gene order.

Several hydrophobic regions have been identified in the predicted amino acid sequences of the precursor proteins, as shown in Fig. 1. A hydrophobic region at the amino-terminal end, which precedes the first glycoprotein, is likely to represent a signal peptide sequence. A hydrophobic region at the carboxy-terminal end may function as a membrane-spanning anchor domain of the second glycoprotein. In addition to these terminal hydrophobic sequences, the polyprotein product of the M segment contains several internal hydrophobic regions as well. A hydrophobic region near the carboxy terminus of the first glycoprotein and one preceding the amino terminus of the second glycoprotein could serve as a membrane anchor sequence and as a signal sequence, respectively. The presence of functional internal hydrophobic domains is well

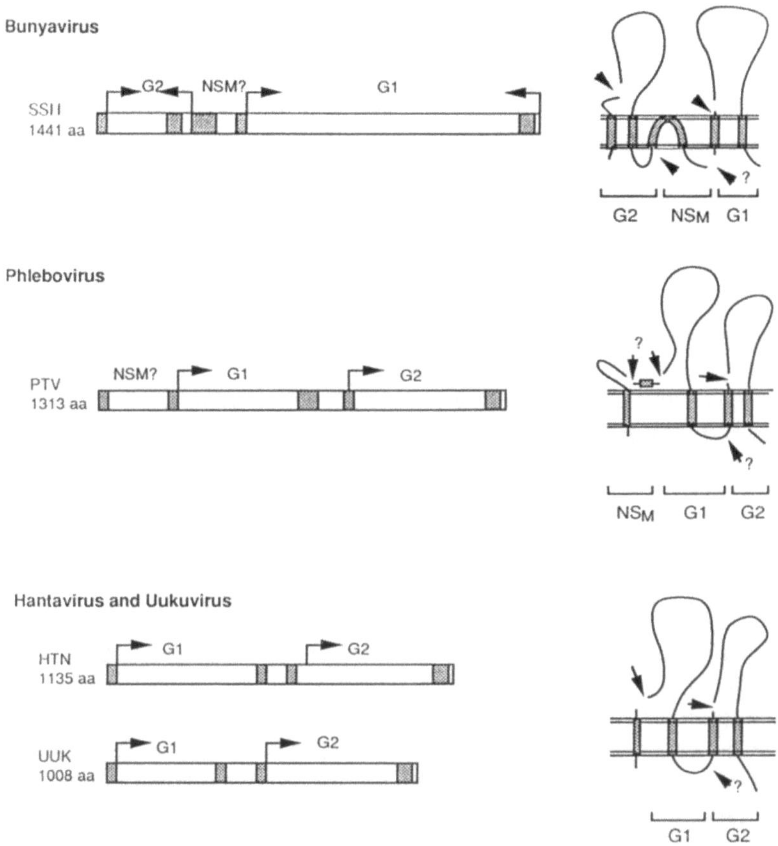

Fig. 1. Gene organization of M genome segments of bunyaviruses and proposed topology of their protein products. The ORF for the precursor protein is shown with hydrophobic regions (■) and known amino termini (↱) and carboxy termini (↰) of the G1 and G2 glycoproteins. The predicted transmembrane topology of proteins based on the hydrophobicity profiles is also shown. *Arrows*, polypeptide cleavage sites. References: SSH (ESHITA and BISHOP 1984; FAZAKERLEY et al. 1988), PTV (IHARA et al. 1985), HTN (SCHMALJOHN et al. 1987), UUK (RONNHOLM and PETTERSSON 1987)

documented with the alphavirus precursor polyproteins (HASHIMOTO et al. 1981; RICE and STRAUSS 1981; WIRTH et al. 1977). It is likely that the amino termini of the first glycoprotein are exposed in the lumen of the endoplasmic reticulum (ER) and outside the viral membrane since the N termini follow a typical signal peptide cleavage site and are followed by most of the potential glycosylation sites before another potential transmembrane sequence is found. It was also suggested by a trypsin digestion study of snowshoe hare virus (SSH) that the amino terminus of the second glycoprotein is external (FAZAKERLEY et al. 1988). A model for the transmembrane topology of the glycoproteins of the Bunyaviridae family can be predicted (Fig. 1). Although the exact carboxy terminus of the first glycoprotein for most viruses has not been identified, except for SSH, it is probably located somewhere between the predicted carboxy-terminal anchoring domain of the first glycoprotein and the signal sequence for the second glycoprotein. The proteolytic cleavage events which generate the two glycoproteins could thus also produce a small peptide and probably a larger nonstructural NS_M molecule for the genus *Bunyavirus*, which may be analogous to the 6-kDa protein found in Semliki Forest virus-infected cells (WELCH and SEFTON 1980). In SSH-infected cells, two small proteins, of 11 and 10 kDa, were identified, and both proteins reacted with antibodies raised against peptides within the predicted NS_M coding region, which is sufficient to code for 19 kDa (FAZAKERLEY et al. 1988). However the exact relationship of these proteins to the NS_M open reading frame (ORF) is not clear.

It was, however, suggested that an alternative translational strategy may exist for the M genome segment of the *Phlebovirus* RVFV. Four polypeptides, 78-kDa, 14-kDa, and the G1 and G2 glycoproteins, were identified as protein products specific for the RVFV M genome segment in RVFV-infected cells as well as in a cell-free transcriptional and translational system (KAKACH et al. 1988, 1989; SUZICH and COLLETT 1988). The biogenesis of these polypeptides was examined by using recombinant vaccinia viruses carrying truncated M genes or M genes with altered potential translational initiation codons (KAKACH et al. 1989; SUZICH et al. 1990). The results suggested that the first and second ATG codons in the same ORF are used for synthesis of the 78-kDa protein and 14-kDa protein, respectively, whereas additional internal in-phase methionine codons as well as the first two ATGs may be utilized for synthesis of the G1 and G2 glycoproteins. Although it is not known whether this translational strategy is used in nature, a similar mechanism which produces multiple protein products from single genes by utilizing internal ATG codons has been also reported for other virus families such as paramyxoviruses (CURRAN and KOLAKOFSKY 1988; MCGINNES et al. 1988) and rhabdoviruses (HERMAN 1986).

2.2 Transport and Glycosylation

All secretory proteins and membrane-bound proteins targeted to various organelles are first synthesized in the rough ER and enter the central vacuolar transport pathway. The organelles involved in this transport system include the rough and smooth ER, the *cis*-, medial, and *trans*-Golgi, the *trans*-Golgi network, secretory vesicles and granules, the endosomal system, lysosomes, and the plasma membrane (KLAUSNER 1989). The envelope glycoproteins of all bunyaviruses so far analyzed possess N-linked oligosaccharide side chains (CASH et al. 1980; PESONEN et al. 1982; SCHMALJOHN et al. 1986). Glycosylation of proteins containing N-linked oligosaccharides is initiated by *en bloc* transfer of preformed oligosaccharide chains from an oligosaccharide pyrophosphoryldolichol intermediate to an asparagine residue in the amino acid sequence Asn-X-Ser/Thr in the newly synthesized polypeptide chain (HUBBARD and IVATT 1981; KLENK and ROTT 1980). Subsequently, sugar residues are modified by enzymes that reside in the ER and the Golgi complex as the glycoproteins are transported through these organelles (KORNFELD and KORNFELD 1985). Thus, the intracellular transport of the glycoproteins of bunyaviruses as well as other enveloped viruses has been studied by analyzing the oligosaccharides of glycoproteins. Endoglycosidase H (endo H) specifically cleaves high-mannose type oligosaccharide side chains which are the primary sugar residues attached in the rough ER. After removal of glucose and some mannose residues in the ER, the side chains are converted into complex sugar chains by the attachment of the peripheral sugars in the Golgi complex, and these mature forms of sugar chains are no longer susceptible to endo H. Therefore, it is generally considered that the time required for the acquisition of endo H resistance corresponds to the transport time from the ER to the Golgi complex (STROUS and LODISH 1980).

The G1 glycoprotein of La Crosse virus (LAC) and both G1 and G2 glycoproteins of UUK (PESONEN et al. 1982; KUISMANEN 1984), PTV (MATSUOKA et al. 1988), and Hantaan virus (HTN) (SCHMALJOHN et al. 1986) were reported to acquire endo H resistance, but to different extents. In general, the G1 glycoprotein contains complex-type oligosaccharide chains, whereas sugar chains of the G2 glycoproteins are mostly high-mannose type except for HTN in which both glycoproteins seem to contain mostly high-mannose type chains. The G1 and G2 glycoproteins of UUK acquire endo H resistance with a half-life of 45 min and 90–150 min, respectively (KUISMANEN 1984). Similar kinetics were observed for glycoproteins of PTV (CHEN et al. 1990; MATSUOKA et al. 1988) and the G1 glycoprotein of LAC (MADOFF and LENARD 1982). In contrast, the G glycoprotein of vesicular stomatitis virus (VSV) and the influenza virus hemagglutinin (HA) acquire endo H resistance much more rapidly, within 15–20 min after synthesis (COPELAND et al. 1988; DOMS et al. 1987; STROUS and LODISH 1980). Thus, the transport of bunyavirus glycoproteins from the ER to the Golgi complex is a relatively slow process.

After being transported from the ER to the Golgi complex, glycoproteins of bunyaviruses accumulate within the Golgi apparatus where virus assembly takes place. Studies with UUK-infected cells using immunofluorescence and immunoelectron microscopy demonstrated that viral glycoproteins accumulate in the perinuclear region, coinciding with the major binding site for wheat germ hemagglutinin (WGA) which stains predominantly the Golgi membranes (KUISMANEN et al. 1982, 1984). It has been also shown by using a temperature-sensitive mutants ts 12, which is defective for assembly of virus particles at restrictive temperature (GAHMBERG 1984), that the glycoproteins accumulate intracellularly in the absence of virus maturation (GAHMBERG et al. 1986). Recent evidence that glycoprotein molecules are targeted to the Golgi complex in the absence of other viral components (structural or nonstructural) has been obtained from studies of recombinant glycoproteins of PTV (MATSUOKA et al. 1988), HTN (PENSIERO et al. 1988), and RVFV (WASMOEN et al. 1988) which were expressed from cloned cDNA in the recombinant vaccinia expression system. The glycoproteins produced from recombinant viruses were stably retained in the Golgi membranes and were not transported to the cell surface. Thus, early reports of the cell surface expression of glycoproteins in LAC (MADOFF and

LENARD 1982) and UUK (KUISMANEN et al. 1982) virus-infected cells may actually reflect the presence of virus particles transported to the cell surface rather than expression of individual protein molecules.

The precise subcellular localization of the glycoproteins of PTV has been analyzed (CHEN et al. 1991) by using Brefeldin A (BFA), a fungal antibacterial reagent reported to inhibit the transport of proteins out of the ER and cause redistribution of Golgi component to the ER (DOMS et al. 1989; LIPPINCOTT-

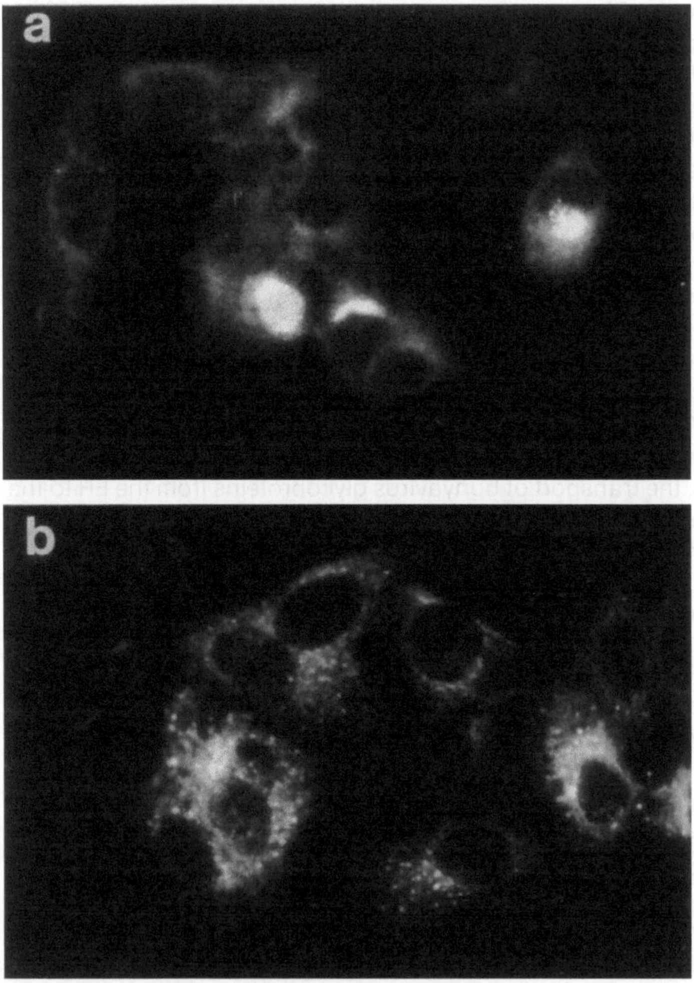

Fig. 2 a, b. Localization of bunyavirus glycoproteins shown by indirect immunofluorescence. **a** PTV glycoproteins expressed from recombinant vaccinia virus showing intracellular retention and accumulation in the Golgi region after 6-h cycloheximide treatment. **b** Effect of BFA treatment showing glycoproteins redistributed throughout the cytoplasm. (From CHEN et al. 1990)

SCHWARTZ et al. 1989; MISUMI et al. 1986; TAKATSUKI and TAMURA 1985). Upon BFA treatment of cells infected with PTV or a recombinant vaccinia virus, PTV glycoproteins which had accumulated in the Golgi region were redistributed throughout the cytoplasm (Fig. 2), whereas little change was observed in the distribution pattern for WGA which has a preferential affinity for the *trans* Golgi network, indicating that PTV glycoproteins specifically localize in the *cis*/medial Golgi membranes (CHEN et al. 1991).

The location and functional roles of other protein products of the M segment are not clear. Two nonstructural proteins are found in SSH-infected cells but do not seem to be glycosylated, and the cellular localization of these proteins has not been reported. The translation products of the predicted NS_M protein coding sequence at the 5′ end of the PTV M segment has not yet been identified either in virions or in infected cells. However, two nonstructural proteins, the glycosylated 78-kDa protein and nonglycosylated 14-kDa protein, are both products from the 5′ end of the coding sequence of the RVFV M genome segment and seem to be localized in the Golgi complex (WASMOEN et al. 1988).

2.3 Possible Mechanisms for Intracellular Transport and Localization

Specific targeting mechanisms must exist for selective transport and localization of proteins in various organelles. The bunyavirus glycoproteins are localized specifically in the membranes of the Golgi complex and thus appear to utilize a similar pathway as that for resident proteins of the Golgi complex. At least two steps which are involved in bunyavirus glycoprotein localization are transport from the ER to the Golgi complex and retention in the Golgi complex. Some proteins remain in the ER as permanent residents, while others are exported to the Golgi complex, most of which are subsequently distributed to various organelles. At least two mechanisms have been demonstrated for retention in the ER. Resident ER proteins may be distinguished from non-resident proteins by the presence of specific signals within the protein sequence. Several soluble proteins including disulfide isomerase, BiP (immunoglobulin heavy-chain binding protein, or grp 78), and others (see review by PELHAM 1989) were found to contain the same C-terminal tetrapeptide sequence, Lys-Asp-Glu-Leu (KDEL). Removal of the KDEL sequence from BiP caused it to be secreted, while addition of this sequence to secretory or lysosomal proteins caused them to accumulate in the ER (MUNRO and PELHAM 1987; PELHAM 1988). Thus, it is evident that the KDEL sequence is responsible for retention of at least some soluble ER specific proteins. Another mechanism to control exit of proteins from the ER has been described as the possession of transport competency. It has become apparent that the correct folding and assembly of proteins is necessary for their efficient transport. A number of mutations intoduced into viral glycoproteins, such as the VSV G protein and influenza HA, were reported to block or slow the export of these proteins from the ER (ROSE and DOMS 1988). Also, for some proteins, including influenza HA and VSV G protein, assembly into oligomers is required for transport from the ER (COPELAND et al. 1986; DOMS et al. 1988; GETHING et al. 1986; HURTLEY and HELENIUS 1989; KREIS and LODISH 1986). Failure to assemble into an oligomer therefore results in ER retention, without requirement for a specific signal.

In virions, the G1 and G2 glycoproteins of UUK exist as heterodimers (PERSSON et al. 1989). Dimerization seems to occur rapidly after protein synthesis; however, each pair of glycoproteins may not be the cleavage products from the same precursor protein since the transport kinetics for the two glycoproteins were reported to be significantly different; the newly synthesized G1 was incorporated into virions 20 min faster than G2 (PERSSON et al. 1989; WIKSTROM et al. 1989). Similar transport kinetics were also obtained for PTV glycoproteins (CHEN et al. 1991). It was also demonstrated from studies using pulse-chase and subcellular fractionation that UUK G1 and G2 glycoproteins moved rapidly from the ER, and that most of the proteins were shifted to lighter density fractions. Furthermore, the glycoproteins appear to remain in this fraction for 15–20 min before entering the Golgi fraction (WIKSTROM et al. 1989). From these results, it was proposed that the dimerization of UUK glycoproteins occurs at an intermediate compartment between the ER and the Golgi complex. The post-ER compartment which lacks ER markers, such as glucose-6-phosphatase, is also distinct from the Golgi complex. At 15 °C, the membrane proteins of Semliki Forest virus were reported to leave the ER but failed to enter the Golgi cisternae and accumulated in the intermediate vacuolar elements (SARASTE and KUISMANEN 1984). A lysosome protein cathepsin D bearing the ER retention signal (KDEL sequence) accumulated within the ER but appeared to be modified by mannose-6-phosphotransferase, suggesting that the post-ER compartment is the site for the sorting proteins into the Golgi or recycling them back to the ER (PELHAM 1988, 1989). It is possible that bunyavirus glycoproteins accumulate at the same post-ER compartment until they obtain a necessary signal, such as dimerization, for entering the Golgi complex. However, it is still not clear why different populations of G1 and G2 glycoproteins, not derived from the same precursor protein, form dimers. Possibly some conformational change after protein synthesis is necessary for heterodimer formation, and the structural change of the G2 may take longer than that of G1 proteins.

Cell fractionation studies have shown that BiP coprecipitates with UUK glycoproteins from the ER fraction (PERSSON et al. 1989). It was shown that unassembled immunoglobulin heavy chains (BOLE et al. 1986; HAAS and WABL 1983), mutant or unglycosylated forms of influenza HA (COPELAND et al. 1988; GETHING et al. 1986), and immature forms of Sendai virus hemagglutinin neuraminidase (HN) protein (ROUX 1990) all associate with BiP in the ER. However, the significance of the association with BiP for bunyavirus glycoprotein transport is not clear at this point. Protein products from recombinant vaccinia viruses containing only G1 or G2 coding sequence of PTV failed to exit from the ER (CHEN et al. 1991; MATSUOKA, unpublished observation). The recombinant proteins did not seem to associate with BiP since anti-BiP antibody failed to coprecipitate these proteins. This result may also suggest the importance of an interaction between the two glycoproteins such as dimer formation.

At least one example of a retention signal for a Golgi-specific protein has been defined. The E1 glycoprotein of a coronavirus which also matures at the Golgi membranes contains three membrane-spanning domains. Deletion of the first and second domains resulted the expression of the protein on the cell surface, whereas a deletion of the second and third transmembrane domains resulted in retention in the Golgi complex (MACHAMER and ROSE 1987). These results suggest that the first transmembrane domain contains a Golgi retention signal.

The situation may be more complicated for the bunyavirus glycoproteins, which are derived from a common precursor protein. Do both G1 and G2 glycoproteins contain their own retention signals, or does only one of them possess such a specific signal and interact with the other to halt both in the Golgi? Are extra sequences such as the NS_M coding sequence involved in retention of glycoproteins in the Golgi? A requirement for an NS_M protein in Golgi retention was ruled out at least for phleboviruses. The G1 and G2 glycoproteins expressed from recombinant vaccinia viruses which contain M genome segment sequences of PTV or RVFV with truncated 5′ ends, and thus lack NS_M coding sequence, were synthesized, processed, and transported to the Golgi complex and were stably retained in the Golgi membranes (MATSUOKA et al. 1988; WASMOEN et al. 1988). Therefore, the region preceding the glycoprotein coding sequence of Phleboviruses does not appear to be responsible for the retention of glycoproteins in the Golgi complex.

Based on the results with the E1 glycoprotein of coronavirus, a logical assumption is that one (or more) of the transmembrane regions of the G1 and/or G2 glycoproteins may play an important role for protein retention. Among several hydrophobic regions illustrated in Fig. 1, two hydrophobic sequences near the carboxy terminal of G1 and G2 are predicted to be membrane anchor domains for each protein and might contain a specific retention signal for bunyavirus glycoproteins in the Golgi complex. Construction of molecules with specific deletions, or chimeras between bunyavirus glycoproteins and other glycoproteins which are expressed on cell surfaces, should enable the identification of the sequences which are important for Golgi retention.

Although it has been possible to identify specific amino acid sequences involved as signals for intracellular retention of some viral proteins, no information is available on the mechanisms by which such retention is effected. Some possibilities which may be considered are the following. (a) The sequences could lead to stable interaction with resident cellular molecules; however, the retention of such cellular molecules would then need to be explained. In addition, cellular proteins are not incorporated into budding virus particles, which argues against such a stable interaction. (b) The sorting signals could interact transiently with components of a recycling pathway which retrieves components from other downstream membranes of the exocytic pathway. (c) The sorting signals could mediate lateral interactions between viral glycoproteins leading to their clustering into localized domains which, because of their curvature or rigidity, may be unable to be incorporated into transport vesicles involved in exit from the Golgi complex.

3 Synthesis and Transport of the Proteins Encoded by the S Genome Segment

The N protein, which is encoded by the S RNA segment, is translated by free ribosomes in the cytoplasm (SCHMALJOHN and PATTERSON 1990). No proteolytic cleavage or posttranslational modification such as phosphorylation has been reported with this protein. The protein is closely associated with the genomic RNAs to form nucleocapsids together with L protein. When and how this interaction occurs is unknown. It was also suggested that N protein controls its own protein synthesis by encapsidating its mRNA in mosquito cells (HACKER et al. 1989). This self-regulation may be related to the establishment of a persistent infection in mosquito cells.

The S segment also encodes a nonstructural polypeptide designated NS_S, although its size and coding strategy varies among the different virus genera (see ELLIOT et al., this volume). It has been reported that the NSs protein of RVFV seems to accumulate in the nucleus in infected cells. High concentrations of NSs were detected in the nuclear fraction from 4 h postinfection (STRUTHERS et al. 1984), and intranuclear fluorescence using immune sheep serum could be observed at 10 h postinfection (STRUTHERS and SWANEPOEL 1982). It is possible that the NSs protein which accumulated in the nucleus may correspond to the intranuclear inclusions in RVFV-infected cells reported previously (SWANEPOEL and BLACKBURN 1977). Neither nuclear inclusions nor intranuclear accumulation of NSs proteins have been reported with other bunyaviruses. The synthesis, processing, and localization of NSs proteins of most bunyaviruses remain to be studied. Monoclonal antibodies against NSs or antibodies raised against NSs-specific polypeptides deduced from known amino acid sequences will be necessary to investigate these processes. The expression of cloned NSs genes, as demonstrated with PTV NSs (OVERTON et al. 1987), will also be a useful for studying the transport and function of this protein.

4 The L Protein

It was shown recently by reassortment that the large (L) segment encodes the L protein (ENDRES et al. 1989). However, little is known about the L protein. Because of its limited quantities both in virions and in infected cells, it is difficult to study the biogenesis of this protein. L protein is believed to be a viral transcriptase which is present in viral nucleocapsids together with viral genomes and N protein (see KOLAKOFSKY and HACKER, this volume). The nucleocapsids accumulate under the Golgi membranes where viral glycoproteins are concentrated and are incorporated into virions upon budding. Although it is unlikely that the L protein has a role in virus assembly, more data are needed about this protein.

5 Assembly and Release of Bunyavirus

The assembly of the bunyaviruses (Fig. 3) is usually restricted to the Golgi apparatus (KUISMANEN et al. 1982; MURPHY et al. 1973; SMITH and PIFAT 1982) although low levels of budding at the cell surface have also been reported (ANDERSON and SMITH 1987). Although the precise mechanisms of budding of enveloped viruses are not fully understood, it has been suggested that budding involves a transmembrane interaction between membrane glycoproteins and the other components of the virus in the cytoplasm, followed by pinching off from the cell surface (LENARD and COMPANS 1974). Upon budding, virions acquire their lipid bilayer from the host cell membrane, whereas most host cell membrane proteins are excluded from the viral particles. Many viruses, such as orthomyxoviruses, paramyxoviruses, rhabdoviruses, and retroviruses, contain a layer of matrix (M) protein between the lipid bilayer and nucleocapsid. It is evident from studies with VSV that the interaction of the M protein with the nucleocapsid is essential for budding (ONO et al. 1987). However, little is known of the interaction between M protein and viral glycoproteins, although it was suggested that the M protein may function as a bridge between the nucleocapsid and the spike glycoproteins (YOSHIDA et al. 1976). Bunyaviruses, as well as arenaviruses and coronaviruses, do not possess an M protein. For these viruses it was proposed that the binding of the cytoplasmic domain of the glycoprotein to the nucleocapsid was the major driving force for budding (GAROFF and SIMONS 1974; STURMAN et al. 1980).

Electron microscopy studies of sandfly fever viruses (*Phlebovirus*) revealed that viral ribonucleoprotein (RNP) condensed at the cytoplasmic face of the Golgi membranes at areas in which viral envelope glycoproteins accumulated at the luminal face (SMITH and PIFAT 1982). Neither RNP nor glycoproteins were observed on the areas which were not directly involved in budding, suggesting that transmembrane interactions between glycoproteins and RNP are essential for budding (SMITH and PIFAT 1982). It was also indicated by enzymatic digestion that approximately 12% of one or both glycoproteins of Karimabad virus remains at the cytoplasmic membrane face. This portion may thus serve to interact with the RNP complex. The G2 glycoprotein of SSH in which the exact amino and carboxy terminals are defined possesses a long external stretch of amino acids (52) after the hydrophobic region presumed to be a membrane anchoring

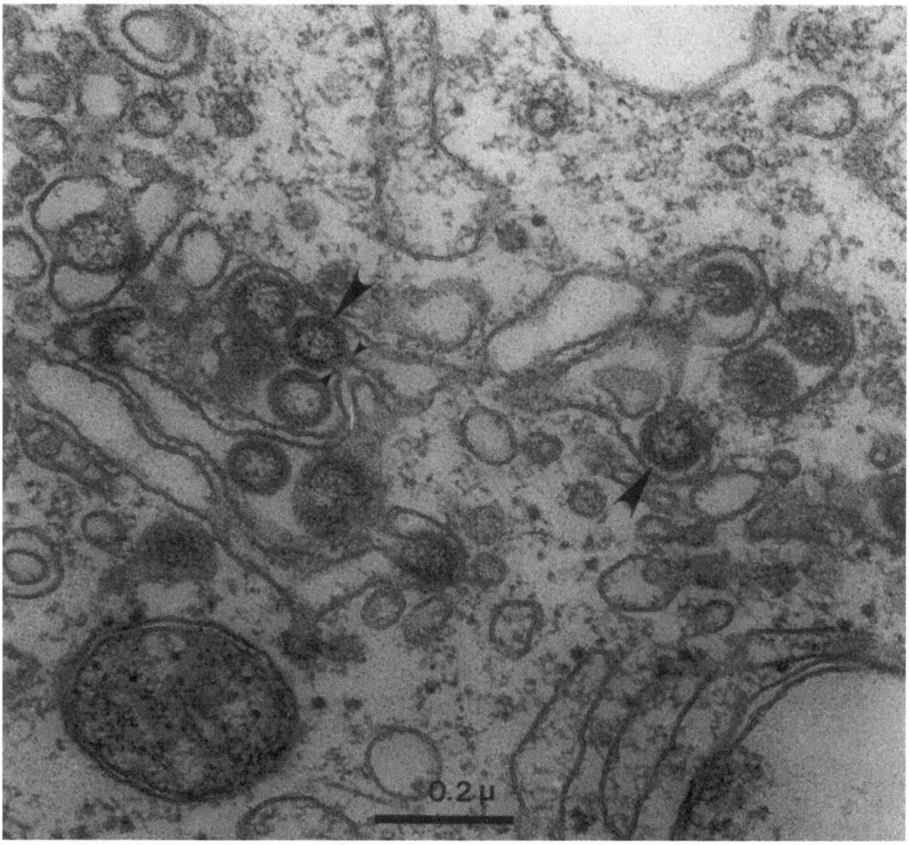

Fig. 3. Assembly of bunyavirus in the Golgi complex. Thin section of PTV-infected Vero cells showing virus particles (arrows) budding from the smooth membrane region of the Golgi apparatus. (From SMITH and PIFAT 1982)

domain, whereas only 20 amino acids are found at the G1 carboxy terminal (FAZAKERLEY et al. 1988). If this 52 amino acid stretch represents the cytoplasmic tail of the G2 glycoprotein of SSH, it may be equivalent to the 12% portion of the Karimabad virus glycoprotein.

Monensin, a carboxylic ionophore which affects the Golgi complex, has been shown to block the transport of membrane as well as secretory proteins from the Golgi complex to the plasma membrane (TARTAKOFF 1983; TARTAKOFF and VASSALLI 1978). Different results have been reported in several studies of the effects of monensin on bunyavirus assembly. In the presence of monensin. Both protein synthesis and virus assembly were inhibited in HTN-infected cells (SCHMALJOHN et al. 1986). In contrast, monensin did not block either virus assembly or the release of SSH bunyavirus particles into culture medium

(ALONSO-CAPLEN et al. 1984). The replication of LAC virus, another member of the *Bunyavirus* genus, was inhibited by monensin, and an accumulation of an incompletely processed form of G1 glycoprotein was observed (CASH 1982; MADOFF and LENARD 1982). This may indicate that proper processing of glycoproteins is required for bunyavirus assembly. The maturation of UUK virus

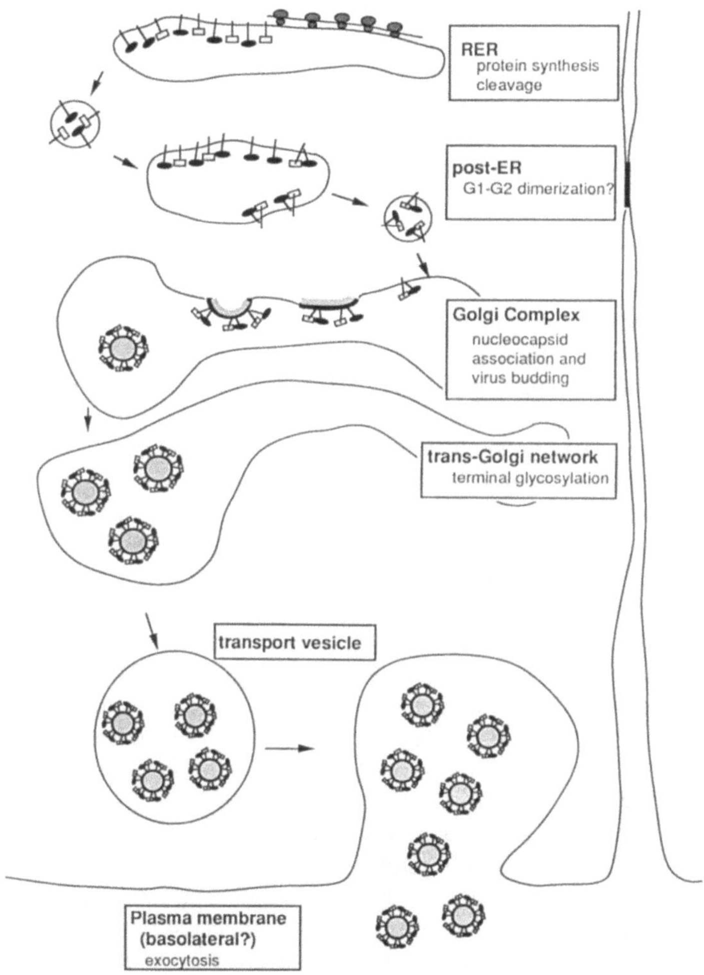

Fig. 4. Model for bunyavirus glycoprotein transport and virus assembly. After translocation into the rough endoplasmic reticulum (*RER*), glycoproteins are transported from the ER to the Golgi complex by acquiring a conformation necessary for the exit from the ER. Virions are assembled by budding at the membranes of the Golgi complex, probably the *cis*/medial Golgi where the glycoproteins accumulate. Virus particles in the Golgi cisternae are further transported through the *trans* Golgi network and released from the cells by exocytosis. In the case of polarized epitherial cells, release appears to occur at the basolateral membranes

was inhibited by monensin at the terminal stage of budding (KUISMANEN et al. 1985). Immunoperoxidase electron microscopy has shown that accumulation of nucleocapsids and viral glycoproteins at the Golgi membranes was not affected by monensin, however, the glycoproteins were assembled into abnormal tubular structures. Surprisingly, monensin did not affect the transport of UUK particles formed before addition of the drug. It is possible that change of the ionic and pH condition of the lumen of the Golgi complex caused by monensin could affect the molecular properties of bunyavirus proteins. In contrast, BFA, which inhibited the transport of proteins out of the ER and caused redistribution of the Golgi-associated proteins, did not affect either protein synthesis or virus maturation in PTV-infected cells (CHEN et al. 1991). During BFA treatment, virions accumulated intracellularly, and as soon as the drug was removed these particles were rapidly transported and released into the medium.

After virus particles bud into the Golgi cisternae, they appear to utilize the exocytic pathway for virus release. The G1 and G2 glycoproteins of mature UUK virions were reported to contain sialic acid (KUISMANEN 1984), indicating that virions assembled in the *cis*/medial Golgi are further transported into the *trans* Golgi where surface glycoproteins are modified by sialyltransferase (KORNFELD and KORNFELD 1985). A schematic pathway of bunyavirus glycoprotein transport and assembly is illustrated in Fig. 4.

The distal side of the *trans* Golgi, designated the *trans* Golgi network, is considered to be a branch point for protein sorting. From here, proteins are sorted into different transport vesicles destined for various organelles (GRIFFITHS and SIMONS 1986). Furthermore, in epithelial cells which maintain apical and basolateral membrane domains separated by tight junctions, it appears that two types of vesicles may exist to transport proteins selectively to the distinct surface domains. Many membrane-associated as well as secretory proteins studied in cultured polarized epithelial cell systems such as Madin-Darby canine kidney cells or human colon carcinoma cells are selectively transported to either or basolateral plasma membrane domains (BURGESS and KELLY 1987; PUGSLEY 1989; SIMONS and FULLER 1985). Assembly of viruses which mature at the plasma membrane is also restricted to the domain to which their glycoproteins are specifically transported; influenza virus and paramyxoviruses are released at apical surfaces and VSV and retroviruses at basolateral surfaces (RODRIGUEZ-BOULAN and SABATINI 1978; ROTH et al. 1983). Bunyavirus particles which mature at the intracellular membranes of the Golgi complex also appear to be transported exclusively to the basolateral domains in polarized epithelial cells after budding. When polarized Vero cells grown on nitrocellulose filters were infected with PTV, the majority of infectious particles were recovered from the basolateral chamber (CHEN et al. 1991). Although the significance of this polarized release is not clear, basolateral release may facilitate spread of the infection to underlying tissues during a systemic infection.

6 Concluding Remarks

The process of virus assembly at intracellular membranes is one of the most interesting features of bunyaviruses. Recent studies have shown that bunyavirus glycoproteins accumulate at the membranes of the *cis*/medial Golgi in the absence of other viral components, clearly indicating their role in determining the intracellular maturation process.

During the past decade, a great deal of information has become available on the molecular biology of bunyaviruses; the sequences of genes, specifically S and M segments, of many bunyaviruses have been determined, and some of the genes have been expressed in various systems. Despite this progress, little is known about certain basic aspects of bunyavirus glycoproteins. Although models for the transmembrane topology of glycoproteins can be predicted from the deduced sequences, direct evidence needs to be obtained concerning the precise configuration of the glycoproteins. In order to determine which portions of the polypeptides are exposed or embedded in the virions (or cell membranes), enzymatic digestion of glycoproteins followed by peptide sequence analysis as well as in vitro analysis of protein synthesis may be necessary. Antibodies against defined peptides of the glycoprotein sequences may also be useful.

The mechanism of intracellular retention and accumulation of bunyavirus glycoproteins is an exciting topic of study. Monoclonal antibodies which can distinguish different maturation stages of the glycoproteins will be powerful tools to study intracellular transport of these proteins. In addition, glycoprotein genes can be dissected at the molecular level by using recombinant DNA approaches, and the exact signals for the retention of these glycoproteins in the Golgi complex may soon be identified. The study of the biogenesis of bunyavirus glycoproteins also provides a useful model system for understanding the mechanisms of biosynthesis and sorting of the Golgi-specific proteins in general.

Acknowledgements. We thank Drs. Jonathan Smith and Dominique Pifat for providing Fig. 3.

References

Alonso-Caplen FV, Matsuoka Y, Compans RW (1984) Transport of viral glycoproteins and its modulation by monensin. In: Compans RW, Bishop DHL (eds) Segmented negative strand viruses. Academic, New York, pp 179–188

Anderson GW Jr, Smith JF (1987) Immunoelectron microscopy of Rift Valley fever viral morphogenesis in primary rat hepatocytes. Virology 161: 91–100

Bole DG, Hendershot LM, Kearney JF (1986) Posttranslational association of immunoglobulin heavy-chain binding-protein with nascent heavy chains in non-secreting and secreting hybridomas. J Cell Biol 102: 1558–1566

Burgess TL, Kelly RB (1987) Constitutive and regulated secretion of proteins. Annu Rev Cell Biol 3: 243–293
Cash P (1982) Inhibition of La Crosse virus replication by monensin, a monovalent ionophore. J Gen Virol 59: 193196
Cash P, Hendershot L, Bishop DHL (1980) The effect of glycosylation inhibitors on the maturation and intracellular polypeptide synthesis induced by snowshoe hare bunyavirus. Virology 103: 235–240
Chen SY, Matsuoka Y, Compans RW (1991) Assembly and polarized release of Punta Toro virus and effect of brefeldin A. J Virol 65: 1427–1439
Collett MS, Purchio AF, Keegan K, Frazier S, Hays W, Anderson DK, Parker MD, Schmaljohn C, Schmidt J, Dalrymple JM (1985) Complete nucleotide sequence of the M RNA segment of Rift Valley fever virus. Virology 144: 228–245
Copeland CS, Doms RW, Bolzau EM, Webster RG, Helenius A (1986) Assembly of influenza hemagglutinin trimers and its role in intracellular transport. J Cell Biol 103: 1179–1191
Copeland CS, Zimmer KP, Wagner KR, Healey GA, Melman I, Helenius A (1988) Folding, trimerization and transport are sequential events in the biogenesis of influenza virus hemagglutinin. Cell 53: 197–209
Curran J, Kolakofsky D (1988) Ribosomal initiation from an ACG codon in the Sendai virus P/C mRNA. EMBO J 7: 245–251
Doms RW, Keller DS, Helenius A, Balch WE (1987) Role for adenosine triphosphate in regulating the assembly and transport of vesicular stomatitis virus G Protein trimers. J Cell Biol 105: 1957–1969
Doms RW, Ruusala A, Machamer C, Helenius A, Rose JK (1988) Differential effects of mutations in three domains on folding, quaternary structure, and intracellular transport of vesicular stomatitis virus G protein. J Cell Biol 107: 89–99
Doms RW, Russ G, Yewdell JW (1989) Brefeldin A redistributes resident and itinerant Golgi proteins to the endoplasmic reticulum. J Cell Biol 109: 61–72
Endres MJ, Jacoby DR, Janssen RS, Gonzalez-Scarano F, Nathanson N (1989) The large viral RNA segment of California serogroup bunyaviruses encodes the large viral protein. J Gen Virol 70: 223–228
Eshita Y, Bishop DHL (1984) The complete sequence of the M RNA of snowshoe hare; bunyavirus reveals the presence of internal hydrophobic domains in the viral glycoprotein. Virology 137: 227–240
Eshita Y, Ericson B, Romanowski V, Bishop DHL (1985) Analysis of the mRNA transcription processes of snowshoe hare bunyavirus S and M RNA species. J Virol 55: 681–689
Fazakerley JK, Gonzalez-Scarano F, Strickler J, Dietzschold B, Karush F, Nathanson N (1988) Organization of the middle RNA segment of snowshoe hare bunyavirus. Virology 167: 422–43
Gahmberg N (1984) Characterization of two recombination-complementation groups of Uukuniemi virus temperature-sensitive mutants. J Gen Virol 65: 1079–1090
Gahmberg N, Kuismanen E, Keranen S, Pettersson RF (1986) Uukuniemi virus glycoproteins accumulate in the cause morphological changes of the Golgi complex in the absence of virus maturation. J Virol 57: 899–906
Garoff H, Simons K (1974) Location of the spike glycoproteins in the Semliki Forest virus membrane. Proc Natl Acad Sci USA 71: 3988–3992
Gething M-J, McCammon K, Sambrook J (1986) Expression of wild-type and mutant forms of influenza hemagglutinin: the role of folding in intracellular transport. Cell 46: 939–950
Grady LJ, Sanders ML, Campbell WP (1987) The sequence of the M RNA of an isolate of La Crosse virus. J Gen Virol 68: 3057–3071
Griffiths G, Simons K (1986) The trans Golgi network: sorting at the exit site of the Golgi complex. Science 234: 438–443
Haas IG, Wabl M (1983) Immunoglobulin heavy chain binding protein. Nature 306: 387–389
Hacker D, Raju R, Kolakofsky D (1989) La Crosse virus nucleopsid protein controls its own synthesis in mosquito cells by encapsidating its mRNA. J Virol 63: 5166–5174
Hashimoto K, Erdei S, Karanen S, Saraste J, Kaarianen L (1981) Evidence for a separate signal sequence for the carboxy-terminal envelope glycoprotein E1 of Semliki Forest virus. J Virol 38: 24–40
Herman RC (1986) Internal initiation of translation of the vesicular stomatitis virus phosphoprotein mRNA yields a second protein. J Virol 58: 797–804
Hubbard SC, Ivatt RJ (1981) Synthesis and processing of asparagine-linked oligosaccharides. Annu Rev Biochem 50: 55–83

Hurtley SM, Helenius A (1989) Control of protein exit from the endoplasmic reticulum. Ann Rev Cell Biol 5: 277–307

Ihara T, Smith J, Dalrymple JM, Bishop DHL (1985) Complete sequence of the glycoproteins and M RNA of Punta Toro phlebovirus compared to those of Rift Valley fever virus. Virology 144: 246–259

Kakach LT, Wasmoen TL, Collett MS (1988) Rift Valley fever virus M segment: use of recombinant vaccinia viruses to study *Phlebovirus* gene expression. J Virol 62: 826–833

Kakach LT, Suzich JA, Colett MS (1989) Rift Valley fever virus M segment: phlebovirus expression strategy and protein glycosylation. Virology 170: 505–510

Klausner RD (1989) Sorting and traffic in the central vacuolar system. Cell 57: 703–706

Klenk HD, Rott R (1980) Cotranslational and posttranslational processing of viral glycoproteins. In: Arber W et al. (eds) Current topics in microbiology and immunology, vol 90. Springer, Berlin Heidelberg New York, pp 19–48

Kornfeld R, Kornfeld S (1985) Assembly of asparagine linked oligosaccharides. Annu Rev Biochem 54: 631–644

Kreis TE, Lodish HF (1986) Oligomerization is essential for transport of vesicular stomatitis virus viral glycoprotein on the cell surface. Cell 46: 929–937

Kuismanen E (1984) Posttranslational processing of Uukuniemi virus glycoproteins G1 and G2. J Virol 51: 806–812

Kuismanen E, Hedman K, Saraste J, Pettersson RF (1982) Uukuniemi virus maturation: accumulation of virus particles and viral antigens in the Golgi complex. Mol Cell Biol 2: 1444–1458

Kuismanen E, Bang B, Hurme M, Pettersson RF (1984) Uukuniemi virus maturation: immunofluorescence microscopy with monoclonal glycoprotein-specific antibodies. J Virol 51: 137–146

Kuismanen E, Saraste J, Pettersson RF (1985) Effect of monensin on the assembly of Uukuniemi virus in the Golgi complex. J Virol 55: 813–822

Lees JF, Pringle CR, Elliott RM (1986) Nucleotide sequence of the bunyamwera virus M RNA segment: conservation of structural feature in the bunyavirus glycoprotein gene product. Virology 148: 1–14

Lenard J, Compans RW (1974) The membrane structure of lipid-containing viruses. Biochim Biophys Acta 334: 51–94

Lippincott-Schwartz J, Yuan LC, Bonifacino JS, Klausner RD (1989) Rapid redistribution of Golgi proteins into the ER in cells treated with Brefeldin A: evidence for membrane cycling from Golgi to ER. Cell 56: 801–813

Machamer CE, Rose JK (1987) A specific transmembrane domain of a coronavirus E1 glycoprotein is required for its retention in the Golgi region. J Cell Biol 105: 1205–1214

Madoff DH, Lenard J (1982) A membrane glycoprotein that accumulates intracellularly: cellular processing of the large glycoprotein of La Crosse virus. Cell 28: 821–829

Matsuoka Y, Ihara T, Bishop DHL, Compans RW (1988) Intracellular accumulation of Punta Toro virus glycoproteins expressed from cloned cDNA. Virology 167: 251–260

McGinnes L, McQuain C, Morrison T (1988) The P protein and the nonstructural 38 K and 29 K proteins of Newcastle disease virus are derived from the same open reading rame. Virology 164: 256–264

Misumi Y, Misumi, Miki K, Takatsuki A, Tamura G, Ikehara Y (1986) Novel blockage by Brefeldin A of intracellular transport of secretory proteins in cultured rat hepatocytes. J Biol Chem 261: 11398–11403

Munro S, Pelham HRB (1987) A C-terminal signal prevents secretion of luminal ER proteins. Cell 48: 899–907

Murphy FA, Harrison AK, Whitfield SG (1973) Bunyaviridae: morphologic and morphogenetic similarities of bunyamwera serologic supergroup viruses and several other arthropod borne viruses. Intervirology 1: 297–316

Ono K, Dubois-Dalcq ME, Schubert M, Lazzarini RA (1987) A mutated membrane protein of vesicular stomatitis virus has an abnormal distribution within the infected cell and causes defective budding. J Virol 61: 1332–1341

Overton HA, Ihara T, Bishop DHL (1987) Identification of the N and NSs proteins coded by the ambisense S RNA of Punta Toro phlebovirus using monospecific antisera raised to baculovirus expressed N and NSs proteins. Virology 157: 338–350

Parker MD, Smith JF, Dalrymple JM (1984) Rift Valley fever virus intracellular RNA: a functional analysis. In: Compans RW, Bishop DHL (eds) Segmented negative strand viruses. Academic, Orlando, pp 21–28

Pelham HRB (1988) Evidence that luminal ER proteins are sorted from secreted proteins in a post-ER compartment. EMBO J 7: 913–918
Pelham HRB (1989) Control of protein exit from the endoplasmic reticulum. Annu Rev Cell Biol 5: 1–23
Pensiero MN, Jennings GB, Schmaljohn CS, Hay J (1988) Expression of the Hantaan virus M genome segment by using a vaccinia virus recombinant. J Virus 62: 696–702
Persson R, Wikstrom L, Pettersson RF (1989) Dimerization and intracellular transport of two Golgi-specific bunyavirus membrane glycoproteins. In: Compans RW, Helenius A, Oldstone MBA (eds) UCLA symposium on cell biology of virus entry, replication and pathogenesis. Liss, New York, pp 229–242
Pesonen M, Kuismanen E, Pettersson RF (1982) Monosaccharide sequence of protein-bound glycans of Uukuniemi virus. J Virol 41: 390–400
Pugsley AP (1989) Protein targeting. Academic, San Diego
Rice CM, Strauss JH (1981) Nucleotide sequence of the 265 mRNA of Sindbis virus and deduced sequence of the encoded virus structural proteins. Proc Natl Acad Sci USA 78: 2062–2066
Rodriguez-Boulan E, Sabaitni DD (1978) Asymmetric budding of viruses in epithelial monolayers: a model system for study of epithelial polarity. Proc Natl Acad Sci USA 75: 5071–5075
Ronnholm R, Pettersson RE (1987) Complete nucleotide sequence of the M RNA segment of Uukuniemi virus encoding the membrane glycoproteins G1 and G2. Virology 160: 191–202
Rose JK, Doms RW (1988) Regulation of protein export from the endoplasmic reticulum. Annu Rev Cell Biol 4: 257–288
Roth MG, Srinivas RV, Compans (1983) Basolateral maturation of retroviruses in polarized epithelial cells. J Virol 45: 1065–1073
Roux L (1990) Selective and transient association of Sendai virus HN glycoprotein with BiP. Virology 175: 161–166
Saraste J, Kuismanen E (1984) Pre- and post-Golgi vacuoles operate in the transport of Semliki Forest virus membrane glycoproteins to the cell surface. Cell 38: 535–549
Schmaljohn CS, Patterson JL (1990) Bunyaviridae and their replication. In: Fields BN et al. (eds) Virology. Raven, New York, pp 1175–1194
Schmaljohn CS, Hasty SE, Rasmussen L, Dalrymple JM (1986) Hantaan virus replication: effects of monensin, tunicamycin and encoglycosidase on the structural glycoproteins. J Gen Virol 67: 707–717
Schmaljohn CS, Schmaljohn AL, Dalrymple JM (1987) Hantaan virus M RNA: coding strategy, nucleotide sequence, and gene order. Virology 157: 31–39
Simons K, Fuller SD (1985) Cell surface polarity in epithelia. Annu Rev Cell Biol 1: 243–288
Smith JF, Pifat DY (1982) Morphogenesis of sandfly fever viruses (Bunyaviridae family). Virology 121: 61–81
Strous GJAM, Lodish HF (1980) Intracellular transport of secretory and membrane proteins in hepatoma cells infected with vesicular stomatitis virus. Cell 22: 709–717
Struthers JK, Swanepoel R (1982) Identification of a major non-structural protein in the nuclei of Rift Valley fever virus-infected cells. J Gen Virol 60: 381–384
Struthers JK, Swanepoel R, Shepherd SP (1984) Protein synthesis in Rift Valley fever virus-infected cells. Virology 134: 118–124
Sturman LS, Holmes KV, Behnke J (1980) Isolation of coronavirus envelope glycoprotein with the viral nucleocapsid. J Virol 33: 449–462
Suzich JA, Collett MS (1988) Rift Valley fever virus M segment: cell-free transcription and translation of virus complementary RNA. Virology 164: 478–486
Suzich JA, Kakach LT, Collett MS (1990) Expression strategy of a phlebovirus: biogenesis of proteins from the Rift Valley fever M segment. J Virol 64: 1549–1555
Swanepoel R, Blackburn NK (1977) Demonstration of nuclear immunofluorescence in Rift Valley fever infected cells. J Gen Virol 34: 557–561
Takatsuki A, Tamura G (1985) Brefeldin A, a specific inhibitor of intracellular translocation of vesicular stomatitis virus G protein: intracellular accumulation of high-mannose type G protein and inhibition of its cell surface expression. Agric Biol Chem 49: 899–902
Tartakoff AM (1983) Perturbation of vesicular traffic with the carboxylic ionophore monensin. Cell 32: 1026–1028
Tartakoff AM, Vassalli D (1978) Comparative studies of intracellular transport of secretory proteins. J Cell Biol 79: 694–707
Wasmoen TL, Kakach LT, Collett MS (1988) Rift Valley fever virus M segment: cellular localization of M segment-encoded proteins. Virology 166: 275–280

Welch WJ, Sefton BM (1980) Characterization of a small, nonstructural viral polypeptide present late during infection of BHK cells by Semliki Forest virus. J Virol 33: 230–237
Wikstrom L, Persson R, Pettersson RF (1989) Intracellular transport of the G1 and G2 membrane glycoproteins of Uukuniemi virus. In: Kolakofsky D, Mahy B (eds) Genetics and pathogenicity of negative strand virus. Elsevier, New York, pp 33–41
Wirth DF, Katz F, Small B, Lodish HF (1977) How a single Sindbis virus mRNA directs the synthesis of one soluble protein and two integral membrane glycoproteins. Cell 10: 253–263
Yoshida J, Nagai U, Yoshii S, Maeno K, Matsumoto T (1976) Membrane (M) protein of HVJ (Sendai virus): its role in virus assembly. Virology 71: 143–161

Antigenic Variance

L. KINGSFORD

1 Introduction

Antigenic variance may be defined as phenotypic expressions that are due to slight differences in gene sequences coding for viral proteins with antigenic reactivity (CALISHER 1988). Evidence for antigenic variance due to mutations and reassortment in the Bunyaviridae family is great and has been obtained by several different approaches. The main intent of this review, however, is not to discuss all the work that has led to the current proposed classification or to review all the papers describing antigenic variance of viruses in the family but to show how the use of monoclonal antibodies, oligonucleotide mapping, sequencing, etc., has led to a better understanding of the relationships of the viruses within the various genera. It is also hoped that one will acquire a better understanding of the need to examine and compare these viruses at all levels from gene coding strategies and replication through structure and function of

Department of Microbiology, California State University, Long Beach, CA 90840, USA

Current Topics in Microbiology and Immunology, Vol. 169

viral proteins. One is referred to articles by BEATY et al. (1988) and by BISHOP and BEATY (1988) for in-depth reviews on arbovirus variation and evolution due to genetic drift (accumulation of point mutations) and shift (RNA segment reassortment), to CALISHER (1988) for evolution and interrelationships among various *Bunyavirus* serogroups, and to ELLIOTT (1990) for the molecular biology of the Bunyaviridae.

2 Classification and Serological Relationships of Members Within the Bunyaviridae Family

Serological tests have been valuable in determining relationships among the more than 300 members of the family Bunyaviridae (KARABATSOS 1985) and have been the primary tools used to study antigenic variance and relatedness. Antigenic relationships among these viruses and thus their classification into genera and serogroups have most commonly been established based on complement fixation (CF), hemagglutination inhibition (HI), indirect immunofluorescent antibody (IFA), and various types of neutralization tests (NT; mouse and cell culture) as well as others (HUNT and CALISHER 1979; BISHOP 1985a). However, each test has its advantages and disadvantages (CASALS and WHITEMAN 1961; HUNT and CALISHER 1979; SWANEPOEL et al. 1986; TESH et al. 1982; BLACKBURN et al. 1987; CASALS and TIGNOR 1974, 1980).

A member of a serogroup is neutralized by homologous antisera and (with different specificities) by antisera specific to other members of that serogroup. With a few exceptions, members of one serogroup are not neutralized by antisera made to members of another serogroup (BISHOP 1985a). Viruses within each serogroup that are easily distinguished by serological assays are considered types (species). Subtypes include viruses which are only slightly different by standard methods; differences are less than those separating serotypes (e.g., a one-way antigenic difference). Those with very slight but significant, repeatable cross-reactions are considered varieties or, if not easily distinguishable, strains (CALISHER 1983; SHOPE 1985).

Relationships among the viral glycoproteins are detected by HI and NT assays while CF tests detect relationships among the more generally conserved nucleocapsid proteins (SHOPE 1985; BISHOP and BEATY 1988). Thus, viruses within a genus have generally been categorized into serogroups based on CF reactions, while HI and NT tests have been used to divide members further within the serogroups. However, the usefulness of various serological assays for the phleboviruses differs as the N protein is conserved to a lesser extent than the G1 and G2 glycoproteins among serogroups of the *Phlebovirus* genus (SHOPE 1985; PIFAT et al. 1988). Results similar to those in the phleboviruses are now being reported for different isolates of tomato spotted wilt virus, a newly proposed member of the family (LAW and MOYER 1990).

For several years now there have been five genera recognized within the family: *Bunyavirus*, *Hantavirus*, *Nairovirus*, *Phlebovirus*, and *Uukuvirus*

(KARABATSOS 1985; BISHOP et al. 1980; BISHOP 1985a,b; MCCORMICK et al. 1982; WHITE et al. 1982; SCHMALJOHN and DALRYMPLE 1983). For the most part, a comparison of RNA sequence data has supported the separation of viruses into these genera (CLERX-VAN HAASTER et al. 1982; ESHITA and BISHOP 1984; IHARA et al. 1985; COLLETT et al. 1985; LEES et al. 1986; SCHMALJOHN et al. 1987; RONNHOLM and PETTERSON 1987; GRADY et al. 1987; KAKACH et al. 1988; PARDIGON et al. 1988; SIMONS et al. 1990). However, it has been suggested that the *Phlebovirus* and *Uukuvirus* genera could be merged into a single genus as a comparison of the replication strategies and protein homology (N and NS_S) indicates that they are evolutionarily related (SIMONS et al. 1990). As noted above, a plant virus with properties of the Bunyaviridae family has been characterized and a new genus *Phythrivirus* has been proposed for it (DE HAAN et al. 1989). In addition to viruses which have been registered and accepted as members of the Bunyaviridae family, there are many which have been proposed as possible members (BISHOP 1985a,b; ZELLER et al. 1989) including new isolates such as Clo Mor in the *Nairovirus* genus and St Abb's Head in the *Uukuvirus* genus (WATRET and ELLIOTT 1985; WATRET et al. 1985); Arboledas (TESH et al. 1986), Corfou (RODHAIN et al. 1985), and five new phlebotomus fever virus serotypes (TESH et al. 1989) in the *Phlebovirus* genus; and Fort Sherman (MANGIAFICO et al. 1988), Brus Laguna (CALISHER et al. 1988a), and Erve (CHASTEL et al. 1989) in the *Bunyavirus* genus. The most recent proposed classification for the Bunyaviridae family is presented in the "Appendix." In addition to what is shown in the "Appendix", the new genus *Phythrivirus*, containing one member, tomato spotted wilt virus, has been proposed. This is a summary of the report submitted by the Bunyaviridae Study Group of the International Committee for Taxonomy of Viruses for review at the International Congress for Virology in Berlin, August 1990 (C.H. CALISHER, Chair of the Study Group, personal communication).

The classification of viruses in the family has been no small feat considering both the number of viruses that have been isolated and the degree of cross-reactivity and interrelationships among them and has taken considerable time, effort, and expense. Classification into the genera seems reasonable for the most part, but when it comes to determining serogroups (groups), complexes, virus species, subtypes, variants, varieties, and strains, the terms have been used rather loosely in the past, and it is very difficult in many instances to know whether the term used to describe the isolate is indeed correct (CALISHER et al. 1988c; CALISHER 1988). Definitive placement in taxa is made by international bodies, such as the International Committee for Taxonomy of Viruses (CALISHER et al. 1988c).

The classification of Bunyaviridae members in itself is in evolution and will continue to be so as long as the viruses continue to evolve, and there are virologists interested in pursuing taxonomy. Serology along with morphological and biochemical characteristics were first used for classification. With the advent of monoclonal antibodies, it would be hoped that the detection of specific shared epitopes would greatly help in sorting out many of the inconsistencies and errors based on conventional serology. However, a panel of monoclonal antibodies may not recognize all epitopes on a given protein, and only certain proteins for a few viruses have been studied. In addition, BISHOP (1985a) also points out that "Because a monoclonal (sic) represents only one of usually several 'neutralizing' epitopes, some monoclonals will be obtained that neutralize viruses that have otherwise diverged beyond the capabilities of recombination inter se." Therefore, cross-neutralization tests and other serological tests can only be used as an indicator of genetic relatedness and are not sufficient per se. Genetic tests in the form of reassortment and complementation analyses can be used to delineate a virus species, and these must also be performed (BISHOP 1985a). BISHOP (1985a) has put forth a rationale for the genetic basis for the classification of Bunyaviridae members.

Except for the rodent-associated hantaviruses, the rest of the members of the family are transmitted by arthropod vectors. These viruses replicate in their invertebrate hosts without causing cytopathology or any overt effects that have been documented. Indeed, often these viruses set up persistent infections in their arthropod hosts (BEATY and BISHOP 1988). These vectors contribute to the dissemination, maintenance, overwintering, and amplification of the virus (BEATY et al. 1988). ELLIOTT and WILKIE (1986) have shown that Bunyamwera virus can establish a persistent infection in mosquito cells (*Aedes albopictus* C6/36) and that the viruses generated have variable genetic and phenotypic characteristics. Thus, the potential is great for emergence of new strains of viruses due to genetic mutations (genetic drift) in a persistently infected arthropod and by reassortment if multiple virus infections occur (see reviews by BEATY and BISHOP 1988; BISHOP and BEATY 1988; CALISHER 1988). Selective pressures brought about by the immune response of the vertebrate host or by "replicative mechanisms congenital to the arthropod vector" may also be factors involved in changes (CALISHER 1988). This evolution has led to different virus types which occupy different ecological niches and to changes in virus coding strategies (BISHOP 1985a).

IROEGBU and PRINGLE (1981) studied the frequency of reassortment between three antigenically distinct and geographically isolated members of the Bunyamwera group. They found that there was no genetic barrier to exchange of genetic material between these three viruses and concluded that "the Bunyamwera complex is potentially a single gene pool if geographical and ecological constraints are discounted." As reviewed by CALISHER (1988), restriction of a virus to vertebrate-vector pairing with defined geographical distribution probably leads to natural isolation and genetic stability.

3 Antigenic Variance as Detected by Polyclonal Antisera

There is a great deal of evidence for natural antigenic variance, and most of it has come from studies such as the following serological ones involving viruses isolated within a geographical area. Two viruses isolated in 1975 in a site in

Ecuador were found to be variants of Pueblo Viejo virus (Gamboa group) which was isolated from the same study site in 1974 (CALISHER et al. 1981). Likewise, Panamanian workers found that not all viruses isolated at a single site were antigenically identical to the prototype Gamboa virus strain MARU 10962 (CALISHER et al. 1988a). Because multiple Gamboa serogroup isolates obtained from a study site possessed correspondingly diverse antigenic characteristics, CALISHER et al. (1988a) proposed that Gamboa virus evolution must be occurring at these sites. In support of this is the fact that of nine Gamboa group viruses that have been adequately tested for identity with other Gamboa group viruses, eight of them are distinct (CALISHER et al. 1988a). The results of the latter work are based on plaque reduction neutralization tests (PRNT) where viruses are considered distinct if titer differences are fourfold or more in both directions as compared with homologous titers.

Maguari virus in the genus *Bunyavirus*, Bunyamwera serogroup, is considered a subtype of Cache Valley virus. Since its isolation in 1957, Maguari virus or strains closely related to it have been found in several South American countries. Many of the Bunyamwera serogroup viruses have also been implicated in diseases of various animals. Thus, a study was undertaken to determine the antigenic relationships between virus strains related to Cache Valley and Maguari viruses. None of the isolates tested was identical to Cache Valley or Maguari virus or to each other. Three were considered to be subtypes of Cache Valley and one a variant of it. Another (Xingu virus) was determined to be a virus distinct from both Cache Valley and Maguari prototype viruses. These results indicates that multiple subtypes of Cache Valley virus exist and suggest that antigenic variation is a phenotypic expression of considerable genetic diversity (CALISHAR et al. 1988c).

A study was carried out in Florida to determine whether Maguari virus was present in the United States. Previously it was not known to occur north of Trinidad. However, all isolates were found to be strains or subtypes of prototype Tensaw virus and not subtypes of Maguari virus, although they cross-reacted with that virus (CALISHER et al. 1988b). In addition, there have been reports of 519 Bunyamwera group isolates from mosquitoes and mammals in Florida. All have been identified as strains of Tensaw virus (see CALISHER et al. 1986). Again on the basis of serological studies, there is clear evidence of genetic drift occurring in Tensaw virus in Florida.

4 Antigenic Variance as Detected by Monoclonal Antibodies

Polyclonal antisera contain antibodies that bind to many epitopes on any or all of the virus structural proteins. Depending on the virus and the way in which the antisera are made, there may be a different set of epitopes recognized by each antiserum preparation. Also, depending on the assay used to measure antigen-antibody reactions, different viral proteins (and epitopes) are detected. Thus, it makes it very difficult to look at antigenic variance and determine what is exactly being examined when polyclonal sera are used in the studies. It also makes it difficult to compare results from various laboratories or results obtained with different serum preparations and assays. One major advance that has led to a better understanding of antigenic variance and to the whole subject of antigen-antibody reactions is the production and use of monoclonal antibodies (MAbs).

MAbs have now been made to three of the four viral structural proteins (G1, G2, and N) and one nonstructural protein. Most have been made with specific goals in mind, such as to study the structure and function of viral glycoproteins (KINGSFORD et al. 1983; KINGSFORD 1984; DANTAS et al. 1986; ARIKAWA et al. 1989; DALRYMPLE et al. 1981; KEEGAN and COLLETT 1986; PIFAT et al. 1988), to determine the phenotype of reassortant viruses (GONZALEZ-SCARANO et al. 1982), to aid in classification of Bunyaviridae members (GONZALEZ-SCARANO et al. 1982; VAN DER GROEN et al. 1987; JAMES and MILLICAN 1986; KINGSFORD and BOUCQUEY 1990; FRANKO et al. 1983; ARIKAWA et al. 1989), to select for variants and to characterize them (SUNDIN et al. 1987; JAMES and MILLICAN 1986), to aid in the identification of viruses associated with specific diseases (VAN DER GROEN et al. 1987; BLACKBURN et al. 1987; SALUZZO et al. 1989a, b), or to screen for new Bunyaviridae isolates (VAN DER GROEN et al. 1987; BLACKBURN et al. 1987). In addition to these primary goals, much has been learned about the number and location of antigenic sites on viral proteins and about the antigenic variance of these viruses.

4.1 Monoclonal Antibodies to Viral Nucleocapsid and Nonstructural Proteins

Of all the MAbs made to California group viruses in the genus *Bunyavirus*, relatively few have been obtained which are specific for nucleocapsid (N) proteins. KINGSFORD et al. (1983) found only one antibody to the La Crosse (LAC) N protein in over 100 clones screened (unpublished results), and GRADY et al. (1983) obtained none. Of 23 MAbs made to Tahyna (TAH) and LAC viruses by GONZALEZ-SCARANO et al. (1982), only eight were specific for N proteins. As expected, they lacked NT and HI activity but could be divided into two groups based on whether they bound preferentially to the homologous virus or cross-reacted with other California group viruses. Two of the antibodies were found to be valuable for detecting reassortant viruses (GONZALEZ-SCARANO et al. 1982).

In contrast to the above studies, BLACKBURN et al. (1987) obtained only MAbs reactive with the N protein of Crimean-Congo hemorrhagic fever (CCHF) virus (a nairovirus). Seven antibodies were obtained, and all were nonneutralizing. None was tested for conventional HI activity. Five of the seven antibodies reacted with high titers to 22 strains of CCHF only and failed to cross-react with closely related Hazara virus or with other nairovirus isolates in Africa. The sixth antibody cross-reacted with Qalyub virus, a member of a different serogroup. The seventh antibody reacted only with CCHF strains but gave a lower IFA titer with one of the strains, indicating that antigenic variance might have occurred in its corresponding epitope.

When 105 hybrid clones were screened for antibodies to Punta Toro (PT) virus antigens (a phlebovirus), 90 were directed against the N protein, while four reacted with G2, one reacted with G1, and ten could not be characterized. None of the anti-N antibodies exhibited NT or HI activities, and none cross-reacted with Rift Valley fever virus (RVFV), an antigenically related virus (DALRYMPLE et al. 1981). In a subsequent study, immunization and screening procedures were designed to increase the number of antibodies reacting with glycoproteins. Although 27 glycoprotein-specific clones were obtained, most were still directed at N (PIFAT et al. 1988). This highly proportionate number of anti-N antibodies has also been observed with RVFV as well as other members of the family (PIFAT et al. 1988). When the MAbs were assayed (IFA) with 35 heterologous phleboviruses, five of nine selected anti-N antibodies reacted with two other viruses (Buenaventura or Karimabad; PIFAT et al. 1988).

Eighteen MAbs mapped to two sites on N and two antibodies reacting with a nonstructural protein (NSP-31) of RVFV (SALUZZO et al. 1989a, b) were used in an IFA assay to compare strains of RVFV. These antibodies (along with ones directed against the glycoproteins) indicated a high degree of epitope conservation. One anti-N MAb could differentiate all of the Egyptian RVFV strains from those isolated in sub-Saharan Africa.

4.2 Monoclonal Antibodies to Virus Glycoproteins

MAbs have been made to glycoproteins of viruses within three genera: *Bunyavirus*, *Hantavirus*, and *Phlebovirus*. Although all viruses in these genera have two glycoproteins, G1 and G2, they cannot be considered as analogous for each genus (see ELLIOTT 1990). The terms G1 and G2 have been assigned on the basis of relative electrophoretic mobilities of the proteins in polyacrylamide gels, G1 having a higher molecular weight. However, the coding order for these two glycoproteins on the M RNAs are not the same for all genera (Fig. 1). Even within one genus (*Phlebovirus*), the G1 glycoproteins are not analogous because, again, name assignments were based on molecular weights. In addition, the sizes of G1 compared with the G2 glycoproteins are very similar for phleboviruses and uukuviruses, while the G1 glycoprotein of bunyaviruses, for example, is a little over three times the size of the G2 (see ELLIOTT 1990). For these reasons, the data available for the glycoproteins are considered separately for viruses of each genus.

4.2.1 *Bunyavirus* Genus

Because the three laboratories that have produced MAbs to LAC have characterized and mapped these antibodies by different approaches, it is not possible to merge the results completely. Also, each laboratory used prototype LAC as antigen, but the virus stocks had different passage histories. These

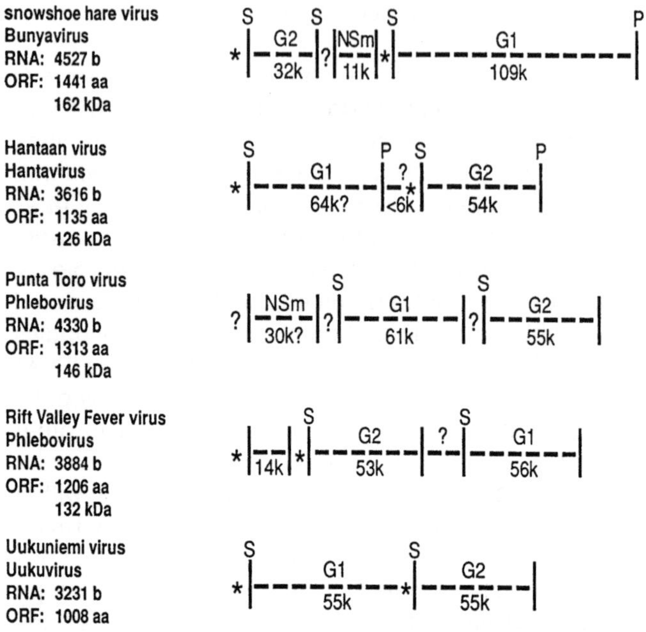

Fig. 1. Comparative organization of the M RNA segment among different genera of the Bunyaviridae (from FAZAKERLEY et al. 1988). Molecular weights are based on the deduced amino acid sequence of the ORF and the amino and carboxy termini of the proteins as given in the appropriate references. *S*, Position determined by protein sequence; *P*, position determined by reactivity of anti-peptide sera; *, signal sequence-like region; ?, number of proteins and/or molecular weight unknown. References: Hantaan virus (SCHMALJOHN et al. 1987, 1986), Punta Toro virus (IHARA et al. 1985; ROBESON et al. 1979), Rift Valley fever virus (COLLETT et al. 1985; KAKACH et al. 1988), Uukuniemi virus (RONNHOLM and PETTERSON 1987; KUISMANEN 1984)

viruses could represent mutants in certain epitopes, as is discussed below (KINGSFORD and BOUCQUEY 1990). In addition, antigens for MAb production were prepared differently by each group. Thus, it is not surprising that antibodies with different specificities, avidities, and characteristics such as the ability to neutralize or inhibit hemagglutination (HA) have been found by each group. However, a considerable amount of work has been done to characterize the LAC G1 glycoprotein with these MAbs, and a picture is emerging as to its function and structure. This information is covered below in context with antigenic variance, as it provides the basis for understanding how various bunyaviruses are serologically related, and where antigenic variance occurs at the glycoprotein level.

Two basic approaches have been used to map epitopes and domains on the G1 glycoproteins of bunyaviruses. The first is competitive binding assays (GRADY et al. 1983b; KINGSFORD et al. 1983), and the second is selection of variant viruses (NAJJAR et al. 1985). Both approaches have limitations but have yielded results for LAC virus that are in good agreement.

MAbs to G1 have been made for two California group viruses, LAC (GONZALEZ-SCARANO et al. 1982; GRADY et al. 1983a; KINGSFORD et al. 1983) and TAH (GONZALEZ-SCARANO et al. 1982). None of the workers obtained antibodies to the G2 glycoprotein. When characterized in PRNT, HI, and enzyme-linked immunosorbent assay (ELISA) tests, a close correlation of HI and NT activities was seen for the MAbs to both LAC and TAH viruses. Almost all antibodies with NT activity also inhibited HA, and there was a good correlation of the titers. This indicates that G1 is the molecule which clearly acts as the viral hemagglutinin, and that the anti-receptor for erythrocytes is the same or close to that which binds to cells in culture (GONZALEZ-SCARANO et al. 1982; GRADY et al. 1983a; KINGSFORD et al. 1983). Two groups (GONZALEZ-SCARANO et al. 1982; GRADY et al. 1983a), however, found a few antibodies with HI but no NT activity, indicating that the binding sites, although overlapping, may differ somewhat in structure or function (GONZALEZ-SCARANO et al. 1982). MAbs were also found that bound to G1 but lacked NT or HI activities (GONZALEZ-SCARANO et al. 1982; GRADY et al. 1983a; KINGSFORD et al. 1983).

For LAC virus, at least, studies with polyclonal antibody indicate that the G1 glycoprotein is the only one involved in NT in BHK cells while polyclonal antisera to both G1 and G2 can inhibit HA (KINGSFORD and HILL 1983). In this latter aspect, LAC glycoproteins are similar to those of hantaviruses and phleboviruses (DALRYMPLE et al. 1981; DANTAS et al. 1986; ARIKAWA et al. 1989). Further elucidation of the role of LAC G2 in HA (and possibly NT) awaits the production of MAbs to G2.

G2, however, has been hypothesized to be involved in receptor binding to midgut cells of mosquitoes (LUDWIG et al. 1989). When intact virus is treated with trypsin, G1 is cleaved leaving G2 intact (KINGSFORD and HILL 1981, 1983; GONZALEZ-SCARANO 1985; LUDWIG et al. 1989). This protease treatment greatly reduces infectivity of LAC in vertebrate cells (KINGSFORD and HILL 1983) because of reduced binding (LUDWIG et al. 1989) while it increases attachment (although not infection rates) to mosquito cells and appears to be a prerequisite for detectable attachment of virus to mosquito midgut cells (LUDWIG et al. 1989). G2 has also been implicated as the actual protein involved in fusion of virus to vertebrate cells (POBJECKY et al. 1989).

The 23 MAbs to the G1 glycoprotein of LAC made by Kingsford and coworkers (KINGSFORD et al. 1983; KINGSFORD and BOUCQUEY 1990) bound to 16 epitopes located within eight antigenic regions A–H. These were determined by competition binding and by assay on serologically related California serogroup viruses. Antibodies to six regions (A, B, D, E, F, G) had HI and NT activities. The single antibody to region C did not inhibit HA and neutralized virus only at high virus and antibody concentrations, possibly due to aggregation (KINGSFORD et al. 1991). The antibodies to epitopes in region H lacked both HI and NT activities (Table 1). There is considerable overlap of antigenic regions A–E and some overlap for regions F–G, with no overlap between regions E and F (KINGSFORD et al. 1983). Thus, these regions may comprise one, but more likely two domains (A–E and F–H). Each domain contains neutralizing and

Table 1. Characteristics of hybridoma cell lines and their antibodies specific for the G1 glycoprotein of La Crosse virus. (Modified from KINGSFORD et al. 1983)

Antigenic[a] region	Epitope[a]	Hybrid cell line	Immunizing agent[b]	Isotype	Binds to trypsinized La Crosse[c]	ELISA titer[d]	Neutralization titer[d]	HI titer[d]	Enhances antibody binding to epitope (s)[e]
A	a_1	6-4B5	LAC	IgG1	–	1:8192	1:104	1:1024	C
	a_2	6-6B4	LAC	IgG1	–	1:4096	1:256	1:256	C
B	b_1	1-6A11	G1	IgG1	–	1:1024	1:16	1:2	G
		1-6A2	G1	IgG1	–	1:256	1:8	1:2	G
	b_2	6-4C5	LAC	IgG2a	–	1:512	1:8	1:4	G
C	c_1	1-4D11	LAC	IgG2a	–	1:256	<1:2	<1:2	A, G
D	d_1	5-10B7	LAC	IgG2a	–	1:4096	1:64	1:32	G
	d_2	3-10B1	LAC	IgG2a	–	1:4096	1:512	1:256	G
E	e_1	6-1A2	LAC	IgG2a	–	1:8192	1:256	1:128	—
		6-1A6	LAC	IgG2a	–	1:8192	1:512	1:128	—
F	f_1	1-4D14	LAC	IgG2a	–	1:8192	1:16,384	1:4096	—
		6-5C4	LAC	ND	–	1:128	ND	1:2	—
	f_2	3-1A7	LAC	IgG2a	–	1:4096	1:1024	1:128	—
		3-2G2	LAC	IgG2b	–	1:8192	1:8192	1:128	—
G	g_1	1-2C3	G1	IgG1	–	1:32	1:4	<1:2	B, C, D
		1-4B5	G1	IgG1	–	1:2048	1:64	1:16	B, C, D
	g_2	1-6C3	LAC	IgG2a	+	1:4096	1:8192	1:512	B, C, D
		1-6B2	G1	IgG1	+	1:4096	1:1024	1:128	B, C, D
	g_3	1-2D1	G1	IgG1	+	1:4096	1:1024	1:128	B, C, D
	g_4	6-5B14	LAC	ND	+	1:32	1:8	1:4	B, C, D
H	h_1	1-4D8	G1	IgG1	+	1:2048	<1:2	1:4	—
		6-4D8	LAC	IgG1	+	1:4096	<1:2	<1:2	—
	h_2	1-5D4	G1	IgG1	+	1:512	1:4	1:32	—

[a] Antigenic region and epitope designations are based on a competition binding assay and serological cross-reactions with other California group viruses

[b] Mice were immunized with infectious LAC virus or purified G1 glycoprotein

[c] Each antibody was assayed on trypsinized LAC by ELISA. A positive reaction (+) indicates the epitope is on the 95- to 100-kDa peptide remaining in the virion after typsinization, while a negative (–) reaction indicates the epitope is located on the 20- to 25-kDa region of G1 removed by the enzyme

[d] Titers in each of these assays were obtained using purified MAbs starting at a protein concentration of 400 μg/ml

[e] Enhancement of binding was determined in a competitive radioimmunoassay and indicates antibody binding at one epitope enhanced binding of antibody at a separate epitope by 50% or greater

nonneutralizing sites. These two domains are close enough that an IgM MAb that was isolated can compete with antibodies binding to sites within each domain (B, C, D, and G). Thus, the surface of the G1 molecule contains a complex array of overlapping antigenic determinants (KINGSFORD et al. 1983; KINGSFORD and BOUCQUEY 1990).

The above results are similar to those reported by NAJJAR et al. (1985) who found that epitopes on the LAC G1 glycoprotein are arranged in five distinct clusters which are interrelated and may be part of a single immunodominant domain. Their results were based on serological cross-reactions with California group viruses and on assays of variant viruses with the selecting neutralizing antibodies.

GRADY et al. (1983b) also examined the spatial relationships of binding sites on G1 of LAC. They found three nonoverlapping antigenic sites. One site contained three overlapping epitopes within it which could be differentiated based on cross-reactions with other California group viruses (see Table 2). However, only five antibodies were employed in this study and thus represent only a few of the G1 epitopes.

The antigenic regions as mapped by KINGSFORD and coworkers (KINGSFORD and HILL 1983; KINGSFORD et al. 1983) are located on the part of the LAC G1 glycoprotein that is distal from the transmembrane region. This must be the aminoterminus, as mapping of the G1 glycoprotein (FAZAKERLEY et al. 1988; D.W. HILL and L. KINGSFORD, unpublished data) to the deduced M polyprotein sequence of LAC (GRADY et al. 1987) and showshoe hare (SSH; ESHITA and BISHOP 1984) viruses indicates that the hydrophobic domain, and thus transmembrane region, of G1 is at the carboxy-terminal end.

When LAC virus is incubated with trypsin at a pH of 6.4 or above, 20–25 kDa of protein is removed leaving a 95- to 100-kDa peptide in the envelope (KINGSFORD and HILL 1983; GONZALEZ-SCARANO 1985). This cleavage site has been mapped directly for SSH virus (FAZAKERLEY et al. 1988). Assay of MAbs on enzyme-treated virus indicates that antigenic regions A through F and epitope g_1 of region G are either located on the 20-kDa region or are altered by trypsin. A peptide containing the epitope for g_1 can be found in the supernatant after trypsinized virus is pelleted. The rest of the G epitopes (g_2, g_3, g_4) as well as antigenic region H are located on the 95- to 100-kDa peptide (see Table 1; KINGSFORD et al. 1983, unpublished data).

As indicated above, when an IgM antibody was used in competition binding, it competed with MAbs to four other antigenic regions (B, C, D, and G; KINGSFORD et al. 1983). Thus, the glycoprotein may be folded back on itself so that the linear amino acid sequence comprising region G is antiparallel with that of regions B, C, and D. When G1 was reduced prior to assay with the MAbs, only antibodies to antigenic region B and epitopes g_1 and g_2 still bound in ELISA. This indicates that the majority of epitopes on G1 are probably discontinuous (L. KINGSFORD and L.D. ISHIZAWA, manuscript in preparation).

G1 of LAC has been shown to be involved in fusion (GONZALEZ-SCARANO et al. 1984) and undergoes an irreversible conformational change at pH 6.3

Table 2. Characteristics of monoclonal antibodies to the G1 glycoprotein of La Crosse virus. (Compiled from GRADY et al. 1983a, b; GRADY and KINCH 1985; by permission of L.J. GRADY)

Antigenic site[a]	Antibody	HI of LAC	NT of LAC on:		ELISA on North American members of California serogroup viruses							
			BHK Cells	*A. albopictus* cells	LAC	LAC (NY isolates)	SSH	CE	SA	JC	KEY	TVT
1	5D4	+	+	+	+	+	+	+	−	+	−	−
	3G8	+	+	−[b]	+	−	−	−	−	−	−	−
	7B11	+	+	+	+	+	−	−	−	−	−	−
2	1A7	−	−	−	+	+	+	−	−	−	+	−
3	2B9	+	−	+	+	+	+	+	+	+	+	+
ND	5D2	−	−	−	+	+	−	+	+	−	+	−

LAC, La Crosse; SSH, snowshoe hare; CE, California encephalitis; SA, San Angelo; JC, Jamestown Canyon; KEY, Keystone; TVT, trivittatus

[a] Three nonoverlapping antigenic sites have been mapped based on competition binding assay. However, designations for these sites have not been made by the authors and are provided here only for convenience. Epitopes in site 1 are overlapping

[b] Antibody did not neutralize LAC when assayed on the designated cell type

(GONZALEZ-SCARANO 1985). The major trypsin cleavage product of virus maintained at neutral pH is the 100-kDa peptide. However, if virus is exposed to a pH of 6.3 or less, a conformational change occurs. If the pH is neutralized and the virus then treated with trypsin, a 72-kDa peptide is left in the viral membrane instead of the 100-kDa peptide. When the MAbs of GONZALEZ-SCARANO et al. (1983) were tested for binding to acidified virus, a group of antibodies forming a distinct cluster (one with high neutralizing ability and two with low neutralizing abilities; NAJJAR et al. 1985) exhibited greatly decreased binding, indicating that a conformational change had occurred (GONZALEZ-SCARANO 1985). In addition, one of these antibodies (high neutralizing) had been used to select variants. These variants demonstrated different pH requirements for the activation of fusion, suggesting that the corresponding region of G1 is involved in the fusion activity (GONZALEZ-SCARANO et al. 1985). Apparently G1 is not the actual fusion protein, however, as G1-containing liposomes cannot induce fusion whereas protease-treated virus (most of G1 but not G2 removed) can fuse liposomes (POBJECKY et al. 1989).

Kingsford and Ishizawa (manuscript in preparation) examined the binding of antibodies to G1 after the virus had been acid-treated (pH 6.3) and trypsinized. Of the antibodies previously binding to trypsinized virus (Table 1), only antibodies to epitope g_2 (one of the continuous epitopes) bound as well as it did to control virus. Antibodies to antigenic region H (nonneutralizing, possibly discontinuous) and epitopes g_3 and g_4 (neutralizing, possibly discontinuous) exhibited decreased binding, indicating that conformational changes may have disrupted these epitopes.

All three laboratories have utilized their MAbs to examine relatedness of the California group viruses. GONZALEZ-SCARANO et al. (1983) used this approach as a first step in characterizing the anti-G1 MAbs to LAC and TAH viruses. Based on HI, NT, and ELISA reactions, they found four groups, designated A, B, C, and D. Groups A and B had high NT and HI titers, with A antibodies being virus specific and B antibodies cross-reactive with heterologous viruses. One exception in group A was an antibody that bound not only to the homologous LAC but also to SSH, a virus known to be closely related antigenically to LAC (BISHOP and SHOPE 1979). Group C and D antibodies had no HI or NT activity with one exception which did not neutralize but had HI activity. Group C antibodies bound preferentially to the homologous virus in an ELISA (usually at least tenfold greater) while the one group D antibody (TAH anti-G1) was cross-reactive and had a high HI titer against all heterologous viruses tested (see Tables 3, 4).

When used in ELISA and HI tests, these MAbs could effectively demonstrate shared antigenic determinants by all 11 members of the California group of viruses tested. NT appeared to be the most useful test for separating and grouping the individual members within the serogroup. The MAbs neutralized the six viruses of the California encephalitis (CE) complex, except that LAC antibodies failed to neutralize CE virus. When tested with viruses belonging to the Melao or trivittatus (TVT) complexes, both LAC and TAH MAbs neutralized Jamestown Canyon (JC) and TVT viruses as frequently as they neutralized

Table 3. Monoclonal antibodies against the G1 proteins of La Crosse and Tahyna viruses: neutralizing activity against 11 California serogroup viruses.[a] (From GONZALEZ-SCARANO et al., Liss Inc., New York, 1983)

Immunizing virus	Group	Clone no.	CE						MEL				TVT
			LAC	SSH	TAH	CE	SA	INK	JC	MEL	KEY	SDN	
LAC	A	807–09	>2.8	—	—	—	—	—	—	—	—	—	—
		807–15	1.6	—	—	—	—	—	—	—	—	—	—
		807–18	>2.8	—	—	—	—	—	—	—	—	—	—
		807–35	>2.8	—	—	—	—	—	—	—	—	—	—
		807–31	>2.5	>2.5	—	—	—	ND	—	ND	—	ND	—
	B	807–12	>2.5	>2.5	>2.5	—	>2.5	ND	>2.5	ND	—	ND	>2.5
		807–22	1.9	1.9	—	—	—	1.3	1.9	—	—	—	1.6
		807–33	>2.8	>2.8	1.3	—	—	>2.8	>2.8	—	—	—	2.5
	C	807–25	—	—	—	—	—	—	—	—	—	—	—
		807–26	—	—	—	—	—	—	—	—	—	—	—
		807–21	—	—	—	—	—	—	—	—	—	—	—
TAH	B	813–13	2.2	>2.8	>2.8	>2.8	>2.8	—	>2.8	1.6	—	>2.8	—
		813–48	1.3	1.9	1.6	1.6	1.3	1.6	1.6	—	—	1.3	1.9
		813–77	1.9	>2.8	>2.8	1.9	>2.8	—	1.6	—	—	2.5	—
	D	814–443	—	—	—	—	—	—	—	—	—	—	—

Titers as $\log_{10}$, —: less than 1.3

LAC, La Crosse; TAH, Tahyna; CE, California encephalitis; MEL, Melao; SSH, snowshoe hare; SA, San Angelo; INK, Inkoo; JC, Jamestown Canyon; KEY, Keystone; SDN, Serra do Navio; TVT, trivittatus

[a] Neutralization titers were determined in plaque reduction assay

Table 4. Monoclonal antibodies against the G1 proteins of La Crosse and Tahyna viruses: hemagglutination inhibition activity against California serogroup viruses.[a] (From GONZALEZ-SCARANO et al. Liss Inc., New York, 1983)

Immunizing virus	Group	Clone no.	CE					MEL		TVT
			LAC	SSH	TAH	CE	SA	JC	KEY	
LAC	A	807–09	2.8	—	—	—	—	—	—	—
		807–15	2.6	—	—	—	—	—	—	—
		807–18	3.2	—	—	—	—	—	—	—
		807–35	2.6	—	—	—	—	—	—	—
		807–31	3.2	3.2	—	—	—	—	—	—
	B	807–12	3.8	4.4	3.8	3.2	2.6	4.4	—	2.6
		807–22	3.2	3.8	3.5	3.8	3.2	3.8	3.2	—
		807–33	3.2	3.8	2.6	2.6	2.3	3.8	—	2.6
	C	807–25	—	2.0	—	—	—	—	—	—
		807–26	—	—	—	—	—	—	—	—
		807–21	—	—	—	—	—	—	—	—
TAH	B	813–13	2.3	3.8	3.8	3.8	3.2	2.6	—	—
		813–48	3.2	3.2	3.2	3.2	3.2	3.2	2.6	2.6
		813–77	2.6	2.6	4.4	2.6	2.6	2.0	—	—
	D	814–443	3.2	3.8	3.2	3.2	3.2	3.2	—	2.6

Titers as $\log_{10}$, —: less than 1.3
CE, California encephalitis; MEL, Melao; LAC, La Crosse; SSH, snowshoe hare; TAH, Tahnya; SA, San Angelo; JC, Jamestown Canyon; KEY, Keystone; TVT, trivittatus
[a] Infected suckling mouse brain antigens were prepared by sucrose-acetone extraction and sonication

viruses within the CE complex (Table 3). The results were essentially the same in an HI assay except that group B and D antibodies inhibited HA of CE virus (Table 4). These data generally support the conventional classification of California group viruses into the three complexes CE, Melao, and TVT with the main exception of JC virus, which appears to be more closely related to the CE complex than to the Melao complex (GONZALEZ-SCARANO et al. 1982, 1983).

With a different panel of MAbs, a different assay, and a different set of data, the same conclusions were made by KINGSFORD and BOUCQUEY (1990). The cross-reactivity of MAbs to G1 of LAC were determined in ELISA with the same 11 California group viruses (Table 5). Both CE and JC viruses exhibited patterns of cross-reactivity somewhat different from other members of their respective complex. Present classification of the CE complex within the group is based on the results of several different serological assays with polyclonal antisera which probably detect antigenic determinants of different viral proteins and lack the sensitivity and specificity of MAbs. Although the data obtained by GONZALEZ-SCARANO et al. (1982, 1983) and KINGSFORD and BOUCQUEY (1990) focus only on G1, they indicate that some reclassification may be in order.

Considerable additional information can be obtained from the information in Table 5 relative to the number of epitopes on G1 and the conservation of them among the various California group viruses. Eight antigenic regions (A through H) on G1 were first mapped with the MAbs by competition binding. When these

Table 5. Reactions of LAC anti-G1 monoclonal antibodies with other viruses of the California group. (From Kingsford and Boucquey 1990)

Antigenic region[a]	Epitope[b]	Monoclonal antibody	California encephalitis[c]						Melao[c]				trivittatus[c]
			LAC	SSH	CE	SA	INK	TAH	MEL	JC	KEY	SDN	TVT
A	a_1	6-4B5	400 000[d]	160 000	–	+	+	+	–	+	–	–	+
	a_2	6-6B4	800 000	4 000	–	–	–	–	–	–	–	–	–
B	b_1	1-6A11	10 000	–	+	–	–	–	–	–	–	–	–
		1-6A2	50 000	–	+	–	–	–	–	–	–	–	–
		6-4C5	10 000	–	+	–	–	–	–	–	–	–	–
C	c_1	1-4D11	50 000	–	–	–	–	–	–	–	–	–	–
D	d_1	5-10B7	1 600 000	–	800	–	–	–	–	1600	+	–	–
	d_2	3-10B1	1 600 000	100 000	800	+	+	+	–	3200	+	+	–
E	e_1	6-1A2	1 600 000	8 000	–	–	–	–	–	–	–	–	–
		6-1A6	800 000	8 000	–	–	–	–	–	–	–	–	–
F	f_1	1-4D14	1 600 000	–	–	+	–	+	–	–	–	–	–
		6-5C4	1 000	–	–	+	–	+	–	–	–	–	–
	f_2	3-2G2	1 600 000	–	–	–	–	–	–	–	–	–	–
		3-1A7	3 200 000	–	–	–	–	–	–	–	–	–	–
G	g_1	1-2C3	10 000	500	400	200	400	100	100	400	400	100	400
		1-4B5	16 000	2 000	400	400	1000	1000	2000	1000	1000	100	800
	g_2	1-6C3	800 000	100	–	+	+	+	–	+	–	–	–
		1-6B2	3 200 000	8 000	–	4000	+	+	–	+	–	–	–
	g_3	1-2D1	200 000	–	–	+	–	–	–	+	–	–	–
	g_4	6-5B14	200	–	–	200	–	200	–	200	–	–	–
H	h_1	1-4D8	1 600 000	8 000	–	–	–	–	–	–	–	–	–
		6-4D8	100 000	100	–	–	–	–	–	–	–	–	–
	h_2	1-5D4	100 000	25 600	–	+	+	+	–	+	+	–	–

Ascitic fluid preparations of each LAC anti-G1 monoclonal antibodies were diluted serially and assayed with all viruses of the California group in ELISA. Comparative reciprocal titers are given for all positive reactions that were greater than fourfold over background. –, Reaction of antibody was negative at the lowest dilution that could be used without binding nonspecifically to the virus (range from 1:20 to 1:100). +, Positive reaction: antibodies positive only at a fourfold dilution over background

LAC, La Crosse; SSH, snowshoe hare; CE, California encephalitis; SA, San Angelo; INK, Inkoo; TAH, Tahyna; MEL, Melao; JC, Jamestown Canyon; KEY, Keystone; SDN, Serra do Navio; TVT, trivittatus

[a] Antigenic region designations are based on a two-way competition binding radioimmunoassay of 23 anti-G1 monoclonal antibodies with LAC virus

[b] Epitope designations within each antigenic region are based on the reactions of LAC MAbs in ELISA with the other California group viruses

[c] Complex groupings of viruses of the California group

antibodies were examined for their reactivity with other California group viruses, at least 12 distinct groups of antibodies were apparent. However, if in addition to the patterns of cross-reactivity, the groupings based on the competition binding assay are considered (regions A to H), there are at least 15 distinct groups of antibodies which define 15 epitopes (Tables 1, 5). When these antibodies were assayed on New York LAC isolates, one additional epitope was detected, giving a total of 16 (KINGSFORD and BOUCQUEY 1990).

Several conclusions regarding the interrelationships of the California group viruses can also be arrived at based on the data in Table 5. The only antibody that reacted equally well (same titer) with other viruses was the antibody to epitope g_4. This epitope is probably the only one detected in these assays that remains unaltered among the California group viruses. Epitope g_1 is the only group-specific epitope detected. However ELISA endpoint titers of antibodies to g_1 varied from 1:100 to 1:16 000, indicating that some genetic drift has occurred in this epitope. The regions most conserved among the members of the group are A (a_1 only), D, G, and H (h_2 only). The greatest conservation seems to be in epitopes g_1 and g_4. These epitopes are minimally involved in NT of LAC (less than 1:8 by PRNT and less than $\log_1$ in kinetic assays) and give little or no enhancement of NT. On the other hand, g_2 and g_3 seem to have undergone a great deal of genetic drift, possibly due to selective pressure by antibody. MAbs to these epitopes exhibit high NT titers on LAC (1:1024 or greater by PRNT and $\log_3$ or more in a kinetic assay) and are involved in enhancement of NT (KINGSFORD 1984; KINGSFORD and BOUCQUEY 1990).

Epitope c_1 was detected in these assays on the LAC G1 glycoprotein only and is possibly virus specific. Although epitope f_2 also appears to be virus specific for LAC, it may not be. The MAbs used in these experiments were made to a cloned stock of prototype LAC. When the antibodies were assayed on uncloned prototype LAC, it was apparent that the cloned stock was a variant in the f_2 epitope. Therefore, antibody to this epitope would not be expected to react with any other California group virus.

SSH seems to be the only group virus closely related to LAC. This supports what has previously been shown (BISHOP and SHOPE 1979). However, the studies with the MAbs indicate that changes have occurred in certain areas of the G1 glycoprotein (GONZALEZ-SCARANO et al. 1982, 1983; GRADY et al. 1983a; KINGSFORD and BOUCQUEY 1990). Thus, although SSH is considered a subtype of LAC (BISHOP 1985a, b). the G1 glycoprotein has undergone major antigenic changes.

In a study by EL SAID et al. (1979), New York isolates of LAC were found to be distinct from the prototype LAC strain by oligonucleotide mapping. However, polyclonal antisera could not consistently differentiate these viruses. When GRADY et al. (1983a) examined several independent isolates made in New York and the prototype LAC with their MAbs, they found at least one epitope involved in both NT and HA on the prototype virus that was either significantly altered or missing in the New York isolates (see Table 2). When two of these New York isolates (provided by L.J. GRADY) were assayed in ELISA with the panel of MAbs

Table 6. Antibodies specific for the G1 glycoprotein of LAC virus that show differences in titer on two New York isolates.[a] (From KINGSFORD and BOUCQUEY 1990)

Antigenic Region	Epitope	Monoclonal antibody	Immunizing antigen	La Crosse isolates		
				LAC	74NY LAC	76NY LAC
B	b_1	1-6A11	G1	512	1024	512
		1-6A2	G1	1024	1024	256
	b_2	6-4C5	LAC	256	16	4
C	c_1	1-4D11	LAC	128	256	1024
D	d_1	5-10B7	LAC	2048	16348	4096
F	f_1	1-4D14	LAC	8192	4	4
	f_2	3-2G2	LAC	4096	4	4
		3-1A7	LAC	16384	2	4
H	h_1	6-4D8	LAC	256	4096	4096

[a] Reciprocal titers are given only for antibodies that exhibited greater than fourfold differences with LAC compared to one or both of the New York isolates (except antibodies 1-6A11 and 1-6A2 which are given for comparison with antibody 6-4C5). Purified MAbs starting at a protein concentration of 400 μg/ml were used in the assay

made by KINGSFORD et al. (1983), it was apparent that variation does occur, and that it is primarily in the F antigenic region (Table 6). When compared with the uncloned reference strain, the cloned LAC used as antigen to make the MAbs is different in the f_2 epitope, and both New York isolates are variants in the f_1 epitope. Because the antibodies used were specific for cloned LAC which has mutated in f_2, it was not possible to determine whether f_2 has also changed on the New York isolates. Of three antibodies which bind to antigenic region B, only two reacted equally well with the New York isolates. This was the basis for determining that there were two overlapping epitopes in region B. This also indicates that there is a second area of the G1 glycoprotein where mutations have occurred in various LAC isolates. In fact, when the data in Table 5 are examined, antibodies to epitopes in antigenic regions B and F (along with C) are the ones least reactive with all the California group viruses, indicating that these two regions may be areas where the greatest number of mutations have occurred (KINGSFORD and BOUCQUEY 1990).

When the neutralizing anti-G1 MAbs to LAC (KINGSFORD et al. 1983) were examined in greater detail in kinetics of neutralization studies, several observations were made, (a) No single antibody neutralized homologous virus to the same extent as did polyclonal antibody, and the degree of NT for the antibodies varied from $\log_{1.5}$ to $\log_3$ in antibody excess. For antibodies of high affinity, this was characteristic of the epitope and not of the antibody (KINGSFORD et al. 1983; KINGSFORD 1984). Further viral infectivity of $\log_2$ to $\log_3$ could be neutralized by the addition of polyclonal anti-G1 antibody or a second MAb (KINGSFORD 1984). (b) NT titers obtained by a PRNT assay did not correlate completely with the rates and extent of NT shown in kinetic reactions. For example, antibody to epitope b_1 in region B had a titer of 1:6 by PRNT. Yet, this antibody neutralized more virus in a kinetics of NT reaction than did antibody to epitope f_1 of region F which had a PRNT titer of 1:16 384. Overall, antibodies to epitopes within regions A, F, and G gave the highest NT titers in a 50% endpoint assay, while antibodies to the a_1 and g_2 epitopes neutralized virus to the greatest extent in a kinetics to NT assay ($\log_3$ to $\log_{3.5}$; KINGSFORD et al. 1983; KINGSFORD 1984; KINGSFORD and ISHIZAWA 1984). (c) Enhanced NT was observed for certain pairs of antibodies. These included antibodies to epitope g_2 plus any other neutralizing antibody outside the G region. One other pair included antibodies to epitopes f_1 and b_1 (KINGSFORD and ISHIZAWA 1984; KINGSFORD 1984; KINGSFORD and BOUCQUEY 1990). Most of the antibody pairs which exhibited enhanced NT were also synergistic in binding in a competitive radioimmunoassay (KINGSFORD 1984). Certain sites seem to be critical for NT, and if antibodies are not bound to them, the degree of NT is considerably less. This synergistic activity has also been shown by HI assays (unpublished data). (d) When Fab fragments were made for each of the MAbs, they were capable of neutralizing virus to the same degree as the bivalent form, although two to three times as much antibody was needed. However, enhancement of NT was not seen for pairs of Fab fragments (or combinations of Fab and bivalent antibodies) indicating that cross-linking by both bivalent antibody molecules is necessary (KINGSFORD et al. 1991).

In kinetics of NT assays, anti-LAC antibody inactivates heterologous California group viruses much less than the homologous virus. This includes SSH virus, which is considered to be a subtype of LAC (KINGSFORD and BOUCQUEY 1990). Figure 2 shows the kinetic reactions for polyclonal anti-LAC mixed with LAC, SSH, JC, San Angelo (SA), or TVT viruses. The SSH, JC, and SA viruses were the three with the most cross-reactive epitopes in ELISA (eight each), while there was only one positive and one weakly positive cross-reactive epitope on TVT (Table 5). Yet there is little difference in the curves of the NT reactions for the four viruses when compared with LAC virus. Considering the number of cross-reactive sites on the various viruses which are recognized by neutralizing MAbs (Table 5; regions A, B, D, E, F, G), it seems that a greater degree of antigenic relatedness should be apparent in view of the results of kinetics of NT assays. This is explained by observation that no single MAb neutralizes LAC to the same extent that polyclonal antibody does and by the conclusion that there are critical sites for NT on LAC. The areas most conserved among the viruses relative to LAC are the a_1, d_2, g_1, g_2, and h_2 epitopes. However, when only the strongly positve reactions are considered, primarily antibodies to g_1 and to a lesser extent those to epitopes d_1, d_2, and g_4 exhibit cross-reactions with many of the California group viruses. Although the epitopes involved in the greatest degree of NT (a_1 and g_2) were detected on several of the viruses, the ELISA titers were very low, indicating minimal cross-reaction. When kinetics of NT was assessed with the heterologous viruses and the neutralizing LAC MAbs that could bind to them, the degree of NT was very minimal (20%–50%) for all the antibodies. The virus that exhibited the greatest degree of NT (up to $\log_1$) was SSH with antibody to epitope d_2.

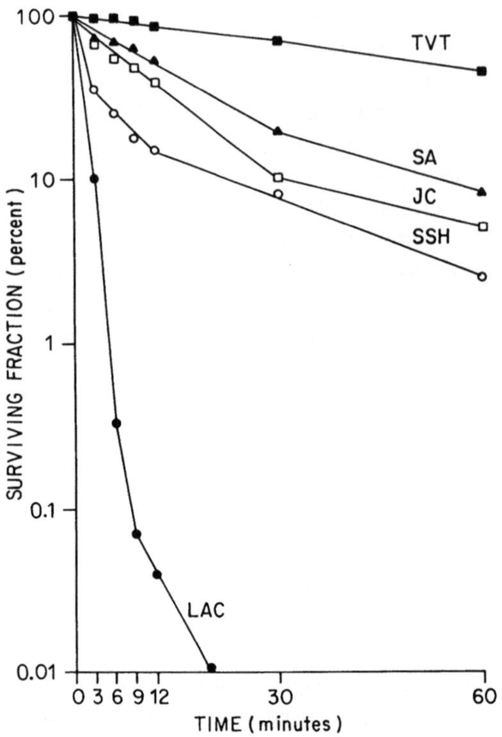

Fig. 2. Neutralization kinetics for trivitattus (*TVT*; ■—■), San Angelo (*SA*; ▲—▲), Jamestown Canyon (*JC*; □—□), snowshoe hare (*SSH*; ○—○), and La Crosse (*LAC*; ●—●) viruses mixed with polyclonal anti-LAC antiserum

As indicated in Table 7, certain pairs of MAbs enhance NT of LAC virus. Of these epitope pairs, 13 were present collectively on the G1 glycoproteins of SSH, SA, Inkoo, TAH, and JC viruses as determined by ELISA. However, when examined in a kinetic reaction for the ability to enhance NT of the heterologous virus, only SSH with one pair of MAbs exhibited any degree of enhancing effect (KINGSFORD 1984).

Thus, based on the above, one explanation for why the California group viruses are not neutralized to a greater degree with the individual G1 MAbs is that changes in the epitopes have occurred such that antibodies have a very low affinity for them. This may be due to electrostatic or conformational changes in the glycoprotein as a direct result of genetic drift in the M RNA gene segment. Thus, antibody may bind (as detected in ELISA) but may not result in NT. However, this does not completely explain why greater degrees of NT did not occur when a pair of LAC MAbs or polyclonal serum was mixed with another California group virus. A second factor involved may be that antibody either does not bind or binds only at a low level to critical sites on the glycoprotein. The key epitopes for enhanced NT of LAC seem to be g_2 and f_1. None of the California group viruses for which the antibodies did bind to these epitopes gave titers greater than 1:4 over background. In NT kinetic assays, none of the California group viruses was neutralized by antibodies to epitope g_2 or f_1 more than $\log_1$,

Table 7. Pairs of LAC anti-G1 monoclonal antibodies that enhance neutralization of LAC virus and bind to epitopes found on other California group viruses. (From KINGSFORD and BOUCQUEY 1990)

Representative antibody pairs which enhance neutralization of LAC virus[a]	Corresponding pairs of epitopes found on the G1 glycoproteins of other California viruses										
	LAC	SSH	CE	SA	INK	TAH	MEL	JC	KEY	SDN	TVT
1-6C3 (g_2) + 1-6A11 (b_1)[b]	+[d]	–[e]	–	–	–	–	–	–	–	–	–
1-6C3 (g_2) + 3-10B1 (d_2)[b]	+	+	–	+	+	+	–	+	–	–	–
1-4D14 (f_1) + 1-6A11 (b_1)[b]	+	–	–	–	–	–	–	–	–	–	–
1-6C3 (g_2) + 6-4B5 (a_1)[c]	+	+	–	+	+	+	–	+	–	–	–
1-6C3 (g_2) + 6-1A6 (e_1)[c]	+	+	–	–	–	–	–	–	–	–	–
1-6C3 (g_2) + 1-4D14 (f_1)[c]	+	–	–	+	–	+	–	–	–	–	–

LAC, La Crosse; SSH, snowshoe hare; CE, California encephalitis; SA, San Angelo; INK, Inkoo; MEL, Melao; JC, Jamestown Canyon; KEY, Keystone; SDN, Serra do Navio; TVT, trivittatus

[a] The epitope for each antibody is given in parentheses after the antibody designation

[b] Pairs of MAbs which bind synergistically in a two-way competition binding assay and enhance neutralization of LAC

[c] Pairs of antibodies which enhance neutralization of LAC but do not exhibit synergistic binding in a two-way competition binding assay

[d] Corresponding pairs of epitopes are located on the virus as demonstrated by ELISA

[e] No reaction of the antibodies with the virus or only one antibody of the pair reacted in ELISA

and in most cases viruses were neutralized less than 50%. In addition, the level of antibody binding to the second epitope in the pair was also low in many cases. Conformational changes induced by antibody binding may be needed for NT. Antibody binding to one site may allosterically change the glycoprotein so that a second antibody binding site becomes more reactive (KINGSFORD et al. 1983; KINGSFORD 1984; LUBECK and GERHARD 1982; EMINI et al. 1983). More antibody could then bind to the virus, which might lead to an increase in neutralization (KINGSFORD 1984). If the affinity of an antibody for the binding site is low, the interaction of antibody and binding site may not be sufficient to induce a conformational change that would allow a greater amount of a second antibody to bind. Alternatively, if one antibody does bind and induce a conformational change, this may not lead to increased antibody binding at the second site because of the low affinity of the second antibody. Thus, without increased antibody binding there would be no increase in the neutralization of the virus (KINGSFORD and BOUCQUEY 1990; KINGSFORD 1984).

When LAC virus is trypsinized and then assayed with polyclonal antibody, the kinetics of NT looks very similar to that observed for untreated LAC assayed with the individual MAbs (KINGSFORD and HILL 1983; KINGSFORD 1983, 1984). Considering that only one antigenic region involved in NT remains on trypsinized LAC, this makes sense. Of particular interest is another observation that when LAC is grown in suckling mouse brain, the resulting virus is neutralized less than $\log_1$ by polyclonal antibody. Thus, its NT reaction is very similar to that of trypsinized virus (KINGSFORD 1977). Whether this is due to proteolytic cleavage of the G1 glycoprotein by protease found in the mouse brain during the disease process or to masking of G1 epitopes by mouse brain proteins that adhere to virus during purification remains to be seen. But considering the sensitivity of the G1 glycoprotein to proteases, it is clearly possible that cleavage could occur in the brain. One should be aware, however, of this reduced neutralizing phenotype with the use of mouse brain derived antigens in serological assays and the implications that this can have on the results.

When LAC virus was incubated with MAbs prepared by GRADY et al. (1983a) and then assayed on BHK versus *Aedes albopictus* cells, two antibodies were found that exhibited host-dependent neutralizing activity (Table 2). One antibody neutralized LAC when assayed on BHK cells but not on mosquito cells while an antibody binding to a different epitope on G1 gave the opposite results. The second antibody bound to an epitope that was detected on all California group viruses and may be an important determinant in the interaction of these viruses with their arthropod vectors. This could indicate that the cell receptor on each cell type recognizes a different part of G1 for attachment (GRADY and KINCH 1985), or it could mean that the virus is handled by different mechanisms during penetration and uncoating in the cell so that antibody has different effects at these levels.

Monoclonal antibodies have also been produced for Bunyamwera (BUN) virus, a member of the Bunyamwera group in the *Bunyavirus* genus (JAMES and MILLICAN 1986). Six of the antibodies with high ELISA titers were tested for

NT of a number of members of the family, including three other Bunyamwera group viruses, four California group viruses, one nairovirus, one uukuvirus, and one phlebovirus. Only one of the antibodies neutralized BUN virus, but all of them neutralized one or more of the other members of the family. This is very surprising because, with the exception of low-level cross-reactions between certain sandfly fever group phleboviruses and certain uukuviruses, viruses in different genera have not previously been shown to be antigenically related (C.H. CALISHER, personal communication).

When many of the studies with MAbs are considered in context with the classification of viruses arrived at with polyclonal sera, one can begin to understand why discrepancies exist between results from different laboratories and with results from different assays and serum preparations within a laboratory. For bunyaviruses, at least, enhancement of NT occurs only if antibody binds to critical sites on G1. If these sites are not found on related viruses, or if the affinity of the antibody for the epitope is low, cross-reactions may not be seen in NT or HI reactions. In addition, NT and HI assays measure only cross-reactive sites which possess these biological activities. Yet there are epitopes detected only by reactivity in ELISA and other assays that are shared by these viruses. Any variance in antigenicity would not be detected in NT or HI assays by antibodies to these epitopes.

4.2.2 *Hantavirus* Genus

Of the MAbs made to hantaviruses, those made to the glycoproteins of Hantaan virus (76–118; DANTAS et al. 1986; ARIKAWA et al. 1989) and to rat-borne hantavirus B1 (serologically related to Hantaan virus; DANTAS et al. 1986) are the best characterized as to functional properties and reactivity with specific topographical sites. Both groups obtained more MAbs to the G2 glycoprotein than to G1 even though different antigenic preparations were used to immunize mice.

In contrast to those observed for LAC and TAH viruses (*Bunyavirus* genus), epitopes involved in HA but not NT seemed to be more characteristic for hantaviruses (DANTAS et al. 1986; ARIKAWA et al. 1989). Only 6 of 20 HI antibodies made to Hantaan by ARIKAWA et al. (1989) showed PRNT activity. HA and NT activities were found on both glycoproteins although HA seemed predominantly associated with G2. Similar results were reported by DANTAS et al. (1986) who found that NT and HI titers correlated only for some of the antibodies produced. Thus, sites involved in NT activity may be close to but separate from those involved in HA. Although both G1 and G2 have neutralizing sites on them, no work has been carried out to characterize them or to determine which might be critical for NT. However, there are at least two critical sites in domains on G2 that are involved in HA (DANTAS et al. 1986). Antibodies to these sites on G2 of Hantaan and B-1 viruses exhibited high HI titers and reacted equally well to each other as well as to Prospect Hill and Puumala viruses in HI assays. Thus, these may be common antigenic sites on hantaviruses. This is in contrast to NT assays with polyvalent sera where virus specificity is shown (SCHMALJOHN et al. 1985). In addition to the common HA sites, one virus-specific HA site was detected on the B-1 G2 glycoprotein. By the IFA test, additional cross-reactive sites could be detected between the hantaviruses.

By a competition binding assay, the 24 MAbs produced by ARIKAWA et al. (1989) were mapped to nine distinct, partially overlapping antigenic sites, two on G1 and seven on G2. These sites could be further divided into 13 based upon cross-reactions in IFA, HI, and PRNT assays with a virus representative of each of the four known antigenic groups within the *Hantavirus* genus, i.e., Hantaan, Seoul, Puumala, and Prospect Hill viruses. Of the six anti-G1 antibodies, only one cross-reacted with Sapporo rat (Seoul) and Puumala viruses, and none of them was cross-reactive with Prospect Hill virus. The G2 antibodies could be divided into four groups based on their cross-reactions. Some were type specific, some reacted with one or two of the other viruses, and some reacted with all four viruses. Based on the various serological reactivity patterns produced with the four groups of viruses, the MAbs could probably be used to assign new isolates to groups within the *Hantavirus* genus (ARIKAWA et al. 1989).

In another study, six MAbs made to Hantaan virus (strain 76–118) were used to examine the antigenic relationships between two Hantaan virus strains and between related hantaviruses (FRANKO et al. 1983). These MAbs were made prior to the characterization of hantavirus proteins, and thus their specificities have not been determined. When used in an IFA assay with Hantaan virus infected cells, five antibodies reacted with antigens on the cell surface and in the cytoplasm of virus-infected cells but gave two different patterns of staining. These antibodies are probably specific for one or both of the envelope glycoproteins. The other antibody reacted with a determinant expressed only in the cytoplasm of virus-infected cells and probably recognizes the viral nucleocapsid protein (FRANKO et al. 1983) or possibly a nonstructural protein. These antibodies could distinguish between the Lee strain (isolated from a human) and the 76–118 strain (isolated from a rodent) of Hantaan virus, which are indistinguishable serologically by rat anti-Hantaan virus antiserum. Only three of the six antibodies recognized antigens expressed by the two viruses, implying that certain epitopes have been conserved during host adaptation while others have changed. Although the MAbs had been prepared in the hope that they could be used to clarify relationships among hantaviruses, none of the antibodies recognized antigenic determinants on serologically related nephropathia epidemica virus or Prospect Hill virus (FRANKO et al. 1983).

Twelve MAbs were prepared by VAN DER GROEN et al. (1987) to Hantaan (76–118) and B-1 viruses. No reports are available on the characteristics or specificities of these antibodies either. However, they were used in IFA assays to determine antigenic relationships among nine hantaviruses which have been isolated from seven different geographical areas worldwide. Two antibodies were found to be group specific for the hantaviruses, while some were strain specific. Six antibodies could be used to differentiate four serotypes: nephropathia epidemica-like viruses, Prospect Hill-like viruses, Hantaan-like viruses, and rat strain-like viruses. Within the nephropathia epidemic group viruses, there were differences between geographical isolates even though they were isolated from the same host and isolated in areas where the mild form of hantavirus disease occurred. There were also minor differences among rat strains although isolates from laboratory rats in Japan and the United Kingdom were very similar. With one exception, the different serotypes of hantaviruses correlated well with the hosts from which they had been isolated.

4.2.3 *Phlebovirus* Genus

DALRYMPLE et al. (1981) characterized four MAbs to the G2 glycoprotein and one to the G1 glycoprotein of PT virus. These antibodies were subsequently included in a study along with a panel of four anti-G1 antibodies and 23 anti-G2 antibodies produced and characterized by PIFAT et al. (1988). The anti-G1 antibodies all exhibited HI and NT activities. None of them cross-reacted in an IFA test with 35 heterologous phleboviruses. The G2 antibodies were varied in their reactions. Although 20 had HI activity, only 10 of them also neutralized in an 80% PRNT test. Twenty of the anti-G2 antibodies were found to cross-react each with two or more (as many as 17) other phleboviruses. None of the 20 antibodies gave the same pattern of reactivity with any of the phleboviruses. The reactions were either strongly positive or negative, with essentially no intermediate reactions. Although considerable homology has been detected between the M RNAs of PT and RVFV viruses (35% for RVFV G2 and PT G1; 49% for RVFV G1 and PT G2; IHARA et al. 1985), only one of the PT G2 antibodies reacted with RVFV. (RVFV G2 corresponds to PT G1; see Fig. 1). Determinant conservation, however, did not necessarily mean functional equivalence, as the one neutralizing antibody to PT that cross-reacted with RVFV by ELISA or IFA failed to neutralize RVFV in vitro or to protect mice challenged with RVFV. The results of the heterologous reactions also indicate that the PT G2 glycoprotein appears to be antigenically more conserved than the G1. This is supported by the sequence homology given above for PT and RVFV glycoproteins (PIFAT et al. 1988).

All of the PT G1-specific MAbs and most of the neutralizing G2-specific antibodies also passively protected hamsters (DALRYMPLE et al. 1981) and mice (PIFAT et al. 1988) from lethal virus challenge. There were a couple of significant exceptions for the G2 antibodies. Based on a competitive binding assay, the protective epitopes were found to be located in at least two distinct sites on G1 and at least three sites on G2 (PIFAT et al. 1988).

The observations that antibodies to both glycoproteins have HI and NT activities, that there is no correlation between HI and NT activities for the G2-specific antibodies, and that there are more epitopes involved in HA than NT indicate that the phlebovirus glycoproteins may be more similar to those of the hantaviruses than the bunyaviruses.

MAbs have also been made against the G1 and G2 glycoproteins of RVFV and characterized (SALUZZO et al. 1989a). The panel included 25 anti-G2 antibodies (seven competition groups) and 14 anti-G1 antibodies (eight competition groups) along with 18 anti-N (two competition groups) and 2 anti-NSP-31 MAbs. MAbs to both glycoproteins are capable of neutralizing RVFV infectivity in vitro by a PRNT test (KEEGAN and COLLETT 1986). SALUZZO et al. (1989a, b) used this panel of MAbs to compare the antigenic relatedness of strains of RVFV isolated from humans, arthropods, and other animals. Fifty-five of the 59 antibodies reacted with all RVFV strains, indicating extensive conservation of epitopes. A single anti-N MAb differentiated the Egyptian RVFV strain isolated during a 1977–1978 epizootic in Egypt from all other isolates in sub-Saharan Africa (with one exception from Zimbabwe). This was the first known widespread involvement of the human population (698 fatal human cases) and indicated the possible emergence of a more pathogenic virus strain. However, the RVFV strains isolated during the 1987 epidemic in Mauritania (severe clinical disease with human fatalities) exhibited antigenic and biological properties in common with the sub-Saharan RVFV strains (SALUZZO et al. 1989a). Only minor differences were detected among the sub-Saharan African RVFV strains with a different anti-N antibody and two anti-glycoprotein antibodies. This indicated a lack of diversity among the Mauritania strains, which might have supported a hypothesis that a more virulent RVFV strain emerged during the 1987 epidemic. Recent studies indicate that stocks of RVFV contain a mixture of variant genomes, which might explain the paradoxical situation in which diverse clinical syndromes are exhibited by different RVFV strains that have similar phenotypes (SALUZZO et al. 1989b). In another study (SALUZZO et al. 1989b), two strains isolated from Madagascar reacted with the MAbs in a pattern indistinguishable from the Egyptian strains. This indicates a possible common ancestor for the Egyptian and Madagascar strains and suggests that epidemic spread could occur if ecological factors were to become favorable.

Three neutralizing antibodies and one nonneutralizing antibody specific for epitopes on the G2 glycoprotein of RVFV have been used to map the antigenic determinants at the amino acid sequence level (KEEGAN and COLLETT 1986). Antibodies to G2 neutralize and protect animals against lethal RVFV challenge. Thus, mapping these epitopes was important in determining antigenic relations, in understanding involvement of G2 in infection and disease, and in designing potential subunit vaccines to protect against RVFV infection. These particular antibodies to G2 were selected because they appeared to recognize epitopes of a linear structure and reacted with a G2 polypeptide β-galactosidase fusion protein to the same extent as the authentic antigen. Plasmid expression libraries were made in which the 3′ end of the G2 protein gene sequence was processively removed. Based on immunoreactivity of the expressed proteins with each MAb, the carboxy-end limit of the respective epitope was defined. Further subcloning of limited G2 polypeptide sequences into the bacterial expression vector defined the maximum bounds of the epitopes. The nonneutralizing antibody recognized a linear epitope located towards the carboxy end of G2. Two of the neutralizing antibodies bound to linear determinants while the third seemed to require some higher order structure, possibly a disulfide bridge between two cysteines in the sequence. Two separate domains are defined by the neutralizing antibodies and are found in a part of the glycoprotein that is predicted to have an overall hydrophilic nature. It was found also that a fusion protein expressed by one of the clones was missing a proline and a leucine in the sequence. The elimination of these amino acids reduced the immunoreactivity of the polypeptide by 80%.

5 Use of Monoclonal Antibodies in the Selection and Study of Variants

One approach to studying antigenic variation, the basis for it at a genetic level and the effect that it has on such things as pathogenicity, is to select for variants and then characterize them. Studies with reassortant California group viruses, for example, have shown that neuroinvasiveness is determined mainly by the M genome segment and its products (GONZALEZ-SCARANO et al. 1985; JANSSEN et al. 1984, 1986). Thus, MAbs to the G1 glycoprotein have been used to select for MAb-resistant variant viruses. When the variants were tested for mouse virulence, one variant had greatly reduced virulence. This variant had been selected with MAb 807-22 (see Tables 3, 4). Variants at a second epitope group were slightly attenuated, while variants at three other epitope groups showed no change from the parent LAC virus (GONZALEZ-SCARANO et al. 1988). The avirulent virus exhibited reduced neuroinvasiveness but minimal loss of neurovirulence in suckling mice. It also required a lower pH for the activation of fusion (pH 5.8 compared with 6.3) and was less effective in cell-to-cell fusion of BHK cells. Thus, an alteration occurred at a specific epitope on G1 that is involved either directly or indirectly in pathogenicity and fusion (GONZALEZ-SCARANO et al. 1985; POBJECKY et al. 1989). After three passages in mosquitoes by the oral route, the resultant virus exhibited an increase in infection rate, was neutralized by the MAb 807-22, and had a restored fusion function. These results suggest that the virus had reverted at the specific epitope (SUNDIN et al. 1987).

A MAb (B16.3) obtained by JAMES and MILLICAN (1986) to BUN virus cross-reacted with several other bunyaviruses by ELISA but effectively neutralized only BUN virus. Because BUN virus grew less well in mosquito cells (C6-36) than Vero cells, it was passed six times in these cells in order to increase its ability to replicate. This host-adapted virus replicated better and was more resistant to NT by MAb B16.3. Passage back into Vero cells did not change this NT phenotype. A similar study was carried out with SSH (JAMES and MILLICAN 1986). When tested with certain MAbs made by GONZALEZ-SCARANO et al. (1982), virus passed in mosquito cells exhibited decreased NT with one antibody (807-12; see Table 3) and increased NT by a second antibody. Thus, host-adaption of two bunyaviruses resulted in antigenic changes in their glycoproteins.

6 Antigenic Variance as Studied at the Genomic Level

Although the term antigenic variance implies detection of differences in proteins by serological means, work performed at the genomic level supports and explains differences detected antigenically. There is evidence available that viruses in the Bunyaviridae family evolve by two major mechanisms, reassortment of RNA genomic segments and genetic drift, the latter being due to accumulation of point mutations and possibly by sequence deletions, duplications, and inversions (BEATY and BISHOP 1988). Natural evolution by genetic drift has been shown for LAC virus by EL SAID et al. (1979), who analyzed the oligonucleotide fingerprints of the three genomic RNA segments of isolates obtained from different geographical areas. In each place where LAC isolates were obtained, variants and varieties were present with related but distinguishable nucleotide sequences for each of their three RNA species. Similar results have been reported by KLIMAS et al. (1981).

EL SAID et al. (1979) proposed the term "virus varieties" for those viruses which "although serologically indistinguishable from a prototype virus are nonetheless radically different at the genome level." This would apply to such isolates as the New York ones when compared with the prototype LAC virus. None of these viruses could be distinguished by polyvalent antisera. Yet, as discussed above, some differences between the G1 glycoproteins of prototype LAC and New York isolates have been detected by MAbs and mapped to two regions on the glycoprotein (GRADY et al. 1983b; KINGSFORD and BOUCQUEY 1990). This undoubtedly accounts for some of the differences in nucleotide sequences detected for the M RNA segment and indicates the value of correlating serological studies utilizing MAbs with methods that detect mutations at the genomic level.

When antigenic and genetic relations among Hantaan-like viruses were examined (SCHMALJOHN et al. 1985), each isolate was antigenically unique yet exhibited a pattern of cross-reactivity (by radioimmunoassay and PRNT) that correlated with host factors rather than geographical location. A comparison of the genome oligonucleotide maps also showed that each virus had a unique RNA fingerprint. This was true even for serologically closely related isolates. In an earlier study, oligonucleotide maps of individual genome segments had been compared for a few of these viruses and found to display considerable diversity. Thus, genetic changes were not confined to any one segment (SCHMALJOHN and DALRYMPLE 1984).

Because of reassortment, viruses with segmented genomes have an even greater evolutionary potential. This is supported by the fact that the majority of arboviruses registered are either members of the Bunyaviridae family or are orbiviruses (KARABATSOS 1985). Reassortment has been shown to occur at a high rate in the laboratory in vivo and in cell culture (BISHOP and SHOPE 1979; IROEGBU and PRINGLE 1981). Reassortant viruses have thus been used for various studies such as to define bunyavirus gene-coding assignments (BISHOP and SHOPE 1979) and to determine the genetic basis for in vivo virulence and tissue tropism (BEATY et al. 1982). Reassortment for the Bunyaviridae is limited, however, to certain members within a serogroup (BISHOP 1985a,b; PRINGLE et al. 1984).

Although the extent to which reasortment occurs in nature is not known, an intertypic reassortment virus was detected by RNA fingerprint analysis among the LAC field isolates examined by KLIMAS et al. (1981). This isolate (from Minnesota) had an L RNA segment identical to that of other Minnesota isolates, while the M RNA was distinguishable but closely related. The S RNA segment, however, was clearly of a different type. Other naturally occuring reassortants have been found by USHIJIMA et al. (1981). Shark River and Pahayokee, members of the Patois serogroup of the *Bunyavirus* genus, were both isolated in the same region of south Florida and from the same vector species but collected at different times. The viruses were found to have almost identical L and S RNA fingerprints but M RNA fingerprints that were different, suggesting that the two viruses may be reassortants. This explains and is supported by serological studies in which they appear to be closely related by CF tests but distinguishable by HI and NT tests (FIELDS et al. 1969).

7 Conclusions

It has been over 40 years now since the first members of the Bunyaviridae family were isolated (such as California encephalitis virus in 1943; HAMMON and REEVES 1952). In this time more than 300 viruses have been isolated and classified. Serological techniques have mostly been used to classify these viruses and study antigenic variance. Based on the number of viruses isolated each year which belong in the family and on the current technology, serology will probably continue to be used for study and comparison of most of them. However, there are many isolates which cannot be characterized this way and remain as unclassified viruses. Because of the number and diversity among family members, other techniques must be developed to study and characterize new isolates.

There is also the need to really understand at a molecular level the basis for genetic diversity and the direct result that it has on antigenic relationships, host susceptibility, pathogenesis, transmission by vectors, etc. Considering the vast number of isolates, one cannot expect to make MAbs to viruses within all the genera and groups or to sequence and map the genomes of all family members. However, this must be done for prototype members within each genus and to some extent within each group. Over the past several years great progress has been made in this area for certain members of the family. Yet it is obvious that more work needs to be done, such as for the uukuviruses, the nairoviruses, and many phleboviruses as well as for the ungrouped viruses.

The dogma of contemporary biological science states that sequencial genomic structure translates to amino acid sequence and hence to higher order structure of encoded gene products. Since the beginning of immunology, detection of antigenic reactivities of certain amino acid sequences (epitopes) of these product proteins has served as the basis for studies on antigenic variation among related molecules. Thus, antigenic variation results from genetic variation and is directly equated with alteration in primary amino acid sequence and subsequent higher order structure of these antigenic determinants. For members of the Bunyaviridae family, reassortment of gene segments also results in the emergence of new viruses and thus antigenic variation as detected serologically. The ultimate goal of biology is the complete understanding of relationships between structure and function, from the ecological and macroscopic to the molecular and atomic levels. Karl Landsteiner, almost 100 years ago, began this inquiry employing the tools of immunology, and it continues today. This review has attempted to further this quest as it relates to an isolated group of viruses and their antigenically related gene products.

Appendix

Bunyavirus genus

Anopheles A group: Anopheles A, Las Maloyas, Lukuni, Tacaiuma, Trombetas, Virgin River, CoAr3624, CoAr1071, CoAr3627, ColAn57389, SpAr2317, H32580

Anopheles B group: Anopheles B, Boraceia

Bakau group: Bakau, Ketapang, Nola, Tanjong Rabok, Telok Forest

Bunyamwera group: Anhembi, Batai, Birao, Bozo, Bunyamwera, Cache Valley, Fort Sherman, Germiston, Guaroa, Iaco, Ilesha, Kairi, Lokern, Macaua, Maguari, Main Drain, Mboke, Ngari, Northway, Playas, Santa Rosa, Shokwe, Sororoca, Taiassui, Tensaw, Tlacotalpan, Tucunduba, Wyeomyia, Xingu, AG83-1746, BeAr328208, CbaAr426

Bwamba group: Bwamba, Pongola

C group: Apeu, Bruconha, Caraparu, Gumbo Limbo, Itaqui, Madrid, Marituba, Murutucu, Nepuyo, Oriboca, Ossa, Restan, Vinces, 63U11

California group: California encephalitis, Inkoo, Jamestown Canyon, Keystone, La Crosse, Melao, San Angelo, Serra do Navio, snowshoe hare, South River, Tahyna, trivittatus, AG83-497

Capim group: Acara, Benevides, Benfica, Bushbush, Capim, Guajara, Juan Diaz, Moriche, GU71u344, GU71u350

Gamboa group: Alajuela, Brus Laguna, Gamboa, Pueblo Viejo, San Juan, 75V-2621, 78V-2441, 75V-2374

Guama group: Ananindeua, Bertioga, Bimiti, Cananeia, Catu, Guama, Guaratuba, Itimirim, Mahogany Hammock, Mirim, Moju, Timboteua

Koongol group: Koongol, Wongal

Minatitlan group: Minatitlan, Palestina

Nyando group: Nyando, Eret-147

Olifantsvlei group: Bobia, Botambi, Olifantsvlei, Oubi, Dabakala

Patois group: Abras, Babahoyo, Estero Real, Pahayokee, Patois, Shark River, Zegla

Simbu group: Aino, Akabane, Buttonwillow, Douglas, Facey's Paddock, Ingwavuma, Inini, Kaikalur, Manzanilla, Mermet, Oropouche, Para, Peaton, Sabo, Sango, Sathuperi, Shamonda, Shuni, Simbu, Thimiri, Tinaroo, Utinga, Utive, YABA-7

Tete group: Bahig, Batama, Matruh, Tete, Tsuruse

Turlock group: Lednice, Turlock, Umbre, Yaba-1

Ungrouped viruses: Kaeng Khoi, Leanyer, Mojui dos Campos, Termeil

Hantavirus genus

Hantaan group: Hantaan, Leaky, Seoul, Prospect Hill, Puumala, Thottapalayam

Nairovirus genus

Crimean-Congo hemorrhagic fever group: Crimean-Congo hemorrhagic fever, Hazara, Khasan

Dera Ghazi Khan group: Abu Hammad, Abu Mina, Dera Ghazi Khan, Kao Shuan, Pathum Thani, Pretoria

Hughes group: Farallon, Fraser Point, Great Saltee, Hughes, Puffin Island, Punta Salinas, Raza, Sapphire II, Soldado, Zirqa

Nairobi sheep disease group: Dugbe, Nairobi sheep disease

Qalyub group: Bandia, Omo, Qalyub

Sakhalin group: Avalon, Clo Mor, Kachemak Bay, Paramushir, Sakhalin, Taggert, Tillamook

Thiafora group: Evre, Thiafora

Phlebovirus genus

Sandfly fever group:

- Sandfly fever Naples complex: Karimabad, sandfly fever Naples, Tehran, Toscana
- Bajaru complex: Bujaru, Munguba
- Candiru complex: Alenquer, Candiru, Itaituba, Nique, Oriximina, Turuna
- Chilibre complex: Cacao, Chilibre
- Frijoles complex: Frijoles, Joa
- Punta Toro complex: Buenaventura, Punta Toro
- Rift Valley fever complex: Belterra, Icoaraci, Rift Valley fever
- Salehebad complex: Arbia, Salehabad
- No complex assigned: Aguacate, Anhanga, Arboledas, Arumowot, Caimito, Chagres, Corfou, Gabek Forest, Gordil, Itaporanga, Odrenisrou, Pacui, Rio Grande, Saint-Floris, Sandfly fever Sicilian, Urucuri
- Uukuniemi group: Grand Arbaud, Manawa, Murre, Oceanside, Ponteves, Precarious Point, St. Abbs Head, Uukuniemi, Zialiv, Terpeniya, EgAn1825-61. Fin V-707, RML 105355

Other possible members of the family Bunyaviridae

Bhanja group: Bhanja, Forecariah, Kismayo

Kaisodi group: Kaisodi, Lanjan, Silverwater

Mapputta group: Gan Gan, Mapputta, Maprik, Trubanaman

Okola group: Okola, Tanga

Resistencia group: Antequera, Barranqueras, Resistencia

Upolu group: Aransas Bay, Upolu

Yogue group: Yogue, Kasokero

Ungrouped viruses: Bangui, Batken, Belem, Belmont, Bobaya, Caddo Canyon, Chim, Enseada, Issyk-Kul (Keterah), Kowanyama, Lone Star, Pacora, Razdan, Salanga, Santarem, Sunday Canyon, Tai, Tamdy, Tataguine, Wanowrie, Witwatersrand, Yacaaba

Proposed classification submitted by the Bunyviridae Study Group to the International Committee for Taxonomy of Viruses for review at the International Congress for Virology, Berlin, August 1990

Addendum in proof

The proposed classification as listed in the "Appendix" was accepted by the International Committee. The name of the proposed genus containing tomato spotted wilt virus (p. 183), however, was changed from *Phythrivirus* to *Tospovirus* (CH Calisher, Chair of the Bunyaviridae Study Group, personal communication). The conclusions of the International Committee will be published in the *Fifth Report of the International Committee for Taxonomy of Viruses* (RIB Franchi, Springer, Vienna, 1991).

References

Arikawa J, Schmaljohn AL, Dalrymple JM, Schmaljohn CS (1989) Characterization of Hantaan virus envelope glycoprotein antigenic determinants defined by monoclonal antibodies. J Gen Virol 70: 615–624

Beaty BJ, Bishop DH (1988) Bunyavirus-vector interactions. Virus Res 10: 289–301

Beaty BJ, Miller BR, Shope RE, Rozhon EJ, Bishop DH (1982) Molecular basis of bunyavirus *per os* infection of mosquitoes: role of the middle-sized RNA segment. Proc Natl Acad Sci USA 79: 1295–1297

Beaty BJ, Trent DW, Roehrig JT (1988) Virus variation and evolution: mechanisms and epidemiological significance. In: Monath TP (ed) The arboviruses, vol 1. CRC Press, Boca, Raton, pp 59–81

Bishop DHL (1985a) The genetic basis for describing viruses as species. Intervirology 24: 79–93

Bishop DHL (1985b) Replication of arenaviruses and bunyaviruses. In: Fields BN, Knipe DM (eds) Virology. Raven, New York, pp 1083–1110

Bishop DHL, Beaty BJ (1988) Molecular and biochemical studies of the evolution, infection and transmission of insect bunyaviruses. Philos Trans R Soc Lond [Biol] 321: 463–483

Bishop DHL, Shope RE (1979) Bunyaviridae. Compr Virol 14: 1–156

Bishop DH, Calisher CH, Casals J, Chumakov MP, Gaidamovich SY, Hannoun C, Lvov DK, Marshall ID, Oker-Blom N, Pettersson RF, Porterfield JS, Russell PK, Shope RE, Westaway EG (1980) Bunyaviridae. Intervirology 14: 125–143

Blackburn NK, Besselaar RG, Shepherd AJ, Swanepoel R (1987) Preparation and use of monoclonal antibodies for identifying Crimean-Congo hemorrhagic fever virus. Am J Trop Med Hyg 37: 392–397

Calisher CH (1983) Taxonomy, classification, and geographic distribution of California serogroup Bunyaviruses. Progr Clin Biol Res 123: 1–16

Calisher CH (1988) Evolutionary significance of the taxonomic data regarding bunyaviruses of the family Bunyaviridae. Intervirology 29: 268–276

Calisher CH, Lazuick JS, Justines G, Francy DB, Monath TP, Gutierrez VE, Sabattini MS, Bowen GS, Jakob WL (1981) Viruses isolated from *Aedeomyia squamipennis* mosquitoes collected in Panama, Ecuador, and Argentina: establishment of the Gamboa serogroup. Am J Trop Med Hyg 30: 219–223

Calisher CH, Francy DB, Smith GC, Muth DJ, Karabatsos N, Jakob WL, McLean RG (1986) Distribution of Bunyamwera serogroup viruses in North America, 1956–1984. Am J Trop Med Hyg 35: 429–443

Calisher CH, Lazuick JS, Sudia WD (1988a) Brus Laguna virus, a Gamboa bunyavirus from *Aedeomyia squamipennis* collected in Honduras. Am J Trop Med Hyg 39: 406–408

Calisher CH, Lazuick JS, Lieb S, Monath TP, Castro KG (1988b) Human infections with Tensaw virus in south Florida: evidence that Tensaw virus subtypes stimulate the production of antibodies reactive with closely related Bunyamwera serogroup viruses. Am J Trop Med Hyg 39: 117–122

Calisher CH, Sabattini MS, Monath TP, Wolff KL (1988c) Cross-neutralization tests among Cache Valley virus isolates revealing the existence of multiple subtypes. Am J Trop Med Hyg 39: 202–205

Casals J, Tignor GH (1974) Neutralization and haemagglutination-inhibition tests with Crimean haemorrhagic fever-Congo virus. Proc Soc Exp Biol Med 145: 960–966

Casals J, Tignor GH (1980) The *Nairovirus* genus: serological relationships. Intervirology 14: 144–147

Casals J, Whitman L (1961) A new serological group of hitherto undescribed arthropod-borne viruses. Immunological studies. Am J Trop Med Hyg 10: 250–258

Chastel C, Main AJ, Richard P, Le Lay G, Legrand-Quillien MC, Beaucournu JC (1989) Erve virus, a probable member of Bunyaviridae family isolated from shrews (*Crocidura russula*) in France. Acta Virol 33: 270–280

Clerx-van Haaster CM, Clerx JPM, Ushijima H, Akashi H, Fuller F, Bishop DHL (1982) The 3′ terminal RNA sequences of bunyaviruses and nairoviruses (Bunyaviridae): evidence of end sequence generic differences within the virus family. J Gen Virol 61: 289–292

Collett MS, Purchio AF, Keegan K, Frazier S, Hays W, Anderson DK, Parker MD, Schmaljohn C, Schmidt J, Dalrymple J (1985) Complete nucleotide sequence of the M RNA segment of Rift Valley fever virus. Virology 144: 228–245

Dalrymple JM, Peters CJ, Smith JF, Gentry MK (1981) Antigenic components of Punta Toro virus. In: Bishop DHL, Compans BW (eds) Replication of negative strand viruses. Elsevier, New York, pp 167–172

Dantas JR, Okuno Y, Asada H, Tamura M, Takahashi M, Tanishi AO, Takahashi U, Furata T, Yamanishi K (1986) Characterization of glycoproteins of viruses causing hemorrhagic fever with renal syndrome (HFRS) using monoclonal antibodies. Virology 151: 379–384

de Haan P, Wagemakers L, Goldbach R, Peters D (1988) Tomato spotted wilt virus: a bunyavirus in plants. Virus Res [Suppl] 2: 83

de Haan P, Wagemakers L, Peters D, Goldbach R (1989) Molecular cloning and terminal sequence determination of the S and M RNAs of tomato spotted wilt virus. J Gen Virol 70: 3569–3473

El Said LH, Vorndam V, Gentsch JR, Clewley JP, Calisher CH, Klimas RA, Thompson WH, Grayson M, Trent DW, Bishop DHL (1979) A comparison of La Crosse virus isolates obtained from different ecological nitches and an analysis of the structural components of California encephalitis serogroup viruses and other bunyaviruses. Am J Trop Med Hyg 28: 364–386

Elliott RM (1990) Molecular biology of the Bunyaviridae. J Gen Virol 71: 501–522

Elliott RM, Wilkie ML (1986) Persistent infection of *Aedes albopictus* C6/36 cells by Bunyamwera virus. Virology 150: 21–32

Emini EA, Shaw-Yi K, Lewis AJ, Crainic R, Wimmer E (1983) Functional basis of poliovirus neutralization determined with monospecific neutralizing antibodies. J Virol 46: 466–474

Eshita Y, Bishop DHL (1984) The complete sequence of the M RNA of snowshoe hare bunyavirus reveals the presence of internal hydrophobic domains in the viral glycoprotein. Virology 137: 227–240

Fazakerley JK, Gonzales–Scarano F, Strickler J, Dietzschold B, Karush F, Nathanson N (1988) Organization of the middle RNA segment of snowshoe hare bunyavirus. Virology 167: 422–432

Fields BN, Henderson BE, Coleman PH, Work TH (1969) Pahayokee and Shark River, two new arboviruses related to Patois and Zelga from the Florida Everglades. Am J Epidemiol 89: 222–226

Franko MC, Gibbs CJ, Lee PW, Gajkusek DC (1983) Monoclonal antibodies specific for Hantaan virus. Proc Natl Acad Sci USA 80: 4149–4153

Gonzalez–Scarano F (1985) La Crosse virus G1 glycoprotein undergoes a conformational change at the pH of fusion. Virology 140: 209–216

Gonzelez-Scarano F, Shope RE, Calisher CH, Nathanson N (1982) Characterization of monoclonal antibodies against the G1 and N protein of La Crosse and Tahyna, two California serogroup bunyaviruses. Virology 120: 42–53

Gonzalez-Scarano F, Shope RE, Calisher CH, Nathanson N (1983) Monoclonal antibodies against the G1 and nucleocapsid proteins of La Crosse and Tahyna viruses. In: Calisher CH, Thompson WH (eds) California Serogroup Viruses, Liss, New York (Progr Clin Biol Res 123: 145–156

Gonzalez-Scarano F, Pobjecky N, Nathanson N (1984) La Crosse bunyavirus can mediate pH-dependent fusion from without. Virology 132: 222–225

Gonzalez-Scarano F, Janssen RS, Najjar JA, Pobjecky N, Nathanson N (1985) An avirulent G1 glycoprotein variant of La Crosse bunyavirus with defective fusion function. J Virol 54: 757–763

Gonzalez-Scarano F, Beaty B, Sundin D, Janssen R, Endres MJ, Nathanson N (1988) Genetic determinants of the virulence and infectivity of La Crosse virus. Microb Path 4: 1–7

Grady LJ, Kinch W (1985) Two monoclonal antibodies against La Crosse virus show host-dependent neutralizing activity. J Gen Virol 66: 2773–2776

Grady LJ, Srihongse S, Grayson MA, Deibel R (1983a) Monoclonal antibodies against La Crosse virus. J Gen Virol 64: 1699–1704
Grady LJ, Sanders ML, Campbell WP (1983b) Evidence for three separate antigenic sites on the G1 protein of La Crosse virus. Virology 126: 395–397
Grady LJ, Sanders ML, Campbell WP (1987) The sequence of the M RNA of an isolate of La Crosse virus. J Gen Virol 68: 3057–3071
Hammon WMcD, Reeves WC (1952) California encephalitis virus, a newly described agent. Calif Med 77: 303–309
Hunt AR, Calisher CH (1979) Relationships of bunyamwera group viruses by neutralization. Am J Trop Med Hyg 28: 740–749
Ihara T, Smith J, Dalrymple JM, Bishop DHL (1985) Complete sequence of the glycoproteins and M RNA of Punta Toro phlebovirus compared to those of Rift Valley fever virus. Virology 144: 246–259
Iroegbu CU, Pringle CR (1981) Genetic interactions among viruses of the Bunyamwera complex. J Virol 37: 383–394
James WS, Millican D (1986) Host-adaptive antigenic variation in bunyaviruses. J Gen Virol 67: 2803–2806
Janssen R, Gonzalez-Scarano F, Nathanson N (1984) Mechanisms of bunyavirus virulence. Comparative pathogenesis of a virulent strain of La Crosse and an avirulent strain of Tahyna virus. Lab Invest 50: 447–455
Janssen RS, Nathanson N, Endres MJ, Gonzalez-Scarano F (1986) Virulence of La Crosse virus is under polygenic control. J Virol 59: 1–7
Kakach LT, Wasmoen TL, Collett MS (1988) Rift Valley fever virus M segment: use of recombinant vaccinia viruses to study phlebovirus gene expression. J Virol 62: 826–833
Karabatsos N (ed) (1985) International catalogue of arboviruses, 3rd edn. American Society of Tropical Medicine and Hygiene, San Antonio
Keegan K, Collett MS (1986) Use of bacterial expression cloning to define the amino acid sequences of antigenic determinants on the G2 glycoprotein of Rift Valley fever virus. J Virol 58: 263–270
Kingsford L (1977) La Crosse virus structural phenotypes induced by host cells. PhD thesis, University of Utah
Kingsford L (1983) Structure and biological function of La Crosse G1 glycoprotein. Prog Clin Biol Res 123: 157–166
Kingsford L (1984) Enhanced neutralization of La Crosse virus by the binding of specific pairs of monoclonal antibodies to the G1 glycoprotein. Virology 136: 265–273
Kingsford L, Boucquey KH (1990) Monoclonal antibodies specific for the G1 glycoprotein of La Crosse virus that react with other California serogroup viruses. J Gen Virol 71: 523–530
Kingsford L, Hill DW (1981) The effects of proteolytic enzymes on structure and function of La Crosse G1 and G2 glycoproteins. In: Bishop DHL, Compans RW (eds) The replication of negative strand viruses. Elsevier, New York, pp 111–116
Kingsford L, Hill, DW (1983) The effect of proteolytic cleavage of La Crosse virus G1 glycoprotein on antibody neutralization. J Gen Virol 64: 2147–2156
Kingsford L, Ishizawa LD (1984) Antigenic sites on the G1 glycoprotein of La Crosse virus that are involved in neutralization. In: Compans RW, Bishop DHL (eds) Segmented negative strand viruses. Academic, Orlando, pp 217–224
Kingsford L, Ishizawa LD, Hill DW (1983) Biological activities of monoclonal antibodies reactive with antigenic sites mapped on the G1 glycoprotein of La Crosse virus. Virology 129: 443–455
Kingsford L, Boucquey KH, Cardozo TP (1991) Effects of specific monoclonal antibodies on La Crosse virus neutralization: aggregation, inactivation by Fab fragments and inhibition of attachment to baby hamster kidney cells. Virology 180: 591–601
Klimas RA, Thompson WH, Calisher CH, Clark GG, Grimstad PR, Bishop DHL (1981) Genotypic varieties of La Crosse virus isolated from different geographic regions of the continental United States and evidence for a naturally occurring intertypic recombinant La Crosse virus. Am J Epidemiol 114: 112–131
Kuismanen K (1984) Posttranslational processing of Uukuniemi virus glycoproteins G1 and G2. J Virol 51: 806–812
Law MD, Moyer JW (1990) A tomato spotted wilt-like virus with a serologically distinct N protein. J Gen Virol 71: 933–938
Lees JF, Pringle CR, Elliott RM (1986) Nucleotide sequence of the Bunyamwera virus M RNA segment: conservation of structural features in the bunyavirus glycoprotein gene product. Virology 148: 1–14

Lubeck M, Gerhard W (1982) Conformational changes at topologically distinct antigenic sites on the influenza A/Pr/8/34 virus HA molecule are induced by the binding of monoclonal antibodies. Virology 118: 1–7

Ludwig GV, Christensen BM, Yuill TM, Schultz KT (1989) Enzyme processing of La Crosse virus glycoprotein G1: a bunyavirus-vector infection model. Virology 171: 108–113

Mangiafico JA, Sanchez JL, Figueiredo LT, LeDuc JW, Peters CJ (1988) Isolation of a newly recognized Bunyamwera serogroup virus from a febrile human in Panama. Am J Trop Med Hyg 39: 593–596

McCormick JG, Palmer EL, Sasso DR, Kiley MP (1982) Morphological identification of the agent of Korean hemorrhagic fever (Hantaan virus) as a member of Bunyaviridae. Lancet i: 765–768

Najjar JA, Gentsch JR, Nathanson N, Gonzalez-Scarano F (1985) Epitopes of the G1 glycoprotein of La Crosse virus form overlapping clusters within a single antigenic site. Virology 144: 426–432

Pardigon N, Vialat P, Gerbaud S, Girard M, Bouloy M (1988) Nucleotide sequence of the M sement of Germiston virus: comparison of the M gene product of several bunyaviruses. Virus Res 11: 73–85

Pifat DY, Osterling MC, Smith JF (1988) Antigenic analysis of Punta Toro virus and identification of protective determinants with monoclonal antibodies. Virology 167: 442–450

Pobjecky N, Nathanson N, Gonzalez-Scarano F (1989) Use of resonance energy transfer assay to investigate the fusion function of La Crosse virus. In: Kolakofsky D, Mahy BWJ (eds) Genetics and pathogenicity of negative strand viruses. Elsevier, Amsterday, pp 24–32

Pringle CR, Lees JF, Clark W, Elliott RM (1984) Genome subunit reassortment among bunyaviruses analysed by dot hybridization using molecularly cloned complementary DNA probes. Virology 135: 244–256

Robeson G, El Said LH, Brandt W, Dalrymple J, Bishop DHK (1979) Biochemical studies on the phlebotomous fever group viruses (Bunyaviridae family). J Virol 30: 339–350

Rodhain F, Madulo-LeBlond G, Hannoun C, Tesh RB (1985) Le virus Corfou: un nouveau *Phlebovirus* isole de phlebotomes en Grece. Ann Inst Pasteur Virol 136E: 161–166

Ronnholm R, Pettersson RF (1987) Complete nucleotide sequence of the M RNA segment of Uukuniemi virus encoding the membrane glycoproteins G1 and G2. Virology 160: 191–202

Saluzzo JF, Anderson GW Jr, Hodgson LA, Digoutte JP, Smith JF (1989a) Antigenic and biological properties of Rift Valley fever virus isolated during the 1987 Mauritanian epidemic. Res Virol 140: 155–164

Saluzzo JF, Anderson GW Jr, Smith JF, Fontenille D, Coulanges P (1989b) Biological and antigenic relationship between Rift Valley fever virus strains isolated in Egypt and Madagascar. Trans R Soc Trop Med Hyg 83: 701

Schmaljohn CS, Dalrymple JM (1983) Analysis of Hantaan virus RNA: evidence for a new genus of Bunyaviridae. Virology 131: 482–491

Schmaljohn CS, Dalrymple JM (1984) Biochemical characterization of Hantaan virus. In: Compans RW, Bishop DHL (eds) Segmented negative strand viruses. Academic, Orlando, pp 117–124

Schmaljohn CS, Hasty SE, Dalrymple JM, LeDuc JW, Lee HW, von Bonsdorff C-H, Brummer-Korvenkontio M, Vaheri A, Tsai TF, Regnery HL, Goldgaber D, Lee PW (1985) Antigenic and genetic properties of viruses linked to hemorrhagic fever with renal syndrome. Science 227: 1041–1044

Schmaljohn CS, Hasty SE, Rasmussen L, Dalrymple JM (1986) Hantaan virus replication: effects of monesin, tunicamycin and endoglycosidases on the structural glycoproteins. J Gen Virol 67: 707–717

Schmaljohn CS, Schmaljohn AL, Dalrymple JM (1987) Hantaan virus M RNA: coding strategy, nucleotide sequence, and gene order. Virology 157: 31–39

Shope RE (1985) Bunyaviruses. In: Fields BN, Knipe DM (eds) Virology. Raven, New York, pp 1055–1082

Simons JF, Hellman U, Pettersson RF (1990) Uukuniemi virus S RNA segment: ambisense coding strategy, packaging of complementary strands into virions, and homology to members of the genus *Phlebovirus*. J Virol 64: 247–255

Sudin, DR, Beaty BJ, Nathanson N, Gonzalez-Scarano F (1987) A G1 glycoprotein epitope of La Crosse virus: a determinant of infection of *Aedes triseriatus*. Science 235: 591–593

Swanepoel R, Struthers JK, Erasmus, MJ, Shepherd SP, McGillivray GM, Shepherd AJ, Hummitzsch DE, Erasmus BJ, Barnard BJH (1986) Comparative pathogenicity and antigenic cross-reactivity of Rift Valley fever and other African phleboviruses in sheep. J Hyg (Lond) 97: 331–346

Tesh RB, Peters CJ, Meegan JM (1982) Studies on the antigenic relationship among phleboviruses. Am J Trop Med Hyg 31: 149–155

Tesh RB, Boshell J, Young DG, Morales A, Corredor A, Modi GB, Ferro De Carrasquilla C, De Rodriquez C, Gaitan MO (1986) Biology of Arboledas virus, a new phlebotomus fever serogroup virus (Bunyaviridae: *Phlebovirus*) isolated from sand flies in Colombia. Am J Trop Med Hyg 35: 1310–1316

Tesh RB, Boshell J, Young DG, Morales A, Ferro de Carrasquilla C, Corredor A, Modi GB, Travassos da Rosa APA, McLean RG, de Rodriguez C, Gaitan MO (1989) Characterization of five new phleboviruses recently isolated from sand flies in tropical America. Am J Trop Med Hyg 40: 529–533

Ushijima J, Clerx-van Hasster CM, Bishop DHL (1981) Analyses of Patois group bunyaviruses: evidence for naturally occurring recombinant bunyaviruses and existence of immune precipitable and nonprecipitable nonvirion proteins induced in bunyavirus-infected cells. Virology 110: 318–332

van der Groen G, Yamanishi K, McGormick J, Lloyd G, Tkachenko EA (1987) Characterization of hantaviruses using monoclonal antibodies. Acta Virol (Praha) 31: 499–503

Watret GE, Elliott RM (1985) The proteins and RNAs specified by Clo Mor virus, a Scottish nairovirus. J Gen Virol 66: 2513–2516

Watret GE, Pringle CR, Elliott RM (1985) Synthesis of bunyavirus-specific proteins in a continuous cell line (XTC-2) derived from *Xenopus laevis*. J Gen Virol 66: 473–482

White JD, Shirey FG, Grench GR, Huggins JW, Brand OM (1982) Hantaan virus aetiological agent of Korean haemorrhagic fever, has Bunyaviridae-like morphology. Lancet i: 768–771

Zeller HG, Karabatsos N, Calisher CH, Digoutte JP, Cropp CB, Murphy FA, Shope RE (1989) Electron microscopic and antigenic studies of uncharacterized viruses. II. Evidence suggesting the placement of viruses in the family Bunyaviridae. Arch Virol 108: 211–227

Pathogenesis

F. GONZALEZ-SCARANO, M. J. ENDRES, and N. NATHANSON

1 Introduction

The family Bunyaviridae consists of over 250 individual virus species grouped into five genera. Most of these viruses rarely infect humans or domestic animals. This chapter focuses on the few viruses that are important pathogens (Table 1)

Departments of Neurclogy and Microbiology, School of Medicine, University of Pennsylvania, Philadelphia, PA 19104, USA

Current Topics in Microbiology and Immunology, Vol. 169

Table 1. Selected bunyaviruses that cause disease in humans or domestic animals

Genus, serogroup	Virus	Host	Disease	Frequency
Bunyavirus				
California	La Crosse	Human	Encephalitis	Endemic
	Snowshoe hare	Human	Encephalitis	Rare
	California encephalitis	Human	Encephalitis	Rare
	Jamestown Canyon	Human	Encephalitis	Sporadic
	Tahyna	Human	Influenzalike	Endemic
Simbu	Oropouche	Human	Acute fever	Epidemic
	Akabane	Ungulates	Fetal malformations	Endemic
Phlebovirus				
RVF	RVF	Sheep	Hepatitis	Epidemic
		Human	Influenzalike	Epidemic
Sandfly fever	Sandfly fever			
	Naples	Human	Acute fever	Endemic
	Sicilian	Human	Acute fever	Endemic
Hantavirus				
Hantaan	Hantaan	Human	HFRS	Endemic
	Seoul	Human	HFRS	Endemic
	Puumala	Human	Renal Syndrome	Endemic
Nairovirus				
NSD	NSD	Sheep	Hepatitis	Endemic
CCHF	CCHF	Human	Hemorrhagic fever	Sporadic

RVF, Rift Valley fever; NSD, Nairobi sheep disease; CCHF, Crimean-Congo hemorrhagic fever, HFRS, hemorrhagic fever with renal syndrome.

Table 2. Animal models for selected pathogenic bunyaviruses

Genus, serogroup	Virus	Model species	Nature of disease	Disease simulation	Containment level
Bunyavirus					
California	La Crosse	Mouse	Encephalitis	Good	II
Simbu	Oropouche	None			III
	Akabane	Ungulate	Fetal malformation	Good	II
Phlebovirus					
RVF	RFV	Rat	Hepatitis encephalitis	Good	IV
		Sheep	Hepatitis	Good	IV
		Monkey	Hemorrhagic fever	Good	IV
Sandfly fever	Sandfly fever	None			III
Hantavirus					
Hantaan		Mouse	Encephalitis	Poor	IV
	Hantaan	Rat	Encephalitis		
Nairovirus					
CCHF	CCHF	None			IV

RVF, Rift Valley fever, NSD, Nairobi sheep disease, CCHF, Crimean-Congo hemorrhagic fever; HFRS, hemorrhagic fever with renal syndrome. Disease simulation refers to disease in humans or domestic animals. Containment level: authors' assessment, based on severity of human disease and transmissibility in the laboratory. Levels: II, laminar flow hood plus standard microbiological safety practices; III, high containment, including externally vented laminar flow hood and other special facilities plus special standard of laboratory practice; IV, maximal containment, requiring elaborate facilities.

and which have been studied in experimental vertebrate hosts. Particular emphasis is given to the California serogroup since this group of agents has received the most detailed analysis. Several other viruses, such as Rift Valley fever virus, Hantaan virus, and Crimean-Congo hemorrhagic fever virus are important causes of disease but are difficult to study because they require special containment facilities and because of a paucity of animal models which mimic the human disease (Table 2). A summary is presented of the information available on these agents.

Additional information on the biological and medical aspects of bunyaviruses can be found in several reviews (GONZALEZ-SCARANO and NATHANSON, 1989, 1990; CALISHER and THOMPSON 1983; PARSONSON and MCPHEE 1985).

2 California Serogroup Viruses

2.1 Background

2.1.1 History

California encephalitis virus was isolated in 1950 by REEVES and HAMMON during their studies of arbovirus encephalitis in Kern County, California (HAMMON and REEVES 1952; REEVES and HAMMON 1962). Tests on sera from patients with encephalitis demonstrated that three cases were due to this virus. Subsequently, California encephalitis virus has been isolated sporadically from mosquitoes but has not been associated with encephalitis (REEVES et al. 1983; LEDUC 1987).

In 1960 a virus, later named La Crosse virus, was isolated from a fatal case of encephalitis in La Crosse, Wisconsin (THOMPSON et al. 1965). La Crosse virus was shown to be closely related to, but distinct from California encephalitis virus, and it was rapidly recognized to cause human infections over a wide area of the midwestern United States (KAPPUS et al. 1983). Extensive serological studies identified several other arboviruses that were related to California encephalitis and La Crosse viruses, and together these were designated the California serogroup (WHITMAN and SHOPE 1962).

The regular reporting of encephalitis due to California serogroup viruses in the United States began in 1963, with the recognition of these agents as an important cause of arboviral encephalitis. Cases reported from 1963 to 1981 (KAPPUS et al. 1983) indicate that the incidence has remained approximately constant, at about 75 cases (range 42–174) annually, which contrasts with the marked epidemicity of St. Louis encephalitis. Circumstantial evidence suggests that encephalitis caused by California serogroup viruses is underreported (KAPPUS et al. 1983).

2.1.2 La Crosse Encephalitis

La Crosse virus produces a classical acute encephalitis in children (CHUN 1983; CHUN et al. 1968; GUNDERSEN and BROWN 1983). The incubation period can be estimated only from studies in experimental animals and is probably about 7 days. The acute illness lasts 10 days or less in most instances. Typically, the first symptoms are nonspecific, last for 1–3 days, and are followed by the appearance of central nervous system (CNS) signs and symptoms such as stiff neck, lethargy, and seizures, which usually abate within 1 week. If cerebrospinal fluid is obtained on initial presentation, meningitis with both polymorphonuclear neutrophil leukocytes and mononuclear cells is found in about 65% of patients, and CSF protein is elevated in about 20% of cases.

The great majority (over 80%) of patients with acute encephalitis recover with no residual evidence of CNS infection. The most important sequela of La Crosse encephalitis is epilepsy, which occurs in about 10% of affected children, almost always in those who have had seizures during the acute illness (CHUN et al. 1968; GUNDERSEN and BROWN 1983). A few patients (estimated 2%) have persistent paresis, and learning disabilities or other objective cognitive deficits have been reported in a small proportion (no more than 2%). Learning performance and behavior of most recovered patients is not distinguishable from comparison groups in the same communities (CHUN et al. 1968; SABATINO and CRAMBLETT 1968).

Acute fatalities are rare (KAPPUS et al. 1983), since only four were recorded among 1348 cases (0.3%). The histologic picture of La Crosse encephalitis is based on the few acutely fatal cases which have come to autopsy (KALFAYAN 1983). The brain lesions are those of a typical acute encephalitis, consisting mainly of scattered glial nodules, perivascular cuffs, mild leptomeningitis, and occasional areas of focal necrosis. Inflammatory lesions are located mainly in the cerebral cortex, with some lesions in the brainstem and very few changes in the medulla and cord.

2.1.3 Epidemiology

La Crosse encephalitis is concentrated geographically in the midwestern United States; Minnesota, Wisconsin, Iowa, Illinois, Indiana, and Ohio report over 90% of all cases. Most cases have onset in July, August, or September. Children and young adults aged 1–19 years, particularly boys, are at greatest risk because of activities such as camping and hiking (BALFOUR et al. 1976; BERRY et al. 1975; MONATH et al. 1970). The distribution and activity of the principal woodland vector, *Aedes triseriatus*, explains these epidemiologic patterns, including factors of geographic localization, seasonality, age, and sex distribution. Humans probably do not transmit the virus to mosquitoes and are therefore considered to be dead-end hosts.

The prevalence of antibodies to California serogroup viruses has been studied by several groups (HENDERSON and COLEMAN 1971; KAPPUS et al. 1983;

MONATH et al. 1970; PARKIN et al. 1972; THOMPSON and EVANS 1965; ROWLEY et al. 1983; THOMPSON and GUNDERSEN 1983). A large survey in Indiana (GRIMSTAD et al. 1984a) showed an age-specific increase in the frequency of antibody to La Crosse virus, reaching about 20% at age 60, consistent with prior studies (HENDERSON and COLEMAN 1971; MONATH et al. 1970). It was estimated that there were about 300 000 new infections each year in the midwest, and that for children under age 16, there are more than 1000 infections per reported case (GRIMSTAD et al. 1984a; KAPPUS et al. 1983).

Extensive testing of mosquitoes in North America from 1960 to the present has shown that a number of California serogroup viruses are enzootic in North America. Several of these viruses infect humans and some cause sporadic cases of encephalitis (Table 1). However, most human cases of encephalitis are caused by La Crosse virus, and there has been little recent evidence of human disease due to California encephalitis virus itself (REEVES et al. 1983).

Jamestown Canyon virus has recently been shown (GRIMSTAD et al. 1984b) to infect humans; in Michigan over 25% of unselected premarital sera exhibited antibody. Surveys in midwestern states and in New York State (DIEBEL et al. 1983; SRIHONGSE et al. 1984) showed that 5%–10% of tested sera were positive, with rates higher in adults than in children. In New York State, Jamestown Canyon virus has also been recognized as a cause of human encephalitis, resembling encephalitis caused by La Crosse virus except that most of the cases occurred in adults rather than in children (DIEBEL et al. 1983; SRIHONGSE et al. 1984).

Snowshoe hare virus (an antigenic variant of La Crosse virus) has been isolated from mosquitoes throughout much of Canada (CALISHER 1983; MCLEAN 1983) but rarely in the United States and thus appears to have a geographic distribution distinct from that of La Crosse Virus, which has never been isolated in Canada. Snowshoe hare virus has been implicated as a cause of human encephalitis in Canada, mainly in Nova Scotia, Quebec, and Ontario (ARTSOB 1983; GRIMSTAD et al. 1984a; SRIHONGSE et al. 1984).

Trivittatus virus has been isolated from mosquitoes in essentially all the contiguous United States (CALISHER 1983; ROWLEY et al. 1983). In certain parts of the midwest, a considerable proportion of the population has antibody against trivittatus virus. However, it has not been associated with human illness (GRIMSTAD et al. 1984b; MONATH et al. 1970; ROWLEY et al. 1983), except for one recently reported case (SRIHONGSE et al. 1984).

2.2 Experimental Infection

2.2.1 Infection of Cell Cultures: ts Mutants, Complementation, and Reassortment

California serogroup viruses cause a productive cytocidal infection in many vertebrate cell lines (BISHOP and SHOPE 1979; DANIELOVA 1975), and they can be grown and plaqued conveniently in BHK-21 or Vero cells (MURPHY et al. 1968,

1975). These viruses also replicate in mosquito cell lines, in which they produce a persistent nonlytic infection (LYONS and HEYDUK 1973; RAJU and KOLAKOFSKY 1989; ROSSIER et al. 1988).

Temperature sensitive (ts) mutants of bunyaviruses are easily obtained by standard methods (BISHOP and SHOPE 1979; VEZZA et al. 1979) and have been used extensively for genetic studies. California serogroup viruses readily complement each other when two ts mutants, with ts lesions in different gene segments, are used to initiate dual infections at the restrictive temperature followed by assay at the permissive temperature (BISHOP 1979). Such dual infections produce virus yields several-fold greater than those obtained when each of the individual ts mutants is grown at the restrictive temperature and assayed at the permissive temperature. The complementation index is the ratio of the yield of the dually infected culture divided by the sum of the yields of the two ts mutants grown separately.

Reassortment of gene segments can occur between many members of the California serogroup with the exception of Guaroa virus (BISHOP 1985a, b; BISHOP et al. 1980; GENTSCH and BISHOP 1976, 1978; GENTSCH et al. 1979). However, California serogroup viruses apparently do not reassort with viruses of other serogroups within the *Bunyavirus* genus, such as members of the Bunyamwera or group C complexes (BISHOP et al. 1984a,b; ELLIOTT et al. 1984; IROEGBU and PRINGLE 1981a,b; PRINGLE and IROEGBU 1982).

Reassortants can be constructed by coinfecting a permissive cell culture with two different viruses and then plaquing the progeny to obtain discrete genetic clones. To facilitate the selection of reassortants, ts mutants may be used as parental viruses, and the progeny can be selected at the restrictive temperature. This method requires that the two parental viruses have ts lesion in different gene segments, so that the parents complement each other. The use of ts mutants of California serogroup viruses has been limited by the fact that it was possible to obtain mutants only in the L and M RNA segments and not in the S RNA segment; thus, this approach could not be used to construct all possible reassortant genotypes (BISHOP 1979; BISHOP and SHOPE 1979). More recently (JANSSEN et al. 1986) reassortants have been constructed using wild-type (i.e. nonmutagenized) parental virus stocks, which yielded all possible segment combinations. However, the latter approach required laborious screening of progeny to identify individual genotypes.

The frequency of reassortment in cell cultures can be estimated by conducting dual infections with two different ts mutants at the permissive temperature and assaying the progeny at the permissive and restrictive temperatures; the yield at the restrictive temperature (as a percent of the yield at the permissive temperature) provides an estimate of the frequency of reassortment. When two California serogroup viruses are crossed (BISHOP and SHOPE 1979), the frequency of reassortment is 5%–50%. Reassortment can occur in mosquitoes that are dually infected (BEATY et al. 1981, 1985), and analysis of field isolates (KLIMAS et al. 1984) suggest that reassortment may also occur in nature.

2.2.2 Infection of Mosquitoes

Infection of *Aedes triseriatus*, the major vector mosquito, with La Crosse virus has been studied in some detail (BEATY and BISHOP 1988; BEATY and THOMPSON 1976; MCLEAN et al. 1975; 1979; PAULSON and GRIMSTAD 1989; TESH and BEATY 1983; TURRELL et al. 1984). Following ingestion of virus, the epithelial cells which line the midgut are the first to become infected. After an eclipse phase during which no infectivity can be detected, virus first appears in the hemocele, just outside the basal lamina of the midgut; from the hemocele virus has access to all body tissues. Many different tissues are permissive, and antigen can be visualized in the heart, neural ganglia, fat body, ovaries, and salivary glands. The salivary glands are one of the last tissues infected, and only a limited number of these cells are antigen positive; infected cells discharge large numbers of virions into the lumen of the gland, whence they can be injected into the vertebrate host during feeding. For La Crosse virus, the extrinsic incubation period, the interval from feeding until the mosquito is infectious, is 7–14 days.

The mosquito midgut may act as a barrier in the stepwise spread of infection following virus feeding. The concept of a midgut barrier arises from the observation that when virus feeding fails to lead to the completion of the extrinsic incubation period, viral antigen may accumulate in the midgut epithelium without dissemination throughout the mosquito (TESH and BEATY 1983). Intrathoracic injection of the same virus may initiate a disseminated infection (MILLER 1983), suggesting that the midgut barrier is the critical block in the infectious cycle. Ingested virus is exposed to proteolytic enzymes in the mosquito midgut, which digest the G1 but not the G2 protein, and increase the infectivity of the virion for mosquito cells (LUDWIG et al. 1989); possibly this phenomenon is involved in the midgut barrier.

Infection of ovaries, including oocytes, occurs as part of systemic infection with California serogroup viruses. This results in transovarial transmission (PAULSON and GRIMSTAD 1989; TESH and BEATY 1983; WATTS et al. 1973) to both female and male offspring. Transovarially infected mosquitoes have widespread infection, including gonadal tissue of both sexes, which provide mechanisms for further vertical transmission of virus and for venereal spread (THOMPSON and BEATY 1978) within the mosquito population. Mosquitoes show no apparent disease as a result of California serogroup virus infection, although subtle reductions in biological viability or breeding efficiency may occur (DEFOLIART 1983; TURELL and HARDY 1980).

It has been possible to identify some of the genetic determinants which control the infectivity of California serogroup viruses for host mosquitoes by the use of reassortants between viruses which differed in their ability to infect mosquitoes. The middle RNA segment appears to be the major genetic determinant of infectivity (BEATY et al. 1982; BEATY and BISHOP 1988). Furthermore, a variant of La Crosse virus, selected as a monoclonal antibody resistant (MAR) clone, exhibited reduced ability to infect mosquitoes. When a revertant was selected by passage of the low-infectivity variant through mosquitoes, the revertant had regained both normal infectivity for mosquitoes and the epitope lost by antibody selection (SUNDIN et al. 1987).

2.2.3 Infection of Vertebrates

La Crosse virus can infect vertebrates of the species traditionally used for experimentation, including rabbits, rats, and mice. However, most investigators have used the laboratory mouse as a model for studying pathogenesis, because many features of the murine infection, such as the pathological process, the variable outcome of infection, and the influence of age, appear to mimic the disease in humans and in natural rodent hosts (JOHNSON and JOHNSON 1968; JOHNSON 1983; JANSSEN et al. 1984; TIGNOR et al. 1983).

Pathogenesis in Mice. Newborn laboratory mice of any strain are highly susceptible to infection with La Crosse virus, regardless of the route of administration, and a fatal infection can be produced with minimal inocula (about 1 PFU) of wild-type La Crosse virus. The virus is much less virulent in adult mice when administered parenterally, but the older animals are still susceptible to small doses of virus injected intracerebrally. This age-dependent variation has been useful in delineating the importance of neuroinvasiveness and neurovirulence to the pathogenesis of the infection (see below).

The course of infection in newborn mice begins with an extraneural phase and progresses to a CNS phase (JOHNSON and JOHNSON 1968; JANSSEN et al. 1984; 1986), as illustrated in Figs. 1 and 2. Following subcutaneous or intraperitoneal injection, the extraneural phase is initiated by replication in muscle tissue adjacent to the site of inoculation. This presumably leads to a low-level systemic viremia that results in further seeding of heart and striated muscle throughout the animal. The plasma viremia is thus markedly amplified, leading to invasion of the CNS. The exact mode of penetration of the virus into the brain has not been determined, but viral antigen has occasionally been seen in or close to blood vessels, making infection of endothelial cells a strong possibility (JOHNSON and JOHNSON 1968; JANSSEN et al. 1984).

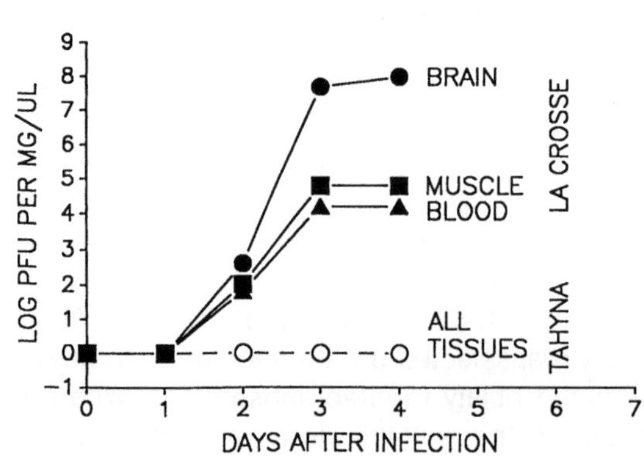

Fig. 1. Replication of La Crosse/original and Tahyna/181-57 viruses after subcutaneous injection of 700 PFU in sucking mice. (From JANSSEN et al. 1984)

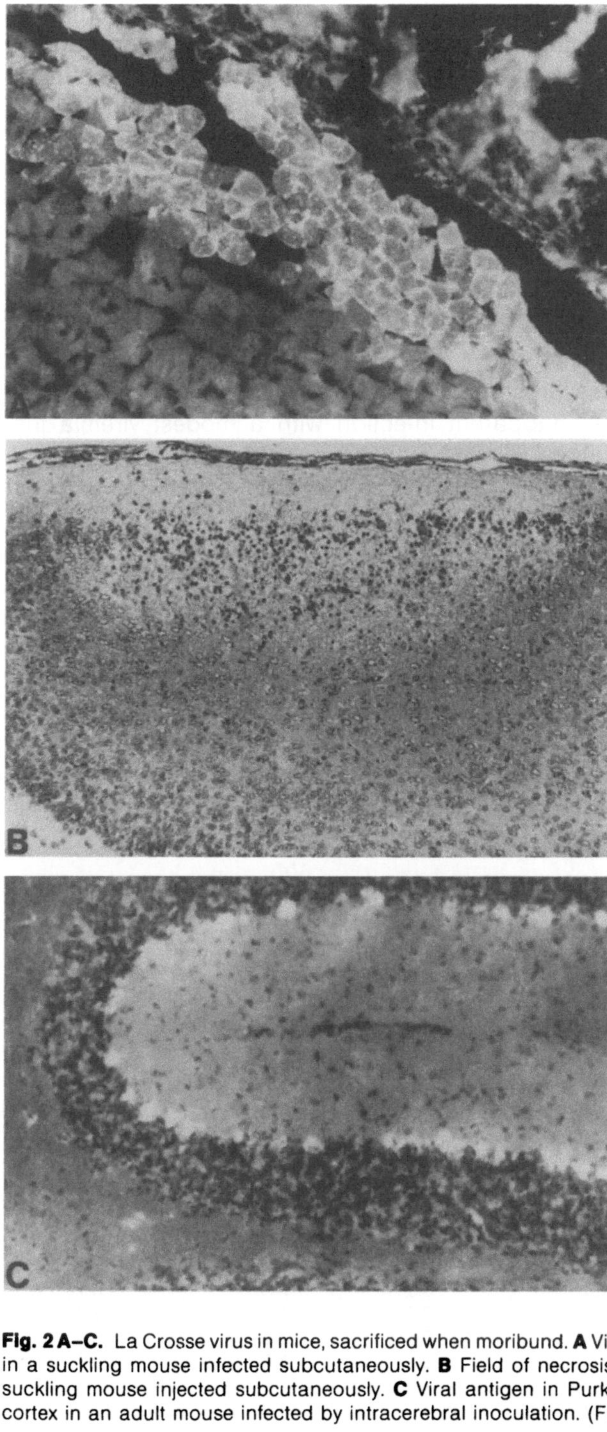

Fig. 2 A–C. La Crosse virus in mice, sacrificed when moribund. **A** Viral antigen in striated muscle in a suckling mouse infected subcutaneously. **B** Field of necrosis in the cerebral cortex of a suckling mouse injected subcutaneously. **C** Viral antigen in Purkinje's cells of the cerebellar cortex in an adult mouse infected by intracerebral inoculation. (From JANSSEN et al. 1984)

Within the CNS of suckling mice, all of the cellular elements are susceptible to infection. As shown in Fig. 2, a severe panencephalitis can be demonstrated by both routine histologic stains, where massive neuronal necrosis is evident, and by immunohistochemistry, which shows robust viral replication in both neurons and glia (JANSSEN et al. 1984). Ultrastructural studies of infected mouse brain showed that California serogroup viruses mature by budding into intracytoplasmic spaces, particularly in the Golgi complex, followed by export of virus in vacuoles migrating to the plasma membrane (MURPHY et al. 1968a, b; LYONS and HEYDUK 1973). In adult mice the virus does not replicate extensively after peripheral inoculation. Intracerebral injection results in a fatal encephalitis following a small virus dose (JANSSEN et al. 1984), with neuronal death being the primary histologic finding.

Wild Rodents. La Crosse virus infection of adult chipmunks, an important natural host, produced an inapparent infection with a modest viremia just adequate to infect mosquitoes (KSIASEK and YUILL 1977, PANTUWATANA et al. 1972; PATRICAN et al. 1985; SEYMOUR et al. 1983). Newborn chipmunks, infected subcutaneously, died of encephalitis (SEYMOUR et al. 1983).

2.2.4 Genetic Determinants of Virulence

Peripheral Virulence. The segmented bunyavirus genome and the construction of reassortant viruses have made it possible to map the determinants of the virulence of California serogroup viruses, following both peripheral and intracerebral routes of infection. An essential prerequisite for this work was the identification of a virus strain, Tahyna/181-57, that is avirulent following peripheral inoculation (Fig. 1), and can reassort with wild-type La Crosse virus

Table 3. Virulence of reassortants between a nonneuroinvasive strain (Tahyna/181-57) and a neuroinvasive strain (La Crosse/original) of California serogroup viruses. (From JANSSEN et al. 1986; GONZALEZ-SCARANO and NATHANSON 1989; ENDRES et al. 1990)

Virus	Genotype	Adult mouse ic $\log_{10}$ PFU/LD_{50}	Suckling mouse ic $\log_{10}$ PFU/LD_{50}	Suckling mouse sc $\log_{10}$ PFU/LD_{50}
Parental viruses				
La Crosse original	LLL	0.1	−0.4	−0.3
Tahyna 181–57	TTT	0.0	−0.5	4.2
Reassortant viruses				
B1-11a	TLT	0.3	ND	−0.3
A1-3a	LLT	0.3	ND	−0.3
P1-2gg	TLL	0.0	ND	0.1
F1-18a	LTL	−0.6	ND	3.2
F1-2a	TTL	−0.5	ND	3.3
B1-1a	LTT	0.3	ND	3.9

Reassortants: intracerebral titrations were done in weanling, not adult mice. ND: not done

(JANSSEN et al. 1986). Reassortants between La Crosse and Tahyna viruses were selected by coinfection in BHK-21 cells, followed by extensive screening of the progeny until clones representing all possible genetic combinations were obtained. The reassortant viruses were then screened for peripheral virulence in suckling mice by subcutaneous injection (Table 3). The results indicated that virulence cosegregated with the middle RNA segment. Thus, the property of peripheral virulence, or neuroinvasiveness, resides within the genomic segment encoding the viral glycoproteins (and at least one nonstructural protein). Similar experiments with reassortants derived from parent viruses with ts mutations have led to the same conclusion (SHOPE et al. 1981).

Neutralizing monoclonal antibodies, recognizing epitopes on the G1 or major glycoprotein, can be used to select MAR variant clones. Some MAR variant viruses demonstrated reduced neuroinvasiveness (GONZALEZ-SCARANO et al. 1985, 1988; GONZALEZ-SCARANO and NATHANSON 1989) and slightly reduced neurovirulence as well. The attenuated variants (but not non-attenuated MAR variants) also showed alterations in the pH profile at which they induced cell-to-cell fusion, suggesting the importance of cell entry in virulence (GONZALEZ-SCARANO et al. 1983, 1985).

Intracerebral Virulence (*Neurovirulence*). All wild-type isolates of California serogroup viruses, as well as the nonneuroinvasive Tahyna/181-57 strain, appear to be virulent after intracerebral injection into adult mice (Table 3). It was therefore necessary to derive a laboratory variant which was neuroattenuated (ENDRES et al. 1990), following a method originally used for Sindbis virus (JOHNSTON and SMITH 1988). A La Crosse–Tahyna reassortant of unexpectedly reduced virulence (presumably a spontaneous mutant) was passaged 25 times in BHK-21 cells, and an attenuated clone, designated B.5, was isolated. Clone B.5 is dramatically reduced in intracerebral virulence, as shown in Fig. 3. Neuroattenuation is a property of this particular clone and not of reassortants of the same genotype (Table 4). B.5 also exhibits markedly reduced neuroinvasiveness when injected subcutaneously in suckling mice, although it retains high virulence following intracerebral injection of suckling mice. Clone B.5 grows

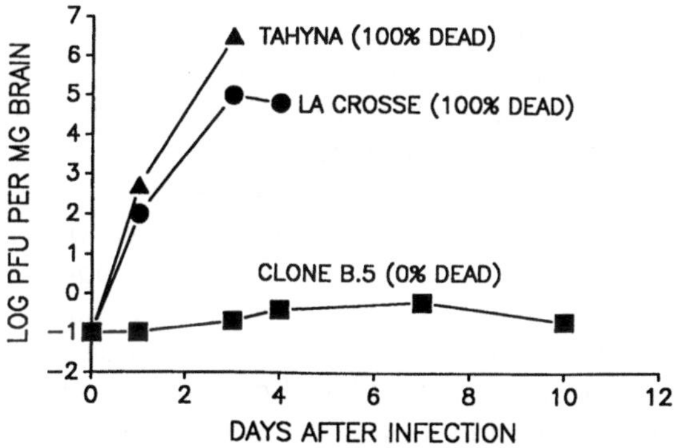

Fig. 3. Brain replication of La Crosse/original and Tahyna/181-57 viruses, and neuroattenuated clone B.5 following intracerebral injection of 700 PFU in adult mice. (From ENDRES et al. 1990)

Table 4. A nonneurovirulent reassortant clone (B. 5) of California serogroup virus compared with parental viruses and other reassortants with similar genotype. (From ENDRES et al. 1990).

Virus	Genotype	Adult mouse ic $\log_{10}$ PFU/LD_{50}	Suckling mouse ic $\log_{10}$ PFU/LD_{50}	Suckling mouse sc $\log_{10}$ PFU/LD_{50}
Reference viruses				
La Crosse original	LLL	0.1	−0.4	−0.3
Tahyna 181-57	TTT	0.0	−0.5	4.2
Reassortant viruses				
P1-2gg	TLL	0.3	0.2	0.1
B. 5	TLL	>6.1	0.4	5.4

Table 5. Neurovirulence of reassortants between a nonneurovirulent clone (B. 5) and a virulent clone. (From ENDRES et al. 1990)

Virus	Genotype	Virulence genotype	Adult mouse ic $\log_{10}$ PFU/LD_{50}[a]	Plaque size
Parental clones				
B. 5	TLL	aaa	>6.1	Small
B1-1a	LTT	vvv	0.0	Large
Reassortant clones				
J-18/6	LLL	vaa	0.8	Large
J-18/33	TTL	ava	>5.7	Small
J-18/38	TLT	aav	4.8	Small

[a] Ratios computed at 14 days after ic injection. For TLT clones, ratio drops from 4.8 to 1.6 when computed at 21 days

well in BHK-21 cells at temperature of 37 °C and lower but replicates very poorly at 39.8 °C (temperature sensitive).

The neuroattenuation of clone B.5 made it possible for the first time to map neurovirulence. For this purpose, reassortants were constructed between clone B.5 and a virulent clone with a reciprocal genotype (Table 5). When reassortants were isolated which carried single gene segments from the virulent parent, it was possible to map neurovirulence primarily to the large (L) RNA segment.

Reassortant viruses have provided an important tool for delineating the properties of neuroinvasiveness and neurovirulence in California serogroup viruses. Peripheral virulence maps to the middle RNA segment, and probably to G1, the major glycoprotein. In contrast, neurovirulence maps to the large RNA segment, which encodes the L protein (ENDRES et al. 1989), thought to represent the viral polymerase (ELLIOTT 1989, 1990), and less to the S segment, which encodes the nucleocapsid protein. Since attenuated clone B.5 replicates well in BHK-21 cells at the body temperature of the mouse, it appears that the viral polymerase and nucleocapsid proteins of this clone function effectively in BHK-21 cells but not in neurons. Thus, both neuroinvasiveness and neurovirulence appear associated with tissue-specific functions of different viral proteins.

In addition to viral genetic determinants, preliminary comparisons of different inbred strains of mice suggest that there are major host determinants of susceptibility (PIFAT and SMITH 1987).

3 Other Members of the *Bunyavirus* Genus

The *Bunyavirus* genus is named after Bunyamwera virus and includes over 150 individual viruses classified in over 15 serogroups (GONZALEZ-SCARANO and NATHANSON 1990). Almost all of these viruses are transmitted by mosquitoes (some members have also been isolated from *Culicoides* midges), but they have a wide variety of vertebrate hosts. A few viruses in the Simbu and Bunyamwera serogroups cause infection or disease in agriculturally important animals, and a number of them have been associated with febrile illness in humans (CALISHER et al. 1988; PARSONSON and MCPHEE 1985; SHOPE 1985). This account is limited to two of these viruses, Akabane and Oropouche, which have caused major epidemics in livestock and humans, respectively. Akabane and Oropouche viruses are both members of the Simbu serogroup, which includes over 20 viruses isolated on several continents in both hemispheres (CALISHER et al. 1969; CAUSEY et al. 1972; DAVID-WEST 1972; HARDY et al. 1970; KINNEY and CALISHER 1981).

3.1 Oropouche Virus

Oropouche virus was initially isolated from a patient with an acute febrile illness on the island of Trinidad off the coast of Brazil (ANDERSON et al. 1961). Since that time it has caused repeated epidemics with thousands of cases in Para province surrounding the Amazon delta in nothern Brazil (DIXON et al. 1981; PINHEIRO et al. 1981a, b; ROBERTS et al. 1981). The major urban vector is the midge *Culicoides paraensis*, and the virus may be maintained in a vector–human cycle during epidemics (ROBERTS et al. 1981). Since epidemics occur only during the rainy season and last no longer than 6 months, it is postulated that the virus is maintained in a sylvatic cycle, which has been shown to involve sloths, monkeys, and probably jungle mosquitoes (PINHEIRO et al. 1982). Oropouche disease is an acute febrile illness of 1–2 weeks' duration with arthralgia, myalgia, headache, and prostration, but there are no deaths and patients recover rapidly.

Oropouche virus kills suckling mice after intracerebral or intraperitoneal injection and adult mice only after intracerebral injection.

3.2 Akabane virus

Epidemics of congenital defects have been observed in cattle, sheep, and goats in Australia since the 1930s (HAUGHEY et al. 1988; PARSONSON and MCPHEE 1985; WHITTEM 1957), and similar outbreaks were reported from Japan (MIURA et al. 1974). Akabane virus was first isolated from mosquitoes in Japan (OYA et al. 1961) and has subsequently been isolated from *Culicoides brevitarsis* midges in Australia (DOHERTY et al. 1972; MURRAY 1987), and various *Culicoides* species have been shown to be competent vectors (JENNINGS and MELLON 1989). The disease potential of Akabane virus was revealed by surveys of sera from calves born during teratogenic epizootics in Japan and Australia, which showed that affected animals had antibody to this virus (HARTLEY et al. 1975; INABA et al. 1975; KUROGI et al. 1975; MIURA et al. 1974). Since neither IgG nor IgM pass the placenta in cattle, antibody in calves implies active infection with the virus. Akabane virus also occurs in Africa and the Middle East, based on serological surveys of ruminants and in occurrence of epizootics of fetal malformation in cattle (DAVIES and JESSETT 1985).

Experimental infection of pregnant cattle confirmed the ability of Akabane virus to produce fetal wastage and anomalies (PARSONSON and MCPHEE 1985; PARSONSON et al. 1975). Under field conditions, serial monitoring of pregnant cows detected new infections, and these were correlated with the occurrence of fetal defects. Subsequent virus isolation from affected calves provided definitive evidence of the cause of natural epizootics (KUROGI et al. 1976). Further studies in cattle have correlated the gestational age of infection with the resulting defects; infection from days 76 to 104 causes hydrancephaly; from days 104 to 173, arthrogryposis; and after day 173, poliomyelitis (HARTLEY et al. 1975; KONNO et al. 1975; KUROGI et al. 1977; PARSONSON and MCPHEE 1985). Similar observations on experimentally infected pregnant ewes showed that there was an initial viremia in the ewe, following which the virus grew in the placenta and fetal membranes and then replicated in those fetal tissues involved in anomalies (PARSONSON and MCPHEE 1985; PARSONSON et al. 1977).

Akabane virus infects laboratory rodents. Adult mice develop lethal encephalitis after intracerebral but not after intraperitoneal injection (PARSONSON and MCPHEE 1985). Infection of pregnant hamsters results in a lethal infection of the fetuses, which are stillborn (ANDERSON and CAMPBELL 1978). Infection of embryonated chicken eggs can be used as a convenient experimental system since it produces a variety of malformations (KONNO et al. 1988; MCPHEE et al. 1984).

In the United States, recent outbreaks of fetal abnormalities in sheep have been associated with Cache Valley virus, another member of the *Bunyavirus* genus (EDWARDS et al. 1989).

4 *Phlebovirus* Genus

The *Phlebovirus* genus comprises over 30 viruses, transmitted either by mosquitoes or by phlebotomine flies (TESH 1988); the type virus is sandfly fever Sicilian virus, which causes phlebotomus fever. The most important pathogens are RVFV (EASTERDAY 1965; MEEGAN and SHOPE 1981; SHIMSONY and BARZILAI 1983) and the sandfly fever viruses (SABIN 1948, 1951).

4.1 Rift Valley Fever Virus (RVFV)

4.1.1 History

RVFV was first isolated in 1930 during an epidemic among sheep on a farm in the Rift Valley in East Africa (DAUBNEY and HUTSON 1931). The disease was transmitted by inoculating blood, plasma, or homogenate of minced liver from an acutely ill sheep into another sheep, and the agent was shown to pass a bacteria-retaining filter. A similar disease occurred in goats and cattle, and humans in the epizootic area developed an acute influenza like illness. The newly isolated agent was named Rift Valley fever virus, and in retrospect it was possible to identify epizootics caused by the virus as far back as 1912 (EDDY and PETERS 1980). For many years, RVFV remained an unclassified arbovirus until it was shown in 1973 to be morphologically similar to the bunyaviruses (MURPHY et al. 1973), and serological tests indicated that it was antigenically related to phlebotomus fever virus and belonged in the *Phlebovirus* genus.

In 1977–1978, RVFV appeared for the first time in Egypt in a epizootic-epidemic of unprecedented size (MEEGAN 1979). The epizootic was very intense in several discrete areas along the Nile; within these areas 25%–50% of sheep and cattle were infected, and infection was frequent in a number of other domestic animals. In the same areas it was estimated that there were as many as 200 000 human cases and at least 600 deaths. In the epidemic areas, human infection rates were as high as 35%. The 1977 epizootic in Egypt began in early September and ended in December with a seasonal drop in temperature and in mosquito populations. *Culex pipiens* was incriminated as the major vector during the epidemic (HOOGSTRAAL 1979).

Epizootics of Rift Valley Fever have been reported from many sub-Saharan countries since 1950, and these have occurred as far north as Egypt and Sudan (EASTERDAY 1965; EDDY and PETERS 1980; MEEGAN 1979; TOMORI 1980) and as far west as Mauritania (KSIASEK et al. 1989; WALSH 1988).

4.1.2 Disease in Animals and Humans

RVFV produces severe disease in domestic animals, sheep being more susceptible than cattle, while goats are least susceptible. Lambs experience over 90% mortality, adult sheep about 25%, and pregnant ewes usually abort (DAUBNEY and HUDSON 1931; MEEGAN and SHOPE 1981).

In humans infected with RVFV, there is an incubation period of 2–6 days, and the influenza like illness begins with abrupt onset of fever and associated symptoms. The acute disease lasts 2–5 days, followed by convalescence which may be prolonged (EDDY and PETERS 1980; LAUGHLIN et al. 1979; MEEGAN 1979; MEEGAN and SHOPE 1981). The viremia lasts for the duration of the acute illness and antibody appears at about the time of termination of viremia. A small proportion of cases (less than 5%) may be more serious or fatal, and these fall into several distinct categories: liver necrosis with hemorrhagic phenomena, retinitis with visual impairment, and meningoencephalitis (MEEGAN and SHOPE 1981; SIAM et al. 1980). The case fatality rate from these serious complications is probably less than 1% (LAUGHLIN et al. 1979).

4.1.3 Ecology

RVFV can be transmitted by a variety of mosquito species, including *Culex pipiens, Culex theileri*, and *Aedes* species (EDDY and PETERS 1980; HOOGSTRAAL et al. 1979; TURELL et al. 1984), and the virus infects many large domestic animals. As a result, epizootics can occur in diverse ecological settings. Outbreaks are characterically sporadic in any one area, and interepizootic intervals may last many years (EDDY and PETERS 1980). Epizootics are usually associated with a particularly wet rainy season and consequent high mosquito density and terminate with the end of the rains (DAVIES 1975; DAVIES et al. 1985). It has been postulated that there is a wildlife–mosquito cycle that maintains the virus in interepizootic periods, but this cycle has not been defined (EDDY and PETERS 1980). In addition, there is evidence that transovarial transmission might constitute an alternative mechanism for interepizootic perpetuation (LINTHICUM et al. 1985).

In an outbreak, disease is usually seen first in animals and then in humans (MEEGAN 1979). Human infection occurs mainly among farmers and others in close contact with animals, and RVFV can be transmitted by contact with tissues or blood of infected animals, during care, autopsy, slaughter, or disposal of infected animals. Infection is postulated to result from transcutaneous or aerosol exposure (CHAMBERS and SWANEPOEL 1980; HOOGSTRAAL et al. 1979; MEEGAN and SHOPE 1981; SHOPE et al. 1982). During an epizootic, when there are many infected and dying animals, contact transmission may be a more important source of human infection than is mosquito transmission (JOUBERT et al. 1951). In the absence of an epizootic, sporadic mosquito-transmitted infections of humans are occasionally reported. Laboratory-acquired infections, probably due to aerosols, have long been recognized (SABIN and BLUMBERG 1947), and extreme precautions should be taken by those working with the virus.

4.1.4 Experimental Infection

Vertebrates. RVFV infects a wide variety of laboratory and domestic animals, replicates to high titer, and is often lethal (DAUBNEY and HUDSON 1931; EASTERDAY 1965; MEEGAN and SHOPE 1981; MIMS1956a–d; MITTEN et al. 1970). The two principal target organs are the liver (focal necrosis) and the brain (necrotic encephalitis). In sheep, the usual epizootic host, there is an incubation period of 2–4 days followed by an acute illness with death from hepatic necrosis (DAUBNEY and HUDSON 1931; EASTERDAY 1965).

Laboratory rodents are also susceptible. Mice develop an acutely fatal hepatic infection that has been used as a model of the pathogenesis of RVFV in larger animals (BROWN et al. 1981; MIMS 1956a–d; MCGAVRAN and EASTERDAY 1963; PETERS and ANDERSON 1981; PIFAT and SMITH 1987), while gerbils undergo an acute encephalitis without extraneural lesions (ANDERSON et al. 1988). Different strains of inbred rats exhibit strikingly different responses; some are resistant and undergo a subclinical immunizing infection, some develop fatal fulminant hepatitis, and some develop encephalitis (ANDERSON et al. 1987; PETERS and ANDERSON 1981). Host resistance segregates as a single Mendelian dominant locus (ANDERSON et al. 1987; PETERS and ANDERSON 1981). Resistant rats appear to contain the initial replication within the liver, while in susceptible animals the infection spreads thoughout the hepatic parenchyma. Encephalitis develops 2–3 weeks after infection and is necrotic and inflammatory.

Experimental infection of rhesus macaques with RVFV produces a severe hemorrhagic fever syndrome in about 20% of animals, similar to some of the complications seen in humans (COSGRIFF et al. 1989; MORRILL et al. 1990; PETERS et al. 1988, 1989). Delayed interferon responses were correlated with prolonged viremia and severe disease (MORRILL et al. 1990), and exogenous interferon alpha, administered at the time of virus challenge, suppressed viremia and reduced mortality (MORRILL et al. 1989).

The pathogenesis of RVFV infection following natural transmission by mosquito bite can be reconstructed from experimental studies (MIMS 1956a–c; PETERS and ANDERSON 1981; PETERS et al. 1989). Inoculated virus presumably spreads from skin to draining lymph nodes and probably replicates there, perhaps in macrophages (LEWIS et al. 1987), before spreading by efferent lymphatics to the circulation. The liver is rapidly invaded, with immediate involvement of hepatocytes; it is unclear whether the initial site of entry is Kupffer's cells, with subsequent spread to the hepatic parenchyma (MCGAVRAN and EASTERDAY 1963). Replication in the liver produces large amounts of progeny virus and is probably a major source of the high-titer plasma viremia. Hepatic necrosis and disseminated intravascular coagulation probably both contribute to the hemorrhagic syndrome seen in monkeys (COSGRIFF et al. 1989; PETERS et al. 1989). Virus may also cross the blood-brain barrier and infect neurons and glia. However, the meningoencephalitis and retinitis that develop 2–3 weeks after infection are highly inflammatory and may be mediated, at least in part, by immune mechanisms (PETERS and ANDERSON 1981).

Different field isolates of RVFV vary dramatically in their virulence for laboratory rodents (PETERS and ANDERSON 1981). Isolates from the 1977 epizootic in Egypt were virulent, while isolates from sub-Saharan Africa were relatively avirulent in rats. Plaque-purified clones derived from a single Egyptian isolate differed in virulence, as assayed by peripheral in suckling mice (PETERS and ANDERSON 1981; MOUSSA et al. 1982) Virulent virus clones were less sensitive to rat interferon and formed plaques on isolated rat hepatocytes (ANDERSON and PETERS 1988).

RVFV can be neuroadapted by serial intracerebral passage in mice (MACKENZIE et al. 1936; SMITHBURN 1949). Neuroadapted strains show reduced hepatotropism and reduced lethality when injected by peripheral routes but are highly virulent after intracerebral injection. RVFV can also be altered by chemical mutagenesis, and a mutagenized clone of RVFV was markedly attenuated when used to infect sheep, by comparison with wild-type virus (CAPLEN et al. 1985; MORRILL et al. 1987; TAKEHARA et al. 1989).

Mosquitoes. In *Culex pipiens* mosquitoes the sequential steps in infection have been studied following ingestion of RVFV. The virus first appears in the midgut, followed by hemolymph, salivary glands, and ovaries (FARAN et al. 1987, 1988). Infection may reduce mosquito viability.

4.2 Sandfly Fever Viruses

Phlebotomus fever viruses are transmitted by phlebotomus flies and produce an acute, nonfatal, influenza like illness in humans (phlebotomus fever or sandfly fever). The vectors are small flies the size of midges which are members of the *Phlebotomus, Sergentomyia*, and *Lutzomyia* genera.

History. Descriptions of an acute illness which probably represented phlebotomus fever date back to the time of the Napoleonic wars, when a similar disease was reported as Mediterranean fever (SABIN 1948). Interest in phlebotomus fever increased during World War II due to the occurrence of epidemics of the disease in Allied troops in Italy in 1943–1944 (SABIN 1948, 1951). Three virus isolates from Sicily, Naples, and the Middle East were made by inoculating acute-phase serum intracerebrally into suckling mice, followed by serial blind passage in mice. Cross-challenge experiments in human volunteers (SABIN 1948) showed that the Naples and Sicilian isolates were antigenically distinct, while a Middle East isolate was identical to the Sicilian strain. These studies also demonstrated that a single infection gave solid protection against the same antigenic type. Phlebotomus fever was controlled in the Mediterranean by the use of DDT, to which sandflies are sensitive.

Sandfly fever viruses (Sicilian and Naples) continue to cause human infections in North Africa and the Middle East and are a potential risk for travelers (BALDUCCI et al. 1985). In addition, Toscana virus, transmitted by *P. perniciosus* in the foothills of the Tuscany region of Italy, causes human infections and sporadic cases of aseptic meningitis (PACI et al. 1985; TESH and MODI 1987; VERANI et al. 1984).

Human Disease and Ecology. Sabin studied over 100 experimentally infected human volunteers, providing a detailed description of the illness (SABIN 1948). After an incubation period of 2–6 days there is sudden onset of fever of up to 40°C and associated severe malaise, which lasts 2–4 days, with complete recovery. *Phlebotomus papatasii* and other *Phlebotomus* and *Sergentomyia* species are the principal vectors around the Mediterranean, in the Middle East, and across the Arabian peninsula and Iran into Pakistan and northwest India (BARNETT and SUYEMOTO 1961). *P. papatasii* breeds in dry sandy areas, near ground level, and feeds in the early evening when it may enter dwellings in search of a blood meal; the flies can even penetrate mosquito netting due to their size. Presumably, sandfly fever viruses are maintained in a wildlife cycle but this has not been identified. During intense epidemics, humans may act as a major host in the vertebrate–insect–vertebrate cycle. In addition, transstadial and transovarial transmission probably serve as alternative mechanisms of virus perpetuation (ENDRIS et al. 1983; TESH and MODI 1984; TESH et al. 1974).

Experimental Infection. Field isolates of sandfly fever viruses have repeatedly been reported to be poorly infectious for laboratory animals and not to kill suckling mice even after intracerebral injection, requiring multiple blind passages for adaptation (BARNETT and SUYEMOTO 1961; SABIN 1948; TESH et al. 1974). The viruses replicate well in Vero cell cultures, are cytocidal, and produce plaques (TESH et al. 1974). Sandfly fever viruses replicate in phlebotomines, with an extrinsic incubation period of 7–10 days (JENNINGS and BOORMAN 1983; SABIN 1948). Sandflies may be infected by feeding or by intrathoracic injection, and virus is transovarially transmitted to a proportion of offspring after experimental infection (TESH and MODI 1984). Phlebotomus fever viruses also replicate in cultured mosquito cells and produce persistent infections (NICOLETTI and VERANI 1985).

5 *Hantavirus* Genus

Hantaviruses are enzootic viruses of wild rodents which cause persistent infections in their natural hosts. They are spread by aerosolized excreta, and human infections occur under conditions of close contact with the enzootic host.

Table 6. Selected biological features of the major *Hantavirus* complexes

Virus complex	Natural host	Geographic distribution	Human disease
Hantaan	*Apodemus agrarius* (striped field mouse) Seoul	Eastern Asia, eastern Europe	Korean hemorrhagic fever, severe
Seoul	*Rattus* species (wild/laboratory rat)	Eastern Asia; seaports, worldwide	Hemorrhagic fever, mild to severe
Puumala	*Clethrionomys glareolus* (bank vole)	Scandinavia, Europe, western USSR	Nephropathia epidemica, mild
Prospect Hill	*Microtus pennsylvanicus* (meadow vole)	United States	Infection only

Diseases caused by Hantaan and Seoul viruses are included in hemorrhagic fever with renal syndrome (HFRS)

Members of the *Hantavirus* genus are the only Bunyaviridae which are not arboviruses. Currently, four distinct complexes of viruses are recognized within the *Hantavirus* genus, as summarized in Table 6 (Sugiyama et al. 1987).

5.1 Hantaan Virus

History. Epidemic hemorrhagic fever was recognized among United Nations soldiers in Korea in 1951 and was called Korean hemorrhagic fever with renal syndrome (EARLE 1954). A similar clinical entity was described in the eastern Soviet Union, Manchuria, and China (CASALS et al. 1970). The transmissible nature of the disease was demonstrated by inoculating serum and urine from patients into human volunteers, and there was strong circumstantial evidence that the infection was contracted from wild rodents or their ectoparasites (LEE et al. 1982; LEE 1989). The occurrence of cases among research workers exposed in the laboratory to wild rodents collected in epidemic areas provided a dramatic demonstration of the possible mode of transmission. Numerous attempts to isolate or propagate the causal agent were unsuccessful (LEE 1982) until the agent was finally identified by the demonstration of immunofluorescent antigen in the lungs of the field mouse *Apodemus agrarius* using sera from patients convalescent from Korean hemorrhagic fever (LEE et al. 1978; French et al. 1981). The virus was passed from infected to uninfected *A. agrarius* but was not readily transmitted to laboratory rodents, and a survey of other Korean rodents showed that infection was confined to *A. agrarius* (LEE et al. 1981a).

Human Disease. Korean hemorrhagic fever is an acutely prostrating fever, in which about one-third of the patients develop hemorrhagic manifestations (the other two-thirds have an influenza-like illness), with 10%–15% developing shock, and a mortality of 5%–10% (EARLE 1954; LEE 1989). The major pathological lesions in fatal cases are disseminated hemorrhages and microscopic abnormalities in the kidneys (OLIVER and MCDOWELL 1957). The renal tubules are particularly affected and exhibit congestion, desquamation, and necrosis. The pathogenesis of hemorrhagic fever is not well understood. Several studies indicate that disseminated intravascular coagulation and platelet abnormalities play a role (M. LEE et al. 1989; PENTTINEN et al. 1981).

Epidemiology and Ecology. There are 300–700 hospitalized cases of Korean hemorrhagic fever annually in South Korea (LEE 1982, 1989). Over 100 000 cases of hemorrhagic fever with renal syndrome are currently reported each year in eastern China, and the disease also occurs in the eastern Soviet Union (LEDUC 1987). *A. agrarius* appears to be the sole natural host of Hantaan virus (LEE et al. 1981a). This rodent ranges over eastern China, Manchuria, and Korea, and its range determines the distribution of Hantaan virus. *A. agrarius* develops an asymptomatic persistent infection, and virus is excreted in the urine for months to years (LEE et al. 1981a).

Experimental Infection. Experimental infections are readily initiated in colonized *A. agrarius*, the natural host, and these mice develop persistent asymptomatic infections with widespread viral antigen, most commonly seen in lungs and kidneys (LEE et al. 1981a). Infectious virus is excreted in saliva, feces, and urine over a long period. However, the period of viremia is brief, 5–10 days after infection, and serum antibody appears about 10 days after infection. Experimental transmission between *Apodemus* mice can be easily demonstrated when uninoculated mice are housed with experimentally infected mice (LEE et al. 1981b). Mice are highly infectious for the first 40 days after their infection, that is, when their virus titers are highest, but are still capable of transmitting infection at low efficiency for as long as 1 year after infection.

Laboratory mice can be infected with Hantaan virus (KIM and MCKEE 1985; KURATA et al. 1983; MCKEE et al. 1985; NAKAMURA et al. 1985a, b; TSAI et al. 1982). Newborn animals have a fatal illness, but resistance to disease increases with age so that 3-week-old mice uniformly survive infection. Injection of newborn mice by any route causes a widespread infection with high titers of virus in brain, lung, and kidney and lower titers in other tissues; capillary endothelium is a prominent site of replication. Viremia is transient, ending at 2 weeks. Mice develop inflammatory and destructive lesions in brain, lung, and liver and die in 2–3 weeks. There are some inconsistencies in reported observations on laboratory mice, and these may reflect the use of virus which was mouse-adapted or tissue-culture passed (MCKEE et al. 1985).

Infected mice develop cytolytic T cells and a proliferative response to viral antigens (ASADA et al. 1988, 1989). Experiments with nude mice suggest that virus clearance is dependent upon both antibody and T cell responses (ASADA et al. 1987; NAKAMURA et al. 1985a, b; MCKEE et al. 1985).

Hantaan virus has been adapted to laboratory rats by blind passage of lung homogenates (P.W. LEE et al. 1981). Adapted virus produces persistent infection of lung and other tissues, similar to that seen in *Apodemus* mice. Rats are readily infected by exposure to virus aerosols, although the ID_{50} is about 100-fold lower than is the intramuscular ID_{50}.

A comparison of an attenuated and a virulent clone of Hantaan virus in suckling mice showed that the attenuated clone was much less efficient in inducing cell-to-cell fusion (TAMARA et al. 1989), similar to an attenuated variant of La Crosse virus (GONZALEZ-SCARANO et al. 1985).

5.2 Seoul Virus

Seoul virus was recognized during clinical studies of Korean hemorrhagic fever as a related but less severe disease in residents of Seoul and other urban areas where *Apodemus* mice were not found (LEE et al. 1987). Beginning in 1979, a similar disease was reported in personnel associated with laboratory rats (DESMYTER et al. 1983; LEE and JOHNSON 1982).

Ecology and Epidemiology. Seoul virus is widespread among urban rats, particularly in seaports (LEDUC et al. 1984, 1986; LEDUC 1987; CHILDS et al. 1987). Infected laboratory rats have been identified mainly in Japan and Korea or in animals distributed from these areas (LEDUC 1987). Screening of rat cell lines and laboratory rat suppliers indicates that Seoul virus is absent from most laboratory rat populations in the United States. Serological surveys suggest that human infections are acquired both from wild and laboratory rats; there is no evidence of person-to-person spread of the virus.

Experimental Infection. There are limited data on the pathogenesis of Seoul virus in experimentally infected rats (TANISHITA et al. 1986). When infected, both newborn and young adult rats undergo a persistent infection, and virus can be isolated from many tissues. Newborn rats exhibit higher virus titers and longer periods of persistence than do adult rats. Infected animals develop neutralizing serum antibody, which appears more rapidly in adults than in newborns, but the coexistence of neutralizing antibody and infectious virus has not been explained. About 50% of infected newborn rats die about 30 days after infection, while adult rats remain healthy. Different isolates of Seoul virus vary in their lethality for rats; an avirulent isolate exhibited reduced neuroinvasiveness and neurovirulence in suckling rats (ZHANG et al. 1989). Passive antibody given prior to challenge effectively protects against potentially lethal challenge in newborn rats (ZHANG et al. 1989a).

5.3 Puumala Virus

Nephropathia epidemica, an acute fever with renal involvement but with a relatively benign outcome, has been recognized for many years in Sweden, Norway, and Finland (MYRHMAN 1951; LAHDEVIRTA 1982; SETTERGREN et al. 1989). Following the isolation of Hantaan virus in Korea, it was found that patients who had recovered from nephropathia epidemica had antibodies to Hantaan virus. However, the absence of *Apodemus* from endemic areas in Scandinavia indicated that there was another wildlife host, and the association of cases in Finland with the bank vole, *Clethrionomys glareolus*, suggested that this rodent was the zoonotic reservoir. Antigen was detected in the lungs of bank voles, and it was then shown that the virus could be isolated and propagated in Vero E-6 cells (BRUMMER-KORVENKONTIO et al. 1980; YANAGIHARA et al. 1984). The virus was called Puumala, after one of the original bank vole collection sites in Finland. Puumala virus is maintained as a natural infection of bank voles and other wild rodents in Scandinavia and parts of Europe (BRUMMER-KORVENKONTIO et al. 1982; SOMMER et al. 1985; NICKLASSON and LEDUC 1987). Bank voles develop persistent infections with continuous virus shedding in urine and, presumably, aerosolized excreta are the source of human infection. Most cases occur during the fall and winter when voles often move from the field into man-made structures.

Experimental Infection. Colonized bank voles can be infected with Puumala virus, which produces an asymptomatic infection (BRUMMER-KORVENKONTIO et al. 1982; GAVRILOVSKAYA et al. 1983; OKER-BLOM et al. 1979; YANIGAHARA et al. 1984, 1985). The virus produces persistent infections in gerbils but not in laboratory mice and rats.

5.4 Prospect Hill Virus

Following the isolation of Hantaan virus, GAJDUSEK and colleagues conducted a serosurvey of microtine rodents in Maryland and found that a considerable proportion of sera from the meadow vole *Microtus pennsylvanicus* had antibody against Hantaan virus. Prospect Hill virus, subsequently isolated in Vero-E cells, can be distinguished antigenically from other hantaviruses and is enzootic in microtine rodents across the United States (LEE et al. 1985; YANAGIHARA et al. 1987). However, the virus rarely if ever infects humans and has not been associated with human disease. Experimental infection of subhuman primates with Prospect Hill virus caused a mild transient renal syndrome (YANAGIHARA et al. 1988).

6 *Nairovirus* Genus

The *Nairovirus* genus, named after Nairobi sheep disease virus (MONTGOMERY 1917), is distinguished by the fact that its members are all tick borne.

6.1 Crimean-Congo Hemorrhagic Fever (CCHF)

History. In the 1940s and 1950s agricultural workers in the Crimean peninsula experienced a newly recognized disease characterized by fever, prostration, and severe hemorrhages, with a mortality of about 10%. The disease was frequently associated with bites of Ixodid ticks of the genus *Hyalomma* (CHUMAKOV 1975). A number of isolates of a virus were made from the blood of acutely ill patients, and many of them showed serological conversion to the same agent (CHUMAKOV et al. 1968). During the 1960s, a virus was isolated from several patients with severe febrile disease in Zaire (formerly, Belgian Congo) and Uganda, and was named Congo virus (SIMPSON et al. 1967). Subsequent laboratory comparisons showed the two viruses to be identical (CASALS 1969; CHUMAKOV et al. 1970; TIGNOR et al. 1980), and they were eventually given the name Crimean-Congo hemorrhagic Fever (CCHF) virus.

Pathogenesis in Humans. The incubation period is estimated at 3–6 days, with abrupt onset of an acute febrile illness (SHEPHERD et al. 1987; SWANEPOEL et al. 1987, 1989). Initially, patients manifest severe influenza-like symptoms. Hemorrhagic manifestations usually occur after several days of illness and include a petechial rash, ecchymoses and bruises, hematemesis, and melena, accompanied by thrombocytopenia and leukopenia. Most cases show signs of hepatitis, with jaundice, hepatomegaly, and elevated serum levels of liver enzymes, and laboratory evidence of disseminated intravascular coagulopathy (Swanepoel et al. 1989). The majority of patients exhibit hemorrhagic signs, and 10%–50% mortality has been seen in different outbreaks. Deaths usually occur during the 2nd week of illness and are often preceded by severe hemorrhages, shock, and renal failure. Patients who survive the acute illness usually recover without sequelae. Subclinical infections also occur, since a number of seropositive humans, who give no history of typical illness, can be found in endemic areas (AL-TIKRITI et al. 1981a, b).

CCHF virus is distributed over the wide geographic range where Hyalomma ticks are found, including sub-Saharan Africa, eastern Europe, the Middle East, and areas in central Asia (AL-TIKRITI et al. 1981a, b; BURNEY et al. 1980; HOOGSTRAAL 1979; SHEPHERD et al. 1987; SULEIMAN et al. 1980; SWANEPOEL et al. 1983; YU-CHEN et al. 1985). CCHF virus is probably maintained in nature in both a transovarial/transstadial cycle in ticks and a tick–vertebrate cycle (HOOGSTRAAL 1979). In South Africa, where the most detailed studies have been conducted, CCHF virus is highly endemic in cattle and large wild herbivores and is also found in small mammals and birds (SHEPHERD et al. 1987; SWANEPOEL et al. 1987).

Transmission to humans occurs by tick bite and also directly by contact with infected animals or with their tissues. Agricultural workers, veterinarians, abbatoir workers, and other persons with exposure to cattle and ticks are at risk (SWANEPOEL et al. 1987). Human infection is relatively rare and most humans are seronegative even in endemic areas. In addition to tick and zoonotic transmission, person-to-person spread can also occur and has been well documented in many small nosocomial outbreaks (BURNEY et al. 1980; SHEPHERD et al. 1985; VAN EEDEN et al. 1985). Transmission to hospital personnel occurs through exposure to infected blood, respiratory secretions and aerosols, and excreta. The intensity of exposure appears to determine the frequency of infections and the severity of illness among contacts, which ranges up to 100% infections and 50% fatalities.

Experimental Infection. The modest published information on the nairoviruses indicates that they kill mice after intracerebral injection; high virus titers can be found in brain and liver (COATES and SWEET 1990; SMIRNOVA 1979). There are no animal models which reproduce the hemorrhagic fever syndrome seen in humans (J.B. MCCORMICK, personal communication, 1989).

References

Al-Tikriti SK, Al-Ani F, Jurji FJ (1981a) Congo-Crimean hemorrhagic fever in Iraq. Bull WHO 59: 85–90

Al-Tikriti SK, Hassan FK, Moslih IM, Jurji F, Mahmud MI, Tantawi HH (1981b) Congo-Crimean hemorrhagic fever in Iraq: a seroepidemiological survey. J Trop Med Hyg 84: 117–120

Anderson AA, Campbell CH (1978) Experimental placental transfer of Akabane virus in the hamster. Am J Vet Res 39: 301–304

Anderson GW, Peters CJ (1988) Viral determinants of virulence for Rift Valley fever (RVF) in rats. Microb Path 5: 241–250

Anderson CR, Spence L, Downs WG, Aitken TH (1961) Oropouche virus: a new human disease agent from Trinidad. Am J Trop Med Hyg 10: 574–578

Anderson GW, Slone TW, Peters CJ (1987) Pathogenesis of Rift Valley fever virus (RVFV) in inbred rats. Microb Path 2: 283–293

Anderson GW, Slone SW, Peters CJ (1988) The gerbil, *Meriones unguiculatus*, a model for Rift Valley fever virus encephalitis. Arch Virol 102: 187–196

Artsob H (1983) Distribution of California serogroup viruses and virus infection in Canada. In: Calisher CH, Thompson WH (eds) California serogroup viruses. Liss, New York, pp 277–292
Asada H, Tamura M, Kondo K, Okuno Y, Takahashi Y, Dohi Y (1987) Role of T lymphocytes subsets in protection and recovery from Hantaan virus infection in mice. J Gen Virol 68: 1961–1969
Asada H, Tamura M, Kondo K, Dohi Y, Yamanishi K (1988) Cell-mediated immunity to virus causing haemorrhagic fever with renal syndrome: generation of cytotoxic T lymphocytes. J Gen Virol 69: 2179–2188
Asada H, Balachandra K, Tamura M, Kondo K, Yamanishi K (1989) Cross-reactive immunity among different serotypes of virus causing haemorrhagic fever with renal syndrome. J Gen Virol 70: 819–825
Balducci M, Fausto AM, Verani P (1985) Phlebotomus-transmitted viruses in Europe. In: Proceedings of the international congress for infectious diseases. Edizione Luigi Pozzi, Rome, pp 101–104
Balfour HH, Edelman CK, Cook FE (1976) California arbovirus (La Crosse) infections. III. Epidemiology of California encephalitis in Minnesota. J Infect Dis 133: 293–301
Barnett IC, Suyemoto W (1961) Field studies on sandfly fever and kala-azar in Pakistan, in Iran, and in Baltistan (Little Tibet) Kashmir. Trans NY Acad Sci 23: 609–617
Beaty BJ, Bishop DH (1988) Bunyavirus-vector interaction. Virus Res 10: 289–301
Beaty BJ, Thompson WH (1976) Delineation of La Crosse virus in developmental stages of transovarially infected *Aedes triseriatus*. Am J Trop Med Hyg 25: 685–691
Beaty BJ, Rohzon EJ, Gensemer P, Bishop DH (1981) Formation of reassortant bunyaviruses in dually infected mosquitoes. Virology 111: 662–665
Beaty BJ, Miller BR, Shope RE, Rohzon EJ, Bishop DH (1982) Molecular basis of bunyavirus per os infection of mosquitoes: role of the middle-sized RNA segment. Proc Nat Acad Sci USA 79: 1295–1297
Beaty BJ, Sundin DR, Chandler LJ, Bishop DH (1985) Evolution of bunyaviruses by genome reassortment in dually infected mosquitoes (*Aedes triseriatus*). Science 230: 548–550
Berry RL, Parsons MA, Restifo RA (1975) Studies of the epidemiology of California encephalitis in an endemic area in Ohio in 1971. Am J Trop Med Hyg 24: 992–998
Bishop DH (1985a) Ambisense RNA genomes of arenaviruses and phleboviruses. Adv Virus Res 31: 1–51
Bishop DH (1985b) The genetic basis for describing viruses as species. Intervirology 24: 79–93
Bishop DH, Shope RE (1979) Bunyaviridae. Comprehensive Virology. Virology 14: 1–156
Bishop DH, Calisher CH, Casals J (1980) Bunyaviridae. Intervirology 14: 125–143
Bishop DH, Fuller F, Akashi H (1984a) The use of reassortant bunyaviruses to deduce their coding and pathogenic potentials. In: Kohn A, Fuchs P (eds) Mechanisms of virus pathogenesis. Nijhoff, Boston, pp 49–60
Bishop DH, Rud E, Belloncik S (1984b) Coding analyses of bunyavirus RNA species. In: Compans RW, Bishop DH (eds) Segmented negative strand viruses. Academic, Orlando, pp 3–11
Brown JL, Dominik JW, Morrissey RL (1981) Respiratory infectivity of a recently isolated Egyptian strain of Rift Valley fever virus. Infect Immun 33: 848–853
Brummer-Korvenkontio M, Vaheri A, Hovi T (1980) Nephropathia epidemica: detection of antigen in bank voles and serological diagnosis of human infection. J Infect Dis 141: 131–134
Brummer-Korvenkontio M, Hentonnen H, Vaheri A, (1982) Hemorrhagic fever with renal syndrome in Finland: ecology and virology of nephropathia epidemica. Scand J Infect Dis 36: 88–91
Burney MI, Ghafoor A, Saleen M, Webb PA, Casals J (1980) Nosocomial outbreak of viral hemorrhagic fever caused by Crimean hemorrhagic fever-Congo virus in Pakistan, Jan 1976. Am J Trop Med Hyg 29: 941–947
Calisher CH (1983) Taxonomy, classification, and geographic distribution of California serogroup bunyaviruses. In: Calisher CH, Thompson WH (eds) California serogroup viruses. Liss, New York, pp 1–18
Calisher CH and Thompson WH (1983) California serogroup viruses. Liss, New York
Calisher CH, Kokernot RH, de Moore JF, Boyd KR, Hayes J, Chappell WA (1969) Arbovirus studies in the Ohio–Mississippi basin, 1964–1967. VI. Mermet: a Simbu group arbovirus. Am J Trop Med Hyg 18: 779–792
Calisher CH, Lazuick JS, Liev S, Monath TP, Castro KG (1988) Human infections with Tensaw virus in South Florida: evidence that Tensaw virus subtypes stimulate the production of antibodies reactive with closely related Bunyamwera serogroup viruses. Am J Trop Med Hyg 39: 117–122
Caplen H, Peters CJ, Bishop DHL (1985) Mutagen-directed attenuation of Rift Valley fevr virus as a method for vaccine development. J Gen Virol 66: 2271–2277

Casals J (1969) Antigenic similarity between the virus causing Crimean hemorrhagic fever and Congo virus. Proc Soc Exp Biol Med 131: 233–236
Casals J, Henderson B, Hoogstraal H, Johnson KM, Shelokov A (1970) A review of Soviet viral hemorrhagic fevers, 1969. J Infect Dis 122: 437–453
Causey OR, Kemp GE, Causey CE, Lee VH (1972) Isolations of Simbu-group viruses in Ibadan, Nigeria 1964–1969, including the new types Sango, Shamonda, Sabo, and Shuni. Am J Trop Med Hyg 66: 357–363
Chambers PG, Swanepoel R (1980) Rift Valley fever in abattoir workers. Cent Afr J Med 26: 122–126
Childs JE, Korch GW, Glass GE, LeDuc JW, Shah KV (1987) Epizootology of hantavirus infections in Baltimore, isolation of a virus from Norway rats, and characteristics of infected rat populations. Am J Epidemiol 126: 55–68
Chumakov MP (1963) Study of viral haemorrhagic fevers. J Hyg Epidemiol 7: 125–140
Chumakov MP, Butenko AM, Shalunova NV (1968) New data on the virus causing Crimean haemorrhagic fever (in Russian). Vopr Virusol 13: 377–390
Chumakov MP, Smirnova SE, Tkachenko EA (1970) Relationship between strains of Crimean hemorrhagic fever and Congo viruses. Acta Virol 14: 82–85
Chun RW (1983) Clinical aspects of La Crosse encephalitis: neurological and psychological sequelae. In: Calisher CH, Thompson WH (eds) California serogroup viruses. Liss, New York, pp 193–201
Chun RW, Thompson WH, Grawbow JD (1968) California arbovirus encephalitis in children. Neurology 18: 369–375
Coates DM, Sweet C (1990) Studies on the pathogenicity of a nairovirus, Dugbe virus, in normal and immunosuppressed mice. J Gen Virol 71: 325–332
Cosgriff TM, Morrill JC, Jennings GB, Hodgson LA, Slayter MV, Gibbs PH, Peters CJ (1989) Hemostatic derangement produced by Rift Valley fever virus in rhesus monkeys. Rev Infect Dis (S4): 807–814
Danielova V (1975) Growth of Tahyna virus at low temperatures. Acta Virol 19: 327–332
Daubney R, Hudson JR (1931) Enzootic hepatitis or Rift Valley fever: an undescribed virus disease of sheep, cattle and man from East Africa. J Pathol Bacteriol 34: 545–557
David-West TS (1972) World distribution and antigenic variation of Simbu arboviruses. Microbios 5: 213–217
Davies FG (1975) Observations on the epidemiology of Rift Valley fever in Kenya. J Hyg (Cambr) 75: 219–230
Davies FG, Jessett DM (1985) A study of the host range and distribution of antibody to Akabane virus (genus *Bunyavirus*, family Bunyaviridae) in Kenya. J Hyg (Cambr) 95: 191–196
Davies FG, Linthicum KJ, James AD (1985) Rainfall and epizootic Rift Valley fever. Bull WHO 63: 941–943
DeFoliart GR (1983) *Aedes triseriatus*: vector biology in relationship to the persistence of La Crosse virus in endemic foci. In: Calisher CH, Thompson WH (eds) California serogroup viruses. Liss, New York, pp 89–105
Desmyter J, Johnson KM, Deckers C, LeDuc JW, Brasseru F, van Ypersele de Strihou C (1983) Laboratory rat associated outbreak of haemorrhagic fever with renal syndrome due to Hantaan-like virus in Belgium. Lancet i: 1445–1448
Diebel R, Srihongse S, Grayson MA (1983) Jamestown Canyon virus: the etiologic agent of an emerging human disease? In: Calisher CH, Thompson WH (eds) California serogroup viruses. Liss, New York, pp 313–328
Dixon KE, Travassos da Rosa AP, Travassos da Rosa JF, Llewellyn CH (1981) Oropouche virus II. Epidemiological observations during an epidemic in Santaram, Para, Brazil in 1975. Am J Trop Med Hyg 30: 161–171
Doherty RL, Carley JG, Standfast A, Dyce AL, Snowdon WA (1972) Virus strains isolated from arthropods during an epizootic of bovine ephemeral fever in queensland. Aust Vet J 48: 81–86
Earle DP (1954) Symposium on epidemic hemorrhagic fever. Am J Med 16: 619–709
Easterday BC (1965) Rift Valley fever Adv Vet Sci 10: 65–127
Eddy GA, Peters CJ (1980) The extended horizons of Rift Valley fever: current and projected immunogens. New developments with human and veterinary vaccines. Liss, New York, pp 179–191
Edwards JF, Livingston CW, Chung SI, Collisson EC (1989) Ovine arthrogryposis and central nervous system malformations associated with in utero Cache Valley virus infection: spontaneous disease. Vet Pathol 26: 33–39
Elliott RM (1989) Nucleotide sequence analysis of the large (L) genomic RNA segment of Bunyamwera virus, the prototype of the family Bunyaviridae. Virology 173: 426–436

Elliott RM (1990) Molecular biology of the Bunyaviridae. J Gen Virol 71: 501–522
Elliott RM, Lees JF, Watret GE (1984) Genetic diversity of bunyaviruses and the mechanism of genetic variation, In: Kohn A, Fuchs P (eds) Mechanisms of virus pathogenesis Nijhoff, Boston, pp 61–75
Endres MJ, Jacoby DR, Janssen RS, Gonzalez-Scarano F, Nathanson N (1989) The large viral RNA segment of bunyaviruses encodes the large viral protein. J Gen Virol 70: 223–228
Endres MJ, Valsamakis, A, Gonzalez-Scarano F, Nathanson N (1990) A neuroattenuated clone of California bunyavirus: derivation, characterization, and revertants. J Virol 64: 1927–1933
Endris RG, Tesh RB, Young DG (1983) Transovarial transmission of Rio Grande virus (Bunyaviridae: *Phlebovirus*) by the sand fly, *Luztomyia anthophora*. Am J Trop Med Hyg 32: 862–864
Faran ME, Turell MJ, Romoser WS, Routier RG, Gibbs PH, Cannon TL, Bailey CL (1987) Reduced survival of adult *Culex pipiens* infected with Rift Valley fever virus. Am J Trop Med Hyg 37: 403–409
Faran ME, Romoser WS, Routier RG, Bailey CL (1988) The distribution of Rift Valley fever virus in the mosquito *Culex pipiens* as revealed by viral titration of dissected organs and tissues. Am J Trop Med Hyg 39: 206–213
French GR, Fulke RS, Brand OA, Eddy GA, Lee HW, Lee PW (1981) Korean hemorrhagic fever: propagation of the etiologic agent in a cell line of human origin. Science 211: 1046–1048
Gavrilovskaya IN, Apenkina NS, Myasnikov YA (1983) Features of circulation of hemorrhagic fever with renal syndrome (HFRS) virus among small mammals in the European U.S.S.R. Arch Virol 75: 313–316
Gentsch J, Bishop DH (1976) Recombination and complementation between temperature-sensitive mutants of a bunyavirus, snowshoe hare virus. J Virol 20: 351–354
Gentsch J, Bishop DH (1978) Small viral RNA segment of bunyaviruses codes for viral nucleocapsid protein. J Virol 28: 417–419
Gentsch J, Robeson G, Bishop DH (1979) Recombination between snowshoe hare and La Crosse bunyaviruses. J Virol 31: 707–717
Gonzalez-Scarano F, Shope RE, Calisher CH (1983) Monoclonal antibodies against the G1 and nucleocapsid proteins of La Crosse and Tahyna viruses. In: Calisher CH, Thompson WH (eds) California serogroup viruses. Liss, New York, pp. 145–156
Gonzalez-Scarano F, Janssen RS, Najjar JA, Pobjecky N, Nathanson N (1985) An avirulent G1 glycoprotein variant of La Crosse bunyavirus with a defective fusion function. J Virol 54: 757–763
Gonzalez-Scarano F, Beaty BJ, Sundin DR, Janssen R, Endres MJ, Nathanson N (1988) Genetic determinants of the virulence and infectivity of La Crosse virus. Microb Path 4: 1–7
Gonzalez-Scarano F, Nathanson N (1989) California serogroup viruses. In: Gilden DH, Lipton HL (eds) Clinical and molecular aspects of neurotropic virus infection. Kluwer, New York, pp 43–68
Gonzalez-Scarano F, Nathanson N (1990) Bunyaviruses. In: Fields BN, Knipe DM (eds) Virology. Raven, New York, pp 1195–1228
Gonzalez-Scarano F, Nathanson N (1989) California serogroup viruses. In: Gilden DH, Lipton HL (eds) Clinical and molecular aspects of neurotropic virus infection. Kluwer, Boston, pp 43–68
Grimstad PR, Barrett RL, Humphrey RL (1984a) Serologic evidence for widespread infection with La Crosse and St. Louis encephalitis viruses in the Indiana human population. Am J Epidemiol 119: 913–930
Grimstad PR, Calisher CH, Haroff RN (1984b) Jamestown Canyon virus (California serogroup) is the etiologic agent of widespread infection in Michigan humans. Am J Trop Med Hyg 35: 376–386
Gundersen CB, Brown KL (1983) Clinical aspects of La Crosse encephalitis: preliminary report. In: Calisher CH, Thompson WH (eds) California serogroup viruses. Liss, New York, pp 169–177
Hammon WM, Reeves WC (1952) California encephalitis virus—a newly described agent. I. Evidence of natural infection in man and other animals. Calif Med 77: 303–309
Hardy JL, Scrivani RP, Lyness RP, Nelson RL, Roberts DR (1970) Ecologic studies on Buttonwillow virus in Kern County, California, 1961–1968. Am J Trop Med Hyg 19: 552–561
Hartley WJ, Wanner RA, Della-Porta AJ, Snowdon WA (1975) Serological evidence for the association of Akabane virus with epizootic bovine congenital arthrogryposis and hydranencephaly syndromes in New South Wales. Aust Vet J 51: 103–104
Haughey KG, Hartley WJ, Della-Porta AJ, Murray MD (1988) Akabane disease in sheep. Aust Vet J 65: 136–140
Henderson BE, Coleman PH (1971) The growing importance of California arboviruses in the etiology of human disease. Prog Med Virol 13: 404–461
Hoogstraal H (1979) The epidemiology of tick-borne Crimean-Congo hemorrhagic fever in Asia, Europe, and Africa. J Med Entomol 15: 307–417
Hoogstraal H, Meegan JM, Khalil GM, Adham FK (1979) The Rift Valley fever epizootic in Egypt 1977–1978. 2. Ecological and entomological studies. Trans R Soc Trop Med Hyg 73: 624–629

Inaba Y, Kurogi H, Omori T (1975) Akabane disease: epizootic abortion, premature birth, stillbirth, and congenital arthrogryposis-hydranencephaly in cattle, sheep, and goats caused by Akabane virus. Aust Vet J 51: 584–585

Iroegbu CU, Pringle CR (1981a) Genetics of the Bunyamwera complex. In: Bishop DH, Compans R (eds) The replication of negative strand RNA viruses. Elsevier, New York, pp 159–165

Iroegbu CU and Pringle CR (1981b) Genetic interactions among viruses of the Bunyavirus complex. J Virol 37: 383–394

Janssen R, Gonzalez-Scarano F, Nathanson N (1984) Mechanisms of bunyavirus virulence. Comparative pathogenesis of a virulent strain of La Crosse virus and an avirulent strain of Tahyna virus. Lab Invest 50: 447–455

Janssen RS, Nathanson N, Endres MJ, Gonzalez-Scarano F (1986) Virulence of La Crosse virus is under polygenic control. J Virol 59: 1–7

Jennings M, Boorman J (1983) Laboratory infection of the sandfly *Phlebotomus papatasii* Scopoli (Diptera, Psychodidae) with three phleboviruses. Trans R Soc Trop Med Hyg 77: 62–64

Jennings M, Mellor PS (1987) *Culicoides*: biological vectors of Akabane virus. Vet Microbiol 21: 125–131

Johnson KP, Johnson RT (1968) California encephalitis. II. Studies of experimental infection in the mouse. J Neuropathol Exp Neurol 27: 390–400

Johnson RT (1983) Pathogenesis of La Crosse virus in mice. In: Calisher CH, Thompson WH (eds) California serogroup viruses. Liss, New York, pp 139–144

Johnston RE, Smith JF (1988) Selection for accelerated penetration in cell culture coselects for attenuated mutants of Venezuelan equine encephalitis virus. Virology 162: 437–443

Joubert JD, Ferguson AL, Gear J (1951) Rift Valley fever in South Africa 2. The occurrence of human cases in the Orange Free State, the North-Western Cape Province, the Western and Southern Transvaal. A. epidemiological and clinical findings. S Afr Med J 25: 890–891

Kalfayan B (1983) Pathology of La Crosse virus infections in humans. Calisher CH, Thompson WH, editors. California serogroup viruses. Liss, New York, pp 179–186

Kappus KD, Monath TP, Kaminski RM (1983) Reported encephalitis associated with California serogroup virus infections in the United States, 1963–1981. In: Calisher CH, Thompson WH (eds) California serogroup viruses. Liss, New York, pp 31–41

Kim GR, McKee KT (1985) Pathogenesis of Hantaan virus infection in suckling mice: clinical, virologic, and serologic observations. Am J Trop Med Hyg 34: 388–395

Kinney RM, Calisher CH (1981) Antigenic relationships among Simbu serogroup (Bunyaviridae) viruses. Am J Trop Med Hyg 30: 1307–1318

Klimas RA, Thompson WH, Calisher CH (1984) Genotypic varieties of La Crosse virus isolated from different geographic regions of the continental United States and evidence for a naturally occurring intertypic recombinant La Crosse virus. Am J Epidemiol 114: 112–131

Konno S, Moriwaki M, Nakagawa M, Uchimura M, Kamimirata M, Tojinbara K (1975) Congenital abnormality of calves with arthrogryposis and hydranencephaly in Japan in 1972–1973. Nat Inst Anim Health Q 15: 52–53

Konno S, Koeda T, Madarame H (1988) Myopathy and encephalopathy in chick embroys experimentally infected with Akabane virus. Vet Pathol 25: 1–8

Ksiazek TG, Yuill TM (1977) Viremia and antibody response to La Crosse virus in sentinel gray squirrels (*Sciuris carolinensis*) and chipmunks (*Tamias striatus*). Am J Trop Med Hyg 26: 815–821

Ksiazek TG, Jouan A, Meegan JM (1989) Rift Valley fever among domestic animals in the recent West African outbreak. Res Virol 140: 67–77

Kurata T, Tsai TF, Bauer SP, McCormick JB (1983) Immunofluorescence studies of disseminated Hantaan virus infection of suckling mice. Infect Immun 41: 391–398

Kurogi H, Inaba Y, Goto Y (1975) Serologic evidence for etiologic role of Akabane virus in epizootic abortion-arthrogryposis-hydranencephaly in cattle in Japan, 1972–1974. Arch Virol 47: 71–83

Kurogi H, Inaba Y, Takahashi E (1976) Epizootic congenital arthrogryposis-hydranencephaly syndrome in cattle: isolation of Akabane virus from affected fetuses. Arch Virol 51: 67–74

Kurogi H, Inaba Y, Tkahashi E (1977) Congenital abnormalities in newborn cattle after inoculation of pregnant cows with Akabane virus. Infect Immun 17: 338–343

Lahdevirta J (1982) Clinical features of HFRS in Scandinavia as compared with East Asia. Scand J Infect Dis 36: 93–95

Laughlin LW, Meegan JM, Strausbuagh LH, Morens DM, Watten RH (1979) Epidemic Rift Valley fever in Egypt: Observations of the spectrum of human illness. Trans R Soc Trop Med Hyg 73: 630–633

LeDuc JW (1987) Epidemiology of Hantaan and related viruses. Lab Anim Sci 37: 413–418

LeDuc JW, Smith GA, Johnson KM (1984) Hantaan-like viruses from domestic rats captured in the United States. Am J Trop Med Hyg 33: 992–998
LeDuc JW, Smith GA, Childs JE (1986) Global survey of antibody to Hantaan-related viruses among peridomestic rodents. Bull WHO 64: 139–144
Lee HW (1982) Korean hemorrhagic fever. Prog Med Virol 28: 96–113
Lee HW (1989) Hemorrhagic fever with renal syndrome. Prog Med Virol 36: 62–102
Lee HW, Johnson KM (1982) Laboratory-acquired infections with Hantaan virus, the etiologic agent of Korean hemorrhagic fever. J Infect Dis 146: 645–651
Lee HW, Lee PW, Johnson KM (1978) Isolation of the etiologic agent of Korean hemorrhagic fever. J Infect Dis 137: 298–308
Lee HW, French GR, Lee PW, Baek LJ, Tsuchiya K, Foulke RS (1981a) Observations on natural and laboratory infection of rodents with the etiologic agent of Korean hemorrhagic fever. Am J Trop Med Hyg 30: 477–482
Lee HW, Lee PW, Baek LJ, Song CK, Seong IW (1981b) Intraspecific transmission of Hantaan virus, the etiologic agent of Korean hemorrhagic fever, in the rodent *Apodemus agragrius*. Am J Trop Med Hyg 30: 1106–1112
Lee HW, Baek LJ, Johnson KM (1982) Isolation of Hantaan virus, the etiologic agent of Korean hemorrhagic fever, from wild urban rats. J Infect Dis 146: 638–644
Lee M, Kim BK, Park S, Han JS, Kim ST, Lee JS (1989) Coagulopathy in hemorrhagic fever with renal syndrome (Korean hemorrhagic fever). Rev Inf Dis 11(SA): 877–883
Lee PW, Amyx HL, Gibbs CJ, Gajdusek DC, Lee HW (1981) Propagation of Korean hemorrhagic fever virus in laboratory rats. Infect Immun 31: 334–338
Lee PW, Amyx HL, Yanagihara R, Gajdusek DC, Goldgaber D, Gibbs CJ (1985) Partial characterization of Prospect Hill virus isolated from meadow voles in the United States. J Infect Dis 152: 826–829
Lewis RM, Cosgriff TM, Peters CJ, Morrill JC (1987) Differentiation of a human monocytic cell line associated with increased production of Rift Valley fever virus by infected cells. J Med Virol 23: 207–215
Linthicum KJ, Davies FG, Kairo A, Bailey CL (1985) Rift Valley fever virus (family Bunyaviriae, genus *Phlevobirus*). Isolations from Diptera collected during an inter-epizootic period in Kenya. J Hyg (Cambr) 95: 197–209
Ludwig GV, Christensen BM, Yuill TM, Schultz KT (1989) Enzyme processing of La Crosse virus glycoprotein G1: a bunyavirus-vector infection model. Virology 17: 108–113
Lyons MJ, Heyduk J (1973) Aspects of the developmental morphology of California encephalitis virus in cultured vertebrate and arthropod cells and in mouse brain. Virology 54: 37–52
Mackenzie RD, Finlay GM, Stern RO (1936) Studies on neurotropic Rift Valley fever virus: the susceptibility of rodents. Br J Exp Pathol 17: 352–361
McGavran MH, Easterday BC (1963) Rift Valley fever virus hepatitis. Am J Pathol 42: 587–607
McKee KT, Lim GR, Green DE, Peters CJ (1985) Hantaan virus infection in suckling mice: virologic and pathologic correlates. J Med Virol 17: 107–117
McLean DM (1983) Yukon isolates of snowshoe hare virus, 1972–1982. In: Calisher CH, Thompson WH (eds) California serogroup viruses. Liss, New York, pp 247–256
McLean DM, Gubash SM, Grass PN, Miller MA, Petric M, Walters TE (1975) California encephalitis development in mosquitoes as revealed by transmission studies, immunoperoxidase staining, and electron microscopy. Can J Microbiol 21: 453–462
McLean DM, Grass PN, Judd BD, Stolz KJ (1979) Bunyavirus development in arctic and *Aedes aegypti* mosquitoes as revealed by glucose oxidase staining and immunofluorescence. Arch Virol 62: 313–322
McPhee DA, Parsonson IM, Della-Porta AJ, Jarrett RG (1984) Teratogenicity of Australian Simbu serogroup and some other Bunyaviridae viruses: the embryonated chicken egg as a model. Infect Immun 43: 413–420
Meegan JM (1979) The Rift Valley fever epizootic in Egypt, 1977–1978. 1. Description of the epizootic and virological studies. Trans R Soc Trop Med Hyg 73: 618–623
Meegan JM, Shope RE (1981) Emerging concepts on Rift Valley fever. Perspect Virol 11: 267–287
Miller BR (1983) A variant of La Crosse virus attenuated for *Aedes triseriatus* mosquitoes. Am J Trop Med Hyg 32: 1422–1428
Mims CA (1956a) Rift Valley fever virus in mice I. General features of the infection. Br J Exp Pathol 37: 99–109
Mims CA (1956b) Rift Valley fever virus in mice. II. Adsorption an multiplication of the virus. Br J Exp Pathol 37: 110–119

Mims CA (1956c) Rift Valley fever virus in mice. III. Further quantitative features of the infective process. Br J Exp Pathol 37: 120–128

Mims CA (1956d) Rift Valley fever virus in mice. IV. Incomplete virus, its production and properties. Br J Exp Pathol 37: 129–143

Mitten JQ, Remmele NS, Walker JS, Carter RC, Stephen EL, Klein F (1970) The clinical aspects of Rift Valley fever virus in household pets. III. Pathologic changes in the dog and cat. J Infect Dis 121: 25–31

Miura Y, Hayashi S, Ishihara T, Inaba Y, Omori T, Matumoto M (1974) Neutralizing antibody against Akabane virus in precolostral sera from calves with congenital arthrogryposis-hydranencephaly syndrome. Arch Gesamte Virusforsch 46: 377–380

Monath TP, Nuckolls JG, Berall J, Bauer H, Chappell WA, Coleman PH (1970) Studies of California encephalitis in Minnesota. Am J Epiemiol 92: 40–50

Montgomery E (1917) On a tick-borne gastro-enteritis of sheep and goats occurring in British East Africa. J Comp Pathol Ther 30: 28–57

Morrill JC, Jennings GB, Caplen H, Turell MJ, Johnson AJ, Peters CJ (1987) Pathogenicity and immunogenicity of a mutagen-attenuated Rift Valley fever virus immunogen in pregnant ewes. Am J Vet Res 48: 1042–1047

Morrill JC, Jennings GB, Cosgriff TM, Gibbs PH, Peters CJ (1989) Prevention of Rift Valley fever in rhesus monkeys with interferon-alpha. Rev Infect Dis 11(54): 815–825

Morrill JC, Jennings GB, Johnson AJ, Cosgriff TM, Gibbs PH, Peters CJ (1990) Pathogenesis of Rift Valley fever in rhesus monkeys: role of interferon response. Arch Virol 110: 195–212

Moussa MI, Wood OL, Abdel-Wahab KS (1982) Reduced pathogenicity associated with a small plaque variant of the Egyptian strain of Rift Valley fever virus (ZH501). Trans R Soc Trop Med Hyg 76: 482–486

Murphy FA, Harrison AK, Tzianabos T (1968a) Electron microscopic observations of mouse brain infected with Bunyamwera serologic supergroup viruses and several other arthropod-borne viruses. J Virol 2: 1315–1325

Murphy FA, Whitfield SG, Coleman PH, Calisher CH (1968b) California group arboviruses: electron microscopic studies. Exp Mol Pathol 9: 44–56

Murphy FA, Harrison AK, Whitfield SG (1973) Bunyaviridae: morphologic and morphogenetic similarities of Bunyamwera serologic supergroup viruses and several other arthropod-borne viruses. Intervirology 1: 297–316

Murphy FA, Whitfield SG, Sudia WD (1975) Interactions of vector with vertebrate pathogenic viruses. Mamamorosch K, Shope RE, editors. Invertebrate immunity. Academic, Oralando, pp 25–37

Murray MD (1987) An outbreak of Akabane disease in calves in New South Wales. Aust Vet J 64: 305–308

Myrhman G (1951) Nephropathia epidemica: a new infectious disease in Northern Scandinavia. Acta Med Scand 140: 52–56

Nakamura T, Yanagihara R, Gibbs CJ, Amyx HL, Gajddusek DC (1985a) Differential susceptibility and resistance of immunocompetent and immunodeficient mice to fatal Hantaan virus infection. Arch Virol 86: 109–120

Nakamura T, Yanagihara R, Gibbs CJ, Gajdusek DC (1985b) Immune spleen cell-mediated protection against fatal Hantaan virus infection in infant mice. J Infect Dis 151: 691–697

Nicoletti L, Verani P (1985) Growth of the phlebovirus Toscana in a mosquito (*Aedes pseudoscutellaris*) cell line (AP-61): establishment of a persistent infection. Arch Virol 85: 35–45

Niklasson B, LeDuc JW (1987) Epidemiology of nephropathia epidemica in Sweden. J Infect Dis 155: 269–276

Oker-Blom N, von Bonsdorff C-H, Korvenkontio M (1979) Nephropathia epidemica in Finland. In: Kurstak E (ed) Arctic and tropical arboviruses. Academic, Orlando, pp 197–209

Oliver J, McDowell M (1957) The renal lesion in epidemic hemorrhagic fever. J Clin Invest 36: 99–222

Oya A, Okino T, Ogata T, Kobayashi I, Matsuyama T (1961) Akabane, a new arbor virus isolated in Japan. Jpn J Med Sci Biol 14: 101–108

Paci P, Balucci M, Verani P (1983) Toscana virus, a new *Phlebotomus*-transmitted virus isolated from Italy. In: Proceedings of the international congress of infectious diseases. Luigi Pozzi, Rome, pp 35–39

Pantuwatana S, Thompson WH, Watts DM (1972) Experimental infection of chipmunks an squirrels with La Crosse and trivittatus viruses an biological transmission of La Crosse virus by *Aedes triseriatus*. Am J Trop Med Hyg 21: 476–481

Parkin WE, Hammon WM, Sather GE (1972) Review of current epidemiologic literature on viruses of the California arbovirus group. Am J Trop Med Hyg 14: 964–978

Parsonson IM, Della-Porta AJ, Snowdon WA (1975) Congenital abnormalities in fetal lambs after inoculation of pregnant ewes with Akabane virus. Aust Vet J 51: 585–586

Parsonson IM, Della-Porta AJ, Snowon WA (1977) Congenital abnormalities in fetal lambs after infection of pregnant sheep with Akabane virus. Infect Immun 15: 254–262

Parsonson IM, McPhee DA (1985) Bunyavirus pathogenesis. Adv Virus Res 30: 279–316

Patrican LA, DeForliart GR, Yuill TM (1985) La Crosse viremia in juvenile, subadult and adult chipmunks (*Tamias striatus*) following feeding by transovarially-infected *Aedes triseriatus*. Am J Trop Med Hyg 34: 596–602

Paulson SL, Grimstad PR (1989) Replication and dissemination of La Crosse virus in the competent vector *Aedes triseriatus* and the incompetent vector *Aedes hendersoni* and evidence for transovarial transmission by *Aedes hendersoni* (Diptera: Culicidae). J Med Entomol 26: 602–609

Penttinen K, Lahdevirta J, Kekomaki R (1981) Circulating immune complexes, immunoconglutinins, and rheumatoid factors in nephrophathia epidemica. J Infect Dis 143: 15–21

Peters CJ, Anderson GW (1981) Pathogenesis of Rift Valley fever. Contrib Epidemiol Biostat 3: 21–41

Peters CJ, Jones D, Trotter R (1988) Experimental Rift Valley fever in rhesus macaques. Arch Virol 99: 31–44

Peters CJ, Lui CT, Anderson GW, Morrill JC, Jahrling PB (1989) Pathogenesis of viral hemorrhagic fevers: Rift Valley fever and Lassa fever contrasted. Rev Infect Dis 11(54): 743–749

Pifat D, Smith JF (1987) Punta Toro virus infection of C57BL/6J mice: a model for phlebovirus-induced disease. Microb Path 3: 409–422

Pinheiro FP, Travassos da Rosa AP, Travassos da Rosa JF (1981a) Oropouche virus. I. A review of clinical, epidemiological, and ecological findings. Am J Trop Med Hyg 30: 149–160

Pinheiro FP, Hoch AL, Gomes MD, Roberts DR (1981b) Oropouche virus. IV. Laboratory transmission by *Culicoides paraensis*. Am J Trop Med Hyg 30: 172–176

Pinheiro FP, Travassos da Rosa AP, Gomes ML, Leduc JW, Hoch HL (1982) Transmission of Oropouche virus from man to hamster by the midge *Culicoides paraensis*. Science 215: 1251–1252

Pringle CR and Iroegbu CU (1982) Mutant identifying a third recombination group in a bunyavirus. J Virol 42: 873–879

Raju R, Kolakofsky D (1989) The translational requirement for complete La Crosse virus mRNA synthesis is cell-type dependent. J Virol 63: 5159–5165

Reeves WC, Hammon WM (1962) Epidemiology of the arthropod-borne viral encephalitides in Kern County, California, 1943–1952. Univ Calif Publ Public Health 4: 1–257

Reeves WC, Emmons RW, Hardy JL (1983) Historical perspective on California encephalitis virus in California. In: Calisher CH, Thompson WH (eds) California serogroup viruses. Liss, New York, pp 19–30

Roberts DR, Hoch AL, Dixon KE, Llewellyn CH (1981) Oropouche virus. III. Entomological observations from three epidemics in Para, Brazil, 1975. Am J Trop Med Hyg 30: 339–350

Rossier C, Raju R, Kolakofsky D (1988) La Crosse virus gene expresion in mammalian and mosquito cells. Virology 165: 539–548

Rowley WA, Wong YW, Dorsey DC, Hunsler WJ, Carrier RW (1988) California serogroup viruses in Iowa. In: Calisher CH and Thompson WH (eds) California serogroup viruses. Liss, New York, pp 237–246

Sabatino DA, Cramblett HG (1968) Behavioral sequelae of California encephalitis virus infection in children. Dev Med Child Neurol 10: 331–337

Sabin AB (1948) Phlebotomus fever. In: Rivers TM (ed) Viral and rickettsial diseases of man. Lippincott, Philadelphia, pp 454–460

Sabin AB (1951) Experimental studies on phlebotomus (pappataci, sandfly) fever during World War II. Arch Virusforsch 4: 367–410

Sabin AB, Blumberg RW (1947) Human infection with Rift Valley fever virus and immunity twelve years after a single attack. Proc Soc Exp Biol Med 64: 385–389

Settergren B, Juto P, Trollfors B, Wadell G, Norrby SR (1989) Clinical characteristics of nephropathia epidemica in Sweden. Rev Infect Dis 11: 921–927

Seymour C, Amundson TE, Yuill TM, Bishop DH (1983) Experimental infection of chipmunks and snowshoe hares with La Crosse and snowshoe hare viruses and four of their reassortants. Am J Trop Med Hyg 32: 1147–1153

Shepherd AJ, Swanepoel R, Leman PA, Blackburn NK, Haliett AF (1985) A nosocomial outbreak of Crimean-Congo hemorrhagic fever at Typerberg Hospital, part V. Virological and serological observations. S Afr Med J 68: 733–736

Shepherd AJ, Swanepoel R, Shepherd SP, McGillivray GM, Searle LA (1987) Antibody to Crimean-Congo hemorrhagic fever virus in wild mammals from southern Africa. Am J Trop Med Hyg 36: 133–142

Shimshony A, Barzilai R (1983) Rift Valley fever. Adv Vet Sci Comp Med 27: 347–425
Shope RE (1985) Bunyaviruses. In: Fields BN (ed) Virology, Raven, New York, pp 1055–1082
Shope RE, Rozhon EJ, Bishop DH (1981) Role of the middle-sized bunyavirus segment in mouse virulence. Virology 114: 273–276
Shope RE, Peters CJ, Dives FG (1982) The spread of Rift Valley fever and approaches to its control. Bull WHO 60: 299–304
Siam AL, Meegan JM, Gharbawi KF (1980) Rift Valley fever ocular manifestations: observations during the 1977 epidemic in Egypt. Br J Ophthalmol 64: 366–374
Simpson DI, Knight EM, Courtois G, Williams MC, Weibren MP, Kibukamosoke JW (1967) Congo virus: a hitherto undescribed virus occurring in Africa. I. Human isolations-clinical notes. East Afr Med J 44: 87–92
Smirnova SE (1979) A comparative study of the Crimean hemorrhagic fever Congo group of viruses. Arch Virol 62: 137–145
Smithburn KC (1949) Rift Valley fever: the neurotropic adaptation of the virus and the experimental use of this modified virus as a vaccine. Br J Exp Pathol 30: 1–16
Sommer A-I, Traavik TM, R., Berdal BP, Dalrymple JM (1985) Reservoir animals for nephropathia epidemica in Norway: indicaton of a major role for the bank vole (*C. glareolus*) in comparison with the woodmouse (*A. sylvaticus*). J Hyg (Cambr) 94: 123–127
Srihongse S, Grayson MA, Diebel R (1984) California serogroup viruses in New York State: the role of subtypes in human infections. Am J Trop Med Hyg 33: 1218–1227
Sugiyama K, Morikawa S, Matsura Y (1987) Four serotypes of hemorrhagic fever with renal syndrome viruses identified by polyclonal and monoclonal antibodies. J Gen Virol 68: 979–987
Suleiman MN, Muscat-Baron JM, Harries JR, Platt GS, Bowen ET, Simpson DI (1980) Congo/Crimean hemorrhagic fever in Dubai. an outbreak at the Rashid hospital. Lancet ii: 939–941
Sundin DR, Beaty BR, Nathanson N, Gonzalez-Scarano F (1987) A G1 glycoprotein epitope of La Crosse virus: a determinant of infection of *Aedes triseriatus*. Science 235: 591–593
Swanepoel R, Struthers JK, Shepherd AJ, McGillivray GM, Nel MJ, Jupp PG (1983) Crimean-Congo hemorrhagic fever in South Africa. Am J Trop Med Hyg 32: 1407–1415
Swanepoel R, Shepherd AJ, Leman PA (1987) Epidemiologic and clinical features of Crimean-Congo hemorrhagic fever in Southern Africa. Am J Trop Med Hyg 36: 120–132
Swanepoel R, Gill DE, Shepherd AJ, Leman PA, Mynhardt JH, Harvey S (1989) The clinical pathology of Crimean-Congo hemorrhagic fever. Rev Infect Dis 11(54): 794–800
Takahara K, Min MN, Battles JK, Sugiyama K, Emery VC, Dalrymple JM, Bishop DH (1989) Identification of mutations in the M RNA segment of a candidate vaccine strain of Rift Valley fever virus. Virology 169: 452–456
Tamara M, Asada H, Kondo K, Tanishita O, Kurata T, Yamanishi K (1989) Pathogenesis of Hantaan virus in mice. J Gen Virol 70: 2897–2906
Tanashita O, Tsakahashi Y, Okuno Y (1986) Persistent infection of rats with hemorrhagic fever with renal syndrome virus and their antibody responses. J Gen Virol 67: 2819–2824
Tesh RB (1988) Phlevovirus and its vectors. Annu Rev Entomol 33: 169–181
Tesh RB, Beaty BR (1983) Localization of California serogroup viruses in mosquitoes. In: Calisher CH, Thompson WH (eds) California serogroup viruses. Liss, New York, pp 67–75
Tesh RB, Modi GB (1984) Studies on the biology of *Phleboviruses* in sand flies (Diptera: Psychodidae). I. Experimental infection of the vector. Am J Trop Med Hyg 33: 1007–1016
Tesh RB, Modi GB (1987) Maintenance of Toscana virus in *Phlebotomus perniciosus* by vertical transmission. Am J Trop Med Hyg 36: 189–193
Tesh RB, Chaniotis BN, Peralta PH, Johnson KM (1974) Ecology of viruses isolated from Panamanian sandflies. Am J Trop Med Hyg 23: 258–266
Thompson WH, Beaty BJ (1978) Venereal transmission of La Crosse virus from male to female *Aedes triseriatus*. Am J Trop Med Hyg 27: 187–195
Thompson WH, Evans AS (1965) California encephalitis virus studies in Wisconsin. Am J Epidemiol 81: 230–244
Thompson WH, Gundersen CB (1983) La Crosse encephalitis: occurrence of disease and control in a suburban area. In: Calisher CH, Thompson WH (eds) California serogroup viruses. Liss, New York, pp 225–236
Thompson WH, Kalfayan B, Anslow RO (1965) Isolation of California encephalitis virus from a fatal human illness. Am J Epidemiol 81: 245–253
Tignor GH, Smith AL, Casals J, Ezekoli CD, Okoli J (1980) Close relationship of Crimean hemorrhagic fever-Congo (CHF-C) virus strains by neutralizing antibody assays. Am J Trop Med Hyg 29: 676–685

Tignor GH, Burrage TG, Smith AL (1983) California serogroup gene structure-function relationships; virulence and tissue tropisms. In: Calisher CH, Thompson WH (eds) California serogroup viruses. Liss, New York, pp 129–138
Tomori O (1980) Rift Valley fever infection in man in Nigeria. J Med Virol 5: 343–350
Tsai TF, Bauer S, McCormick JB, Kurata T (1982) Intracerebral inoculation of suckling mice with Hantaan virus. Lancet i: 503–504
Turell MJ, Hardy JL (1980) Carbon dioxie sensitivity of mosquitoes infected with California encephalitis virus. Science 209: 1029–1030
Turell MJ, Gargan TP, Bailey CL (1984) Replication and dissemination of Rift Valley fever virus in Culex pipiers Am J Trop Med Hyg BB 176–181
Turell MJ, Rossignol PA, Spielman A, Rossi CA, Bailey CL (1984) Enhanced arboviral transmission by mosquitoes that concurrently ingested microfilariae. Science 225: 1039–1041
van Eeden PJ, Joubert RR, van de Wal BW, King JB, de Kock A, Gorenwald JH (1985) A nosocomial outbreak of Crimean-Congo hemorrhagic fever at Tygerberg Hospital, part I. Clinical features. S Afr Med J 68: 711–717
Verani P, Nicoletti L, Ciufolini MG (1984) Antigenic and biological characterization of Toscana virus, a new phleblotomus fever group virus isolated in Italy. Acta Virol 28: 39–47
Vezza AC, Repik DM, Cash P, Bishop DH (1979) In vivo transcription and protein synthesis capabilities of bunyaviruses: wild-type snowshoe hare virus and its temperature-sensitive group I, group II, and group I/II mutants. J Virol 31: 426–436
Walsh J (1988) Rift Valley fever virus rears its head. Science 240: 1397–1399
Watts DM, Panuwatana S, DeFoliart GR, Yuill TM, Thompson WH (1973) Transovarial tramsmission of La Crosse virus (California encephalitis group) in the mosquito, *Aedes triseriatus*. Science 182: 1140–1143
Whitman L, Shope RE (1962) The California complex of arthropod-borne viruses and its relationship to the Bunyamwera serogroup through Guaroa virus. Am J Trop Med Hyg 11: 691–696
Whittem JH (1957) Congenital abnormalities in calves: arthrogryposis and hydranencephally. J Pathol Bacteriol 73: 375–387
Yanagihara R, Svedmyr A, Amyx HL (1984) Isolation and propagation of nephropathia epidemica virus in bank voles. Scand J Infect Dis 16: 225–228
Yanagihara R, Golgaber D, Gajdusek DC (1985) Propagation of nephropathia epidemica virus in Mongolian gerbils. J Virol 53: 973–975
Yanagihara R, Daum CA, Lee PW (1987) Serological servey of Prospect Hill virus infection in indigenous wild rodents in the USA. Trans R Soc Trop Med Hyg 81: 42–45
Yanagihara R, Amyx HL, Lee PW, Asher DM, Gibbs CJ, Gajusek DC (1988) Experimental hantavirus infection in nonhuman primates. Arch Virol 10: 125–130
Yu-Chen Y, Ling-Xiong K, Ling L (1985) Characteristics of Crimean-Congo hemorrhagic fever virus (Xinjiang strain) in China. Am J Trop Med Hyg 34: 1179–1182
Zhang XK, Takashima I, Hashimoto N (1989a) Characteristics of passive immunity against hantavirus infections in rats. Arch Virol 105: 235–246
Zhang XK, Takashima I, Mori F, Hashimoto N (1989b) Comparison of virulence between two strains of Rattus serotype hemorrhagic fever with renal syndrome (HFRS) virus in newborn rats. Microbiol Immunol 33: 195–205

Subject Index

Current Topics in Microbiology and Immunology

Volumes published since 1985 (and still available)

Vol. 122: **Potter, Michael (Ed.):** The BALB/c Mouse. Genetics and Immunology. 1985. 85 figs. XVI, 254 pp.
ISBN 3-540-15834-0

Vol. 124: **Briles, David E. (Ed.):** Genetic Control of the Susceptibility to Bacterial Infection. 1986. 19 figs. XII, 175 pp.
ISBN 3-540-16238-0

Vol. 125: **Wu, Henry C.; Tai, Phang C. (Ed.):** Protein Secretion and Export in Bacteria. 1986. 34 figs. X, 211 pp.
ISBN 3-540-16593-2

Vol. 126: **Fleischer, Bernhard; Reimann, Jörg; Wagner, Hermann (Ed.):** Specificity and Function of Clonally Developing T-Cells. 1986. 60 figs. XV, 316 pp.
ISBN 3-540-16501-0

Vol. 127: **Potter, Michael; Nadeau, Joseph H.; Cancro, Michael P. (Ed.):** The Wild Mouse in Immunology. 1986. 119 figs. XVI, 395 pp. ISBN 3-540-16657-2

Vol. 128: 1986. 12 figs. VII, 122 pp.
ISBN 3-540-16621-1

Vol. 129: 1986. 43 figs., VII, 215 pp.
ISBN 3-540-16834-6

Vol. 130: **Koprowski, Hilary; Melchers, Fritz (Ed.):** Peptides as Immunogens. 1986. 21 figs. X, 86 pp. ISBN 3-540-16892-3

Vol. 131: **Doerfler, Walter; Böhm, Petra (Ed.):** The Molecular Biology of Baculoviruses. 1986. 44 figs. VIII, 169 pp.
ISBN 3-540-17073-1

Vol. 132: **Melchers, Fritz; Potter, Michael (Ed.):** Mechanisms in B-Cell Neoplasia. Workshop at the National Cancer, Institute, National Institutes of Health, Bethesda, MD, USA, March 24–26, 1986. 1986. 156 figs. XII, 374 pp. ISBN 3-540-17048-0

Vol. 133: **Oldstone, Michael B. (Ed.):** Arenaviruses. Genes, Proteins, and Expression. 1987. 39 figs. VII, 116 pp.
ISBN 3-540-17246-7

Vol. 134: **Oldstone, Michael B. (Ed.):** Arenaviruses. Biology and Imminotherapy. 1987. 33 figs. VII, 242 pp.
ISBN 3-540-14322-6

Vol. 135: **Paige, Christopher J.; Gisler, Roland H. (Ed.):** Differentiation of B Lymphocytes. 1987. 25 figs. IX, 150 pp.
ISBN 3-540-17470-2

Vol. 136: **Hobom, Gerd; Rott, Rudolf (Ed.):** The Molecular Biology of Bacterial Virus Systems. 1988. 20 figs. VII, 90 pp.
ISBN 3-540-18513-5

Vol. 137: **Mock, Beverly; Potter, Michael (Ed.):** Genetics of Immunological Diseases. 1988. 88 figs. XI, 335 pp.
ISBN 3-540-19253-0

Vol. 138: **Goebel, Werner (Ed.):** Intracellular Bacteria. 1988. 18 figs. IX, 179 pp.
ISBN 3-540-50001-4

Vol. 139: **Clarke, Adrienne E.; Wilson, Ian A. (Ed.):** Carbohydrate-Protein Interaction. 1988. 35 figs. IX, 152 pp.
ISBN 3-540-19378-2

Vol. 140: **Podack, Eckhard R. (Ed.):** Cytotoxic Effector Mechanisms. 1989. 24 figs. VIII, 126 pp. ISBN 3-540-50057-X

Vol. 141: **Potter, Michael; Melchers, Fritz (Ed.):** Mechanisms in B-Cell Neoplasia 1988. Workshop at the National Cancer Institute, National Institutes of Health, Bethesda, MD, USA, March 23–25, 1988. 1988. 122 figs. XIV, 340 pp.
ISBN 3-540-50212-2

Vol. 142: **Schüpach, Jörg:** Human Retrovirology. Facts and Concepts. 1989. 24 figs. 115 pp. ISBN 3-540-50455-9

Vol. 143: **Haase, Ashley T.; Oldstone Michael B. A. (Ed.):** In Situ Hybridization 1989. 22 figs. XII, 90 pp.
ISBN 3-540-50761-2

Vol. 144: **Knippers, Rolf; Levine, A. J. (Ed.):** Transforming. Proteins of DNA Tumor Viruses. 1989. 85 figs. XIV, 300 pp.
ISBN 3-540-50909-7

Vol. 145: **Oldstone, Michael B. A. (Ed.):** Molecular Mimicry. Cross-Reactivity between Microbes and Host Proteins as a Cause of Autoimmunity. 1989. 28 figs. VII, 141 pp. ISBN 3-540-50929-1

Vol. 146: **Mestecky, Jiri; McGhee, Jerry (Ed.):** New Strategies for Oral Immunization. International Symposium at the University of Alabama at Birmingham and Molecular Engineering Associates, Inc. Birmingham, AL, USA, March 21–22, 1988. 1989. 22 figs. IX, 237 pp. ISBN 3-540-50841-4

Vol. 147: **Vogt, Peter K. (Ed.):** Oncogenes. Selected Reviews. 1989. 8 figs. VII, 172 pp. ISBN 3-540-51050-8

Vol. 148: **Vogt, Peter K. (Ed.):** Oncogenes and Retroviruses. Selected Reviews. 1989. XII, 134 pp. ISBN 3-540-51051-6

Vol. 149: **Shen-Ong, Grace L. C.; Potter, Michael; Copeland, Neal G. (Ed.):** Mechanisms in Myeloid Tumorigenesis. Workshop at the National Cancer Institute, National Institutes of Health, Bethesda, MD, USA, March 22, 1988. 1989. 42 figs. X, 172 pp. ISBN 3-540-50968-2

Vol. 150: **Jann, Klaus; Jann, Barbara (Ed.):** Bacterial Capsules. 1989. 33 figs. XII, 176 pp. ISBN 3-540-51049-4

Vol. 151: **Jann, Klaus; Jann, Barbara (Ed.):** Bacterial Adhesins. 1990. 23 figs. XII, 192 pp. ISBN 3-540-51052-4

Vol. 152: **Bosma, Melvin J.; Phillips, Robert A.; Schuler, Walter (Ed.):** The Scid Mouse. Characterization and Potential Uses. EMBO Workshop held at the Basel Institute for Immunology, Basel, Switzerland, February 20–22, 1989. 1989. 72 figs. XII, 263 pp. ISBN 3-540-51512-7

Vol. 153: **Lambris, John D. (Ed.):** The Third Component of Complement. Chemistry and Biology. 1989. 38 figs. X, 251 pp. ISBN 3-540-51513-5

Vol. 154: **McDougall, James K. (Ed.):** Cytomegaloviruses. 1990. 58 figs. IX, 286 pp. ISBN 3-540-51514-3

Vol. 155: **Kaufmann, Stefan H. E. (Ed.):** T-Cell Paradigms in Parasitic and Bacterial Infections. 1990. 24 figs. IX, 162 pp. ISBN 3-540-51515-1

Vol. 156: **Dyrberg, Thomas (Ed.):** The Role of Viruses and the Immune System in Diabetes Mellitus. 1990. 15 figs. XI, 142 pp. ISBN 3-540-51918-1

Vol. 157: **Swanstrom, Ronald; Vogt, Peter K. (Ed.):** Retroviruses. Strategies of Replication. 1990. 40 figs. XII, 260 pp. ISBN 3-540-51895-9

Vol. 158: **Muzyczka, Nicholas (Ed.):** Viral Expression Vectors. 1991. Approx. 20 figs. Approx. XII, 190 pp. ISBN 3-540-52431-2

Vol. 159: **Gray, David; Sprent, Jonathan (Ed.):** Immunological Memory. 1990. 38 figs. XII, 156 pp. ISBN 3-540-51921-1

Vol. 160: **Oldstone, Michael B. A.; Koprowski, Hilary (Ed.):** Retrovirus Infections of the Nervous System. 1990. 16 figs. XII, 176 pp. ISBN 3-540-51939-4

Vol. 161: **Racaniello, Vincent R. (Ed.):** Picornaviruses. 1990. 12 figs. X, 194 pp. ISBN 3-540-52429-0

Vol. 162: **Roy, Polly; Gorman, Barry M. (Ed.):** Bluetongue Viruses. 1990. 37 figs. X, 200 pp. ISBN 3-540-51922-X

Vol. 163: **Turner, Peter C.; Moyer, Richard W. (Ed.):** Poxviruses. 1990. 23 figs. X, 210 pp. ISBN 3-540-52430-4

Vol. 164: **Bækkeskov, Steinnun; Hansen, Bruno (Ed.):** Human Diabetes. 1990. 9 figs. X, 198 pp. ISBN 3-540-52652-8

Vol. 165: **Bothwell, Mark (Ed.):** Neuronal Growth Factors. 1991. 14 figs. IX, 173 pp. ISBN 3-540-52654-4

Vol. 166: **Potter, Michael; Melchers, Fritz (Ed.):** Mechanisms in B-Cell Neoplasia 1990. 143 figs. XIX, 380 pp. ISBN 3-540-52886-5

Vol. 167: **Kaufmann, Stefan H. E. (Ed.):** Heat Shock Proteins and Immune Response. 1991. 18 figs. IX, 214 pp. ISBN 3-540-52857-1

Vol. 168: **Mason, William S.; Seeger, Christoph (Ed.):** Hepadnaviruses. Molecular Biology and Pathogenesis. 1991. 21 figs. IX, 206 pp. ISBN 3-540-53060-6

MIX
Papier aus verantwortungsvollen Quellen
Paper from responsible sources
FSC® C105338

If you have any concerns about our products,
you can contact us on
ProductSafety@springernature.com

In case Publisher is established outside the EU,
the EU authorized representative is:
Springer Nature Customer Service Center GmbH
Europaplatz 3, 69115 Heidelberg, Germany

Printed by Libri Plureos GmbH
in Hamburg, Germany